Fluorescence Assay
in Biology and Medicine

Molecular Biology

An International Series of Monographs and Textbooks

Edited by

NATHAN O. KAPLAN
Graduate Department of Biochemistry
Brandeis University
Waltham, Massachusetts

HAROLD A. SCHERAGA
Department of Chemistry
Cornell University
Ithaca, New York

Fluorescence Assay
in Biology and Medicine

SIDNEY UDENFRIEND

Laboratory of Clinical Biochemistry, National Heart Institute
Bethesda, Maryland

1962

New York **Academic Press** *London*

Third Printing, 1964 with literature appendix 1962 through 1964

ACADEMIC PRESS INC.
111 Fifth Avenue
New York 3, N. Y.

United Kingdom Edition
Published by
ACADEMIC PRESS INC. (London) Ltd.
Berkeley Square House, London W. 1

Library of Congress Catalog Card Number 61–18301

First Printing, February 1962
Second Printing, August 1962
Third Printing, May 1964

PRINTED IN THE UNITED STATES OF AMERICA

ONULP

PREFACE

The purpose of this book is to make available to both the novice and the expert a reference volume containing mainly information of a practical nature. The changes that have taken place in instrumentation for fluorescence assay have been so rapid that the available texts, some published as late as 1953, although excellent as far as theoretical considerations are concerned, have little practical use for the biologist today.

This volume is intended to serve as a laboratory manual presenting details of many fluorescence procedures which are currently of interest in various fields of biology and medicine. Attempts have been made to bring each of these procedures up to date by including recent modifications and advances and by presenting critical evaluations wherever possible. I have intentionally made use of much material with which I am familiar, either through direct experimental participation or through personal acquaintance with individuals involved in the research, in order to make such evaluations meaningful. The term biology has been used in a broad sense to include applications to public health and sanitation, food inspection including quality control, toxicology, detection of insecticide residues, etc.

The profusion of figures and chemical formulas is to familiarize the reader with the vast amount of recent work in this area and to make apparent the relationships between chemical structure and fluorescence.

A major purpose of this book, other than providing a working knowledge of fluorescence theory and practice, is to convey to the biologist the potentialities of fluorescence assay, to make him aware that applications to structural studies on proteins and other macromolecules, enzyme-coenzyme-substrate interaction, and to immunochemistry will open new vistas and that the development of newer instrumentation and chemical methodology will make more metabolites amenable to microfluorometric assay. An attempt has been made to instill into this book the feeling that this is a versatile and powerful tool and that present applications do not begin to realize its full potential.

My first use of fluorescence assay was as a member of the Antimalarial Program during World War II. With Dr. Bernard B. Brodie, we applied the procedure to the assay of quinine, atabrine, chloroquine, pamaquine, and many other antimalarial drugs in human tissues. The successful application of fluorescence assay to so many drugs led us to make attempts at more general applications only to find the procedure limited to compounds which absorbed at those wavelengths of the mercury arc above 350 mμ) and fluoresced in the visible range. It was indeed fortunate that in 1950, when he joined the staff of the National Institutes of Health, we were able to interest Dr. Robert Bowman in the broader aspects of fluorescence assay. Within a few years he was able to make available to us a spectrophotofluorometer, an instrument which compares with the fluorometer as the spectrophotometer does with the colorimeter. This spectrophotofluorometer (on which later commercial instruments were based) demonstrated convincingly that fluorescence assay could be even more sensitive and specific and more widely applicable than had been imagined.

At present, with spectrophotofluorometer-type instruments commercially available, and with more sensitive and more widely applicable fluorometers also available, the biologist is turning to fluorescence to help solve problems which have previously presented great analytical difficulties. As a result, the number of publications relating to fluorescence assay are now very large and are increasing rapidly. This newly awakened interest in the field means that more and more scientists are seeking to apply this procedure to their particular problems. These individuals require appropriate reference sources to help them learn about fluorometry, about appropriate instrumentation, and about the particular applications to their respective field. All this information can, of course, be found in suitable texts, in various periodicals, and in the literature supplied by instrument manufacturers. It is hoped that this book will consolidate much of this material into a single source.

The encouragement and help of many colleagues made it possible to complete this book within a reasonable time. Many individuals made available unpublished data and manuscripts in advance of publication. Acknowledgements of these are presented in the text. Particular thanks are owed to Hugh Howerton, for putting his excellent files on fluorescence at my disposal. Others who were most helpful during the writing of the manuscript are Mr. Louis Heiss, Mr. David Korman, and Dr. Robert E. Philips. The encouragement and helpful criticism of Dr. Robert Bowman must also be gratefully acknowledged.

Bethesda, Maryland   SIDNEY UDENFRIEND

PREFACE TO THE THIRD PRINTING

The reception of this book has been most gratifying and has made necessary a third printing. I have taken the opportunity to make several corrections, some of which were pointed out to me by readers of the book. Progress in the field of fluorescence analysis has been so rapid that we thought it advisable to include an up-to-date, annotated bibliography. References and new instrumentation which appeared subsequent to the initial printing of the book have been incorporated into an additional section, Appendix IV. It is hoped that the additional references, which include many in early 1964, will increase the usefulness of this book. I want to thank Dr. Robert E. Phillips for making available to me his excellent files on fluorescence.

Bethesda, Maryland   SIDNEY UDENFRIEND
April, 1964

CONTENTS

Fluorescence Assay
in Biology and Medicine

1

INTRODUCTION

I. GENERAL

ADVANCES in biology and medicine are limited by available analytical techniques and the biologist has been in the forefront of development of such procedures as ultraviolet and infrared absorptiometry, radioactive assay, mass spectrometry, and chromatography. The biologist requires not only diverse analytical procedures, but extremely sensitive ones as well, since he is continually examining intracellular mechanisms and investigating the functions of chemical agents which are active in extremely low concentrations in tissues. The term nanogram (one thousandth of a microgram) was introduced by physiologists as a convenient unit for following the actions of agents such as epinephrine and histamine. Biologists have long been aware of the great sensitivity of fluorescence assay and were among the first to utilize it. The fluorescence of quinine, chlorophyll, and other plant materials was known to Stokes when he published his now historic paper on fluorescence in 1852.[1] Vitamins, trace metals, and drugs have been assayed fluorometrically for many years. However, when one realizes that almost all molecules are capable of fluorescing, either directly or after suitable chemical reaction, it is surprising that this procedure has not had greater application before now.

The biologist is really just beginning to use fluorescence as an analytical tool and has only recently discovered that it is not limited to the visible range. In his enthusiastic explorations of this phenomenon, he frequently takes liberties with some of the principles of fluorescence spectroscopy. The physicist, who has studied this phenomenon as an exact science, is highly critical of biologists who use instruments which yield uncorrected excitation and fluorescence spectra and who report these uncorrected values in the literature. These criticisms are certainly justifiable, but the physicist has been of little direct assistance to the biologist in his attempts to develop fluorescence assay as a useful and practical tool. The biologist has had to develop and modify instruments and procedures largely through trial and error. Studies on the mechanism of fluorescence and its relationship to molecular structure have occupied physicists for many years. However, the information gathered by them has penetrated only slowly

1

into biological areas. This may be due to poor communication between the two sciences. However, biologists have not followed the literature concerning fluorescence as a physical phenomenon for other reasons. Many of the physical studies on fluorescence were carried out in the solid or gaseous state. Those studies which were carried out in solution usually utilized compounds which were of little interest to the biologist and measurements were made at temperatures far from the physiologic range. Furthermore, the physicist made no attempt to standardize or simplify the instruments with which he made his measurements, each investigator using an individually designed and complicated instrument. Consequently, we find that biologists have started to use fluorescence on a wide scale without a comprehensive understanding of the phenomenon or its limitations. This means that some misuse it and that some of the information gathered with it may prove to be in error. Such errors are unfortunate, but they are unavoidable whenever new procedures are introduced. However, as with other techniques, it should not take long for the biologist to master its intricacies and to use fluorescence not only in a "proper manner," but in ways which are not even apparent to today's expert.

The sensitivity of fluorescence detection is so great that it has served many of the functions for which radioactive isotopes are now used. Measurement of the size of various pools and of rates of flow, either in the body or in streams and sewer lines, was estimated with fluorescent indicators and appropriate detecting devices long before radioisotopes were more than a scientific curiosity. With the development of new types of sensitive instrumentation there is no reason why fluorescence should not assume many of the indicator roles now considered to be limited to radioisotopes. In pharmacology and related fields of biology it has become almost a requirement to synthesize a newly discovered metabolite, drug, or insecticide in radioactive form in order to determine its *in vivo* disposition. This procedure is used with great success to determine blood levels and rates of absorption and excretion. Where the compound is fluorescent or can be made to fluoresce it may often be detected at even lower concentrations by fluorescence assay. This can be done without the need of a costly radioactive synthesis and without the hazards of radiation. Of course, fluorescence cannot substitute, as effectively, for isotopes in detecting split products resulting from the metabolic breakdown of a compound. Finally, scintillation counting is an application of fluorescence which is, at present, the most sensitive and stable procedure for detecting radioactivity. The use of compounds which fluoresce on bombardment with beta particles emitted during radioactive decay has revolutionized counting procedures for the soft beta emitters. It is now possible to measure carbon-14, sulfur-35, and tritium in solution, simply and with high efficiency.

The great sensitivity of fluorescence assay and some of the reasons for this sensitivity are best summarized in the following statement from a paper by Lowry[2]:

"Fluorometry is inherently a much more sensitive analytical tool than colorimetry for measuring concentrations of substances. In colorimetric procedures a significant percentage of the light must be absorbed, and this is determined almost entirely by the concentration of the substance and the length of light path. In the measurement of a fluorescent substance the sensitivity is proportional not only to the concentration of the substance and the length of light path, but also to the intensity of illumination and the sensitivity of the photometer. There are limitations to the increase in illumination permissible with light sensitive materials, but the sensitivity of the photometer can be greatly increased. There are few substances which can be measured colorimetrically at concentrations less than 1 part per 10 million with a reasonable light path. This is a concentration 2000 times stronger than that required to measure riboflavin fluorometrically and this vitamin is not an exceptionally fluorescent substance."

Fluorescence assay should now be considered as a part of the biologist's armamentarium of analytical tools. It is actually of more general use to him than all other assay procedures except absorptiometry. The two, fluorometry and absorptiometry, permit the biologist to make precise measurements over an extremely wide range (0.001–1000 μg.). The development of elegant methods for separating complex mixtures (chromatographic and electrophoretic) make it possible to utilize fluorometry in instances where it would formerly have been impossible.

II. TERMINOLOGY

The terminology in the field of fluorescence is not always consistent. For the sake of uniformity and clarity it will, therefore, be necessary to adopt certain terms. Thus, an instrument which is employed to measure fluorescence intensity is called a *fluorometer* in the United States and a *fluorimeter* in the United Kingdom. In this book the former term will be used. The term *photofluorometer* has been used widely to signify instruments which utilize photoelectric detectors but this is a registered trade name in the United States of America. The simpler term fluorometer will be used.

It would be logical to use the term spectrophotofluorometer for instruments designed to determine excitation and fluorescence spectra. The term has been used frequently in the literature. However, the terms spectrophotofluorometer, spectrofluorometer, and spectrofluorimeter have also been used in connection with various commercial instruments. For this reason a simpler and more general term, *fluorescence spectrometer*, has been used to signify an instrument for the measurement of excitation and fluorescence spectra.

As for the fluorescence process itself, this is characterized by two spectra. A fluorescent molecule emits fluorescence when it absorbs radiation anywhere within its absorption spectrum. When measured by fluorometry the absorption spectrum is called an excitation or activation spectrum (see Chapter II). Although both terms have been used, the former is now more generally accepted and is less ambiguous. Fluorescence-excitation spectrum has also been used for this purpose. However, in an article or book devoted to the subject of fluorescence, the term *excitation spectrum* should be sufficiently clear to the reader. Terms which have been used for the emitted fluorescence are emission spectrum, fluorescence-emission spectrum, and *fluorescence spectrum*. The latter is the simplest and most explicit of the three and is used in preference to the others.

REFERENCES

1. Stokes, G. C., *Phil. Trans. Roy. Soc. London* **A142**, 463 (1852).
2. Lowry, O. H., *J. Biol. Chem.* **173**, 677 (1948).

2

PRINCIPLES
OF FLUORESCENCE

I. INTRODUCTORY REMARKS

As WITH many other analytical procedures, application of fluorescence assay does not require knowledge of the theoretical background of the phenomenon. However, some information concerning the physical principles of fluorescence is helpful in understanding and extending its use. An oversimplified definition of fluorescence is that it is the immediate emission of light from a molecule or atom following the absorption of radiation. In solution, at normal temperatures, fluorescence occurs at a longer wavelength than does the absorbed light. However, in certain gases at elevated temperatures this may be reversed. Since many of the manifestations of fluorescence vary with the physical state of the system, the following discussions will be limited to absorption and fluorescence in solution at ordinary temperatures, conditions which generally prevail in biological studies. Before discussing fluorescence it is necessary to review, in brief, our present knowledge of the nature of light itself, and the ways in which it interacts with matter. The explanations presented in the following discussion may be considered as oversimplifications. However, they are intended to summarize many physical-chemical concepts in a manner that will be readily followed by investigators and students of biology and medicine.

II. LIGHT AND ITS INTERACTION WITH MATTER

Current theories of the absorption and emission of radiation by matter combine the classical and quantum theories of optics with the quantum mechanical theory of atomic and molecular structure. The terminology in this discussion is taken from both theories.

Light is a form of electromagnetic radiation (energy), the propagation of which is regarded as a wave phenomenon. It is characterized by a frequency (ν), and has a wavelength (λ) and, in a vacuum, a constant velocity (c). These are related in the following manner:

$$\nu = \frac{c}{\lambda}$$

When light enters matter two things may happen to it:

(1) It may pass through the matter with little absorption taking place. The material is then considered as being essentially transparent. In this case there is little loss of energy, but the velocity of the light is diminished to an appreciable extent. The refractive index of a substance is the ratio of the velocity of light in a vacuum to the velocity in that substance. In general, refractive index varies with wavelength so that light is separated into its component parts (spectrum) as it passes through a transparent prism.

(2) On its passage through the matter, the light may be absorbed, either entirely or in part. In either case the absorption involves a transfer of energy to the medium. The absorption itself is a highly specific phenomenon, and radiation of a particular energy can be absorbed only by characteristic molecular structures. According to the quantum theory, energy from light is absorbed in integral units, called quanta. The energy of a quantum is given by the expression

$$E = h\nu \tag{2}$$

or

$$E = h\frac{c}{\lambda} \tag{3}$$

where h is Planck's constant, 6.62×10^{-27} erg second, c the velocity of light, ν the vibration frequency (second^{-1}), and λ the wavelength.

Each molecule possesses a series of closely spaced energy levels and can pass from a lower energy level to a higher one by absorbing an integral quantum of light which is equal in energy to the difference between the two energy states. All of the light which is absorbed, at any one instant, by a liquid solution containing many molecules, is taken up by only a few molecules. Only those few molecules are, therefore, promoted to an "excited" state and are capable of fluorescing or undergoing photochemical change.

The relationship between energy and wavelength of light, in the spectrum corresponding to molecular frequencies, is shown in Table I.* It is

---

* A more detailed tabulation of wavelength, frequency, and chemical bond energy is presented in Appendix II.

TABLE I

WAVELENGTH, FREQUENCY, AND ENERGY RELATIONSHIPS OF RADIATION

Wave-length mμ	Frequency ν second^{-1}	Chemical units of energy kcal./mole	Color	Molecular effects
1000	3×10^{14}	29	Infrared	Stimulation of molecular vibration
800	3.8×10^{14}	36	Visible limit	
700	4.3×10^{14}	41	Red	
620	4.9×10^{14}	46	Orange	
580	5.2×10^{14}	49	Yellow	Absorption in the visible
530	5.8×10^{14}	54	Green	and ultraviolet regions
470	6.4×10^{14}	60	Blue	produces electronic ex-
420	7.2×10^{14}	68	Violet	citation.
400	7.5×10^{14}	71	Visible limit	
300	1×10^{15}	95	Ultraviolet	
200	1.5×10^{15}	143	Ultraviolet	

apparent from this that the spectral regions generally used for measuring absorption and fluorescence (200–800 mμ) correspond to electronic changes in the molecule. In other words, the absorption of a quantum of light by a molecule in this region of the spectrum raises one electron to a higher electronic energy level. The electronic energy levels in a molecule assume discrete or quantized values. Diagrammatic presentations of the manner in which radiation interacts with polyatomic molecules have been made. However, these are rather complicated since they require plotting in several dimensions. For this reason, the following discussions will be based on the simpler diatomic molecule. Curves representing the various potential energy levels of a diatomic molecule are shown in Fig. 1. The abscissa represents the internuclear distance between atoms A and B. The ordinate represents potential energy. The lowest energy level, or ground state, is represented as N, and increased amplitudes of vibrational energy in the lowest electronic state are represented as 0, 1, 2, 3, 4 of the lower curve. At room temperature a combination of thermal agitation and attractive and repulsive forces keep the atoms vibrating along the path labeled N. When a quantum of light impinges on such a molecule, it is absorbed in about 10^{-15} second or within the period of the frequency of the light. This is so rapid that we may consider the internuclear distances as remaining constant during the absorption process.

At ordinary temperatures we may assume that the molecule, when it absorbs the light, is in its lowest electronic level. The absorption of the

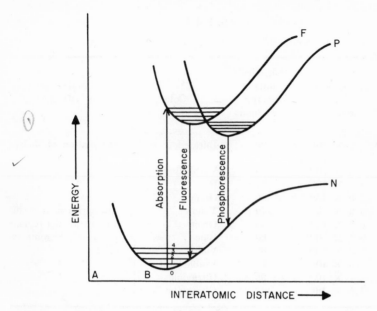

FIG. 1. Potential energy curves of a diatomic molecule at various energy levels (according to Jablonski[2]).

quantum of light will then raise the energy by promoting the electron to a higher energy state. This is represented by F in Fig. 1. Although the electronic transition due to light absorption is almost instantaneous (10^{-15} second), represented by the straight line (absorption), the excited state persists for a finite time (10^{-8} second). Subsequently, there occurs deactivation by collision with other molecules and a return to the lowest vibrational level of the excited electronic state (lower portion of curve F). The probability for return to the ground state is highest at this point and the electron drops back to the normal state, N.

Now, just as light absorption accompanies the transfer of radiant energy to the molecule, so does the reverse transition from the excited state to the ground state lead to release of energy. In fact, several types of energy release may occur.

In some gases at extremely low pressures the reverse transition occurs before vibrational deactivation can occur and has the same energy as the absorption transition (Fig. 1). The emitted light has, therefore, the same wavelength as the absorbed light. This type of emission, which is called *resonance radiation,* does not occur at higher vapor pressures or in solution because vibrational deactivation of the excited molecules to the lowest vibrational state occurs by collision before fluorescence emission takes place. This phenomenon will, therefore, seldom concern the biologist.

Light, of a frequency which is incapable of producing in a molecule a transition to an excited electronic state, is nevertheless capable of being absorbed by the molecule. An absorbed photon of energy excites an electron from its ground state to a higher *vibrational* level. Since there is no electronic transition, energy is entirely conserved and a photon of the same energy is re-emitted within 10^{-15} second, while the electron returns to its original state. Since the absorbed and emitted photons are of the same energy, the emitted light has the same wavelength as the exciting light. Light emitted in this manner is referred to as Rayleigh scattering. It occurs at all wavelengths and its intensity varies inversely with the fourth power of the wavelength. Vibrational excitation, leading to Rayleigh scattering, influences a large proportion of the molecules in a solution as compared to electronic excitation which produces relatively few excited molecules. Light scattering is relatively nonspecific and is exhibited by the container, the solvent, and by many solutes. It is, therefore, a problem with which the biologist must frequently contend in applying fluorescence as an analytical tool. It is particularly a problem when the absorption and fluorescence spectra of a substance are close together and when the intensity of fluorescence is low compared to that of the exciting light. The problem of light scattering will be discussed from a more practical standpoint in subsequent sections.

The *Raman effect* is another form of scattering emission which is related to Rayleigh scattering. As in the latter case, an electron absorbs a photon of energy producing vibrational excitation. Vibrational energy may be added or subtracted to this photon depending upon the characteristic frequency of the electronic state. When re-emission occurs (within 10^{-15} second), the electron may contain more ($+$) or less ($-$) energy. For a given substance there are emitted both ($+$) and ($-$) satellites of the exciting energy, the latter being more prominent. The energy difference is characteristic of a given molecular structure and is independent of the energy of the exciting light. In fluorescence spectrometers Raman scatter appears, therefore, at a lower frequency (higher wavelength) than does the exciting light. The Raman bands appear as satellites of the Rayleigh scatter peak, having a constant frequency difference from the exciting light. This difference is characteristic of the molecular structure of a given substance and is used for structural studies. In liquids, the Raman bands are much fainter than the Rayleigh scatter peak, but become significant in fluorescence studies when high intensity exciting sources are used. It becomes of concern in fluorescence studies because it can occur near a wavelength coinciding with the fluorescence emission band of a given compound. Raman scatter of a solvent is therefore another factor with which the biologist must contend. A simplified diagrammatic presentation showing the relationships between Rayleigh scatter, Raman scatter, and fluorescence is shown in Chapter 4, Fig. 4. The practical aspects of the various forms of scattered light with relation to fluorescence assay will be considered in that section. Conversely, fluorescence is a source of error in studies which utilize light scattering as a tool, as in the molecular weight determination of proteins.[1]

The emission of light following its *selective* absorption is referred to as luminescence. It is the general term for the emission of light from a molecule following its excitation through the absorption of any form of energy. Heat, electricity, and chemical reaction can also bring about molecular excitation leading to luminescence. Chemiluminescence is well known in biology, and the mechanism of fire-fly luminescence has been studied in great detail. Fluorescence and phosphorescence are the two classifications of luminescence which occur following excitation due to the absorption of any form of electromagnetic radiation.

In order to understand some of the distinctions between fluorescence and phosphorescence, we must remember that one of the important concepts in the modern theory of molecular structure is that the electron can spin about its own axis. Electrons are considered as having a spin, S, equal to $\pm\frac{1}{2}$. A normal polyatomic molecule in its ground state usually has an even number of electrons with paired spins; there are as many electrons of $S = +\frac{1}{2}$ as of $S = -\frac{1}{2}$. The resulting spin is therefore 0. *Multiplicity* is a term used to express the orbital angular momentum of a given state of an atom or molecule. It is related to the electron spin, S, through the expression $2S + 1$. In molecules having all electron spins paired, $S = +\frac{1}{2} - \frac{1}{2} = 0$, and the multiplicity is 1. Such molecules are said to have a "singlet" electronic level. When, through some internal energy transition, the spin of a single electron in a polyatomic molecule is reversed the molecule finds itself with two unpaired electrons ($S = \frac{1}{2} + \frac{1}{2} = 1$), each occupying a different orbital. The multiplicity term, $2S + 1$, is therefore equal to 3, and the molecule is said to have a "triplet" electronic level. Such a molecule may be considered to be in a metastable state. It should be pointed out that the oxygen molecule, in its ground state, has a triplet electronic level.

Among other things, interaction with radiation must be accompanied by changes in the electrical center of a molecule. This is termed a dipole moment of transition. The selection rules, for electronic transitions in atoms, state that energy transitions within a molecule which do not yield appreciable dipole moments of transition are highly improbable. Transitions from one singlet level to another or from one triplet level to another are accompanied by large transition moments, those from singlet to triplet or vice versa are relatively weak. Transitions within a given level are therefore considered as *"allowed transitions"* (highly probable), the latter are *"forbidden transitions"* (low probability).

According to Jablonski[2] and Lewis and Kasha,[3] most absorbing molecules have two excited electronic states, F (singlet) and P (triplet), which are related to each other as shown in Fig. 1. Transitions from the ground state, N (singlet), to the excited state F (singlet), constitute normal absorption; the reverse is *fluorescence*. A molecule in state F has a mean lifetime (τ) of the order of 10^{-8} second. *Phosphorescence* involves transitions from the *excited state*, P (triplet). The probability of transitions occurring directly from N to P is very small ($N \rightleftarrows P$ is a *forbidden transition*). However, internal conversion (electron spin reversal) from F to P may occur with some probability since the energy of the lowest vibrational level of P is lower than that of F. Molecules in the excited state, P, can return to the ground state N via F only by acquiring energy from the environment (at elevated temperatures or by a flash of light). Under these conditions the emitted light is the same as that produced by normal fluorescence, but the lifetime of the excited state is longer than 10^{-8} second. At low temperatures, thermal energy is not available and the return to the ground state can proceed only via the forbidden transition $P \rightarrow N$. Since the probability of this is quite low, the excited state persists for a relatively long time; several seconds at ordinary temperatures. It is evident, too, from Fig. 1, that the emitted light will be of a lower frequency (higher wavelength) than the fluorescent light. The excited state, P, apparently represents a tautomeric form of the ground state in which a pair of electron spins are uncoupled (triplet state). Phosphorescence is, therefore, emission produced by the transition from a triplet state (unpaired electrons) to a singlet state (all electron spins are paired). From a practical standpoint, phosphorescence differs from fluorescence mainly insofar as the persistence of the luminescence following excitation by the light source and by the marked enhancement and increase of the emission time with the lowering of temperature. While the biological applications of phosphorescence are not yet large in number, the phenomenon is important to the biologist in explaining energy transfer mechanisms.

Coming back to the phenomenon of fluorescence, how does absorption of light by a molecule lead to a characteristic emission and how is the latter related to the absorbed light? The molecule absorbs a photon of light and an electron is raised to a higher level of energy. The molecule is, therefore, left in an excited state. The electronic change yields a band spectrum of absorption which comprises the whole series of transitions, electronic, vibrational, and rotational. If the molecule does not decompose as a result of the increase in energy and if all the energy is not dissipated by subsequent collisions with other molecules, then after a short period of time, τ, which is characteristic of the atom or molecule (10^{-8}–10^{-7} second), the electron returns to the lower energy level, emitting a photon in the process. The difference between the energy of the initial state (N) and the final state (F) determines the energy of the emitted radiation. It is this radiation which we call fluorescence. The emitted fluorescence has a greater wavelength or lower energy than the light which is absorbed (Stokes law). This implies that the electron, after being raised to a higher energy level by the absorbed radiation, falls first to an intermediate vibrational level, and so emits radiation of lower energy than that used to induce the fluorescence. In other words, a small amount of energy is dissipated as heat (thermal deactivation) in the over-all process. "Anti-Stokes" fluorescence, the emission of light of a lower wavelength or greater energy than the exciting light, implies that the electron transition is accompanied by greater energy than is supplied by the absorbed photon. The surplus energy may be supplied, in such cases, by thermal agitation of the molecules. Sufficient energy to produce anti-Stokes effects can be attained in polyatomic molecules in the gas phase. However, almost all fluorescence in solution at ordinary temperatures, is of the Stokes type.

III. LIFETIME OF THE EXCITED STATE

Upon withdrawal of the exciting light, the fluorescence intensity of a solution of fluorescent molecules decays with a rate characteristic of a first-order process, exactly like radioactive decay. The relationship between the change in fluorescence with time and the original fluorescence can be expressed in the same exponential manner as radioactive decay:

$$I = I_0 e^{-t/\tau} \tag{4}$$

where I = fluorescence intensity at time t.

I_0 = maximum fluorescence intensity during excitation.

t = time after removing the source of excitation.

τ = average lifetime of the excited state.

In fluorescence studies τ has been taken as the time required for the fluorescence intensity to fall to $1/e$ of its initial value. In radioactivity, on the other hand, the time required for radioactivity to fall to one-half of the initial value is used as an index of the lifetime of the process.

Measurements of τ can be made in a number of ways. For direct measurement, a device such as a Kerr cell placed between crossed Nicol prisms has been used. The cell contains a liquid of high dipole moment and polarizability which becomes doubly refracting in an electric field. Nitrobenzene is one such liquid. When the electric field is applied, the cell transmits light.

However, transmission of light ceases when the field is interrupted. The cell therefore acts as a shutter of extremely low inertia and can be activated in fractions of a microsecond. By using two such shutters, one for the exciting light and one for the emitted fluorescence, and arranging for them to be activated fractions of a microsecond apart, it is possible to measure values of τ in the region from 10^{-10} to 10^{-4} second.

An indirect method for calculation of τ requires the measurement of fluorescence polarization in media of known viscosity. The relationship between τ, polarization, and viscosity (Chapter 6) permits calculation of τ. For small molecules the method usually requires polarization measurements in highly viscous solvents such as glycols. For macromolecules, such as proteins, τ can be measured in dilute aqueous solution. The values shown in Table II indicate the wide range of τ with chemical structure.

TABLE II

AVERAGE LIFETIME OF THE EXCITED STATE (τ) OF SOME COMPOUNDS

Compound	τ seconds	Reference
DPNH	5×10^{-10}	(4)
Fluorescein anion	5.1×10^{-9}	(5)
Quinine	4×10^{-8}	(6)
Chlorophyll	3×10^{-8}	(6)
Anthracene	2.5×10^{-7}	(6)
Naphthalene	3.3×10^{-5}	(7)

IV. RELATIONSHIP BETWEEN CONCENTRATION AND FLUORESCENCE INTENSITY

When monochromatic light passes through a solution containing molecules which can absorb it, the light which is emitted is found to be of lower intensity. The Beer-Lambert law combines the relationship between the concentration of solute (c), its extinction coefficient (K), and the length of the light path (l), into an expression which can be used to calculate the intensity of the transmitted light (I) of a solution:

$$I = I_0 e^{-Kcl} \tag{5}$$

where I_0 is the initial intensity of the incident light. This expression holds only in the absence of fluorescence, and a highly fluorescent substance will give erroneous absorption data when measured in the usual type of photometer. Absorption errors due to fluorescence can be quite large, and procedures for their estimation and correction are important enough to

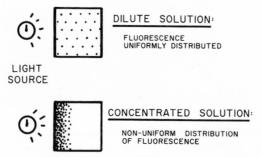

FIG. 2. The distribution of fluorescent molecules in dilute and concentrated solutions. The black dots represent individual absorbing and fluorescing molecules (adapted from Bowen and Wokes[10]).

merit serious consideration.[8] However, because of the low intensity of the light which is generally used in making absorption measurements, fluorescence interferences is not as great a problem as it might otherwise be.

Just as fluorescence creates a problem when measuring the absorption, so absorption raises certain problems during fluorometric assay. Obviously, light must be absorbed before fluorescence can occur. However, it is also apparent that when absorption is so great as to make the solution opaque, no light will pass through to cause excitation. At intermediate concentrations, even though the light does penetrate through the solution, it is not evenly distributed along the light path. The portions nearest the light source absorb much of it so that progressively less and less is available for the remainder of the solution. The result is that much excitation occurs at the entrance, but there is less and less through the remainder of the cell. This is an example of fluorescence loss due to an inner filter effect. The non-

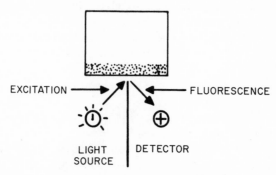

FIG. 3. Excitation and fluorescence detection at a glancing angle from the same surface.

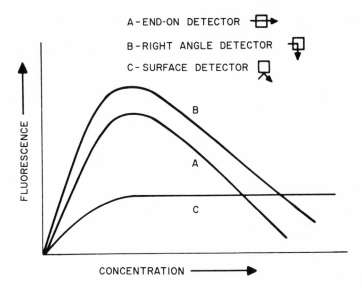

FIG. 4. The effect of fluorophore concentration on the measurement of fluorescence at various angles.

uniform distribution of the fluorescence in a solution of strong absorption presents a problem for detection. This is shown in Fig. 2. In highly concentrated solutions one can measure fluorescence from the surface at a glancing angle just as with crystals (Fig. 3). Plotting fluorescence against concentration yields the following types of data for the different angles of detection (Fig. 4). Although measurement at the surface can be made independent of concentration, it is used by the biologist only for special purposes, such as in studies on whole cells or complex cell extracts (see Chapter 9). The biologist who is mainly an analyst must usually resort to that end of the scale where fluorescence is proportional to concentration.* This occurs in highly dilute solution. Since fewer molecules are

---

* The biologist who is familiar with radioactive isotope procedures will recognize the similarity between the problems involved in fluorescence assay and those inherent in the assay of dried samples of beta emitters with a Geiger counter. In radioassay, the biologist is faced with self-absorption by the sample. To obtain the greatest sensitivity and proportionality between the amount of isotope and counting rate, it is necessary to use essentially weightless samples. This is comparable to the measurement of fluorescence in solutions which absorb little light. When the dried sample has appreciable mass, correction must be made for self-absorption of the beta radiation. Finally, when the sample reaches a certain thickness ("infinitely thick"), only the top layers contribute to the assay. Beyond this point the measured radioactivity is related to the specific activity of the sample and variations in mass or sample thickness do not affect it. This is comparable, in a way, to fluorometric assay of concentrated or highly absorbing solutions at a glancing angle.

present under these conditions, it is necessary to increase the sensitivity of the detecting device and the intensity of the exciting radiation to excite a larger proportion of them. That is why modern fluorometers use mercury arcs, xenon arcs, and other high-intensity lamps for excitation, and utilize the most sensitive electronic detectors. The various types of lamps and detectors will be discussed in the section on instrumentation (Chapter 3).

The intensity of fluorescence (emitted in all directions) is equal to the intensity of the absorbed light multiplied by the quantum efficiency of fluorescence[9]:

$$F = [I_0 (1 - 10^{-\epsilon cd})] [\phi] \tag{6}$$

where F = total fluorescence intensity, quanta per second.

I_0 = intensity of exciting light, quanta per second.

c = concentration of solution.

d = optical depth of solution.

ϵ = molecular extinction coefficient.

ϕ = quantum efficiency (yield) of fluorescence (see below).

When the solutions used are so dilute that the amount of light absorbed is very small, then equation (6) reduces to

$$F = [I_0 (2.3 \epsilon cd)] [\phi] \tag{7}$$

It is apparent that at such low concentrations when the exciting wavelength and intensity are kept constant, the meter response (on any instrument) will be directly proportional to the concentration of fluorophore.

In what region of concentration is fluorescence proportional to concentration? A general rule is that a linear response will be obtained until the concentration of the fluorescent substance is sufficiently great so as to absorb significant amounts of the exciting light. A solution having an absorbency of 0.500 in the spectrophotometer would be expected to lower the apparent fluorescence efficiency to a great extent. Bowen and Wokes[10] state that for a linear response to obtain, the solution must absorb less than 5% of the exciting radiation. Other problems arise when high light intensities are used for excitation to overcome limitations in the detecting mechanism. A limiting factor with high light intensities is, of course, photodecomposition (see below). Light scattering also becomes a problem under such conditions. In the regions of concentration where fluorescence is proportional to concentration, one is essentially measuring fluorescence in the absence of any significant absorption. This represents a limiting case of the Beer-Lambert law where the energy available for excitation is uniformly distributed through the solution.

With the highly sensitive instruments that are now available, fluorescence can be measured at concentrations of 0.0001 gamma/ml. and linearity obtains up to 10 gammas/ml. or higher. It has been found convenient to plot concentration versus fluorescence as a log/log plot in order to cover such a wide range.[11] This is a useful device for the physicist in studying the phenomenon of fluorescence. The biologist, however, will seldom be faced with ranges greater than ten- to fiftyfold and can, if necessary, dilute his sample to give concentrations of the same order of magnitude.

V. QUANTUM YIELD OF FLUORESCENCE

The percentage of the absorbed energy which can be re-emitted as fluorescence is indicated by the term "quantum yield of fluorescence," designated as ϕ, where

$$\phi = \frac{\text{number of quanta emitted}}{\text{number of quanta absorbed}} \tag{8}$$

In the absence of any perturbations the emitted fluorescence is equal to the absorbed radiation and $\phi = 1$. The lifetime of the excited state under these ideal conditions is referred to as τ_0 and can be estimated from absorption data.[6] The actual value of ϕ under practical conditions can be derived from the equation

$$\phi = \tau/\tau_0 \tag{9}$$

where τ has been experimentally determined. Direct measurements of τ present many technical problems as do estimations of τ_0 from absorption data.[6] For this reason more direct measurements of ϕ are utilized frequently.

Many procedures are known for measuring fluorescence efficiency directly in solution. The following comparative procedure for determining the quantum yield of fluorescence, by Weber and Teale,[12] is one of the more recent ones. The intensity of fluorescence emitted at right angles to the direction of excitation is compared with the intensity of the light scattered in the same direction by a glycogen solution. When monochromatic exciting light is used and the apparent optical densities of the fluorescing and scattering solutions (glycogen) are the same, the latter may be considered to be a standard with a quantum yield of 1.0. The ratio of fluorescence intensity to scattered intensity is proportional to the quantum yield, *provided appropriate corrections are made for any unequal spatial distribution of the radiation*. More recently, Parker and Rees[9] have described a comparative procedure for determining fluorescence efficiency. The method is based on the relationship between the intensity of the emitted fluorescence (F), exciting intensity (I_0), and fluorescence efficiency (ϕ), where

$$F \simeq I_0 \, \phi \, \epsilon \, cd \tag{10}$$

If the fluorescence intensities of two solutions are measured at their respective maxima, in the same apparatus, using the same exciting source, then the two intensities* are related in the following manner:

$$\frac{F_2}{F_1} = \frac{\phi_2}{\phi_1} \cdot \frac{\text{optical density of 2}}{\text{optical density of 1}} \tag{11}$$

In order to utilize this method, the fluorescence quantum efficiency of one substance must be known. Parker and Rees[9] utilized as a comparative standard the value of 0.55 obtained by Melhuish[13] for quinine. Needless to say, the values for F_2 and F_1 must be corrected for instrumental variables, as discussed by the authors. Only when the absorption and fluorescence maxima of the two substances coincide can such a simple calculation be made without corrections. This was the case when tryptophan was used as a standard for estimating the quantum efficiencies of adenine and guanine (see Chapter 8). Some fluorescence quantum efficiencies obtained by Parker and Rees[9] and by Weber and Teale[12] are shown in Table III. The agreement is probably as good as that which has been obtained by any two laboratories. Shore and Pardee[14] reported a relatively simple method for determining fluorescence efficiency utilizing a Beckman spectrophotometer to measure both the absorption and fluorescence (with a filter to cut off absorbed light). Again, certain assumptions concerning geometry and photomultiplier response were required. Some of the values obtained agreed with literature reports. However, many of their reported values have been questioned, particularly those for the aromatic amino acids. Because of the complexities involved it is not surprising that the reported values for fluorescence efficiency have sometimes not been too accurate. Fluorescence efficiencies approaching 1.0 have been reported for some dyes in organic solvents. Fluorescein in water and quinine in dilute acid have been reported to have efficiencies ranging from 0.7 to 0.9 and 0.3 to 0.6, respectively. The highest efficiencies are observed in dilute solutions where fluorescence is proportional to concentration, since under these conditions little fluorescent light is lost due to interaction with other molecules. Weber and Teale[15] give values of 4, 21, and 20% for the fluorescence efficiencies of phenylalanine, tyrosine, and tryptophan. In the sections below, fluorescence efficiencies will be reported for other molecules. It is now generally held that the efficiency of fluorescence is independent of

---

* The fluorescence intensity, F, is the entire emission representing the whole spectral output. With recording spectrometers the area under the fluorescence spectrum can be used as F in the comparative method for determining quantum yield.

TABLE III

QUANTUM YIELDS OF FLUORESCENCE (ϕ) IN SOLUTION

Compound	Solvent	ϕ^a
Fluorescein	Water, pH 7	0.65
Fluorescein	0.1 N NaOH	0.92
Fluorescein	0.1 N NaOH	0.85*
Eosin	0.1 N NaOH	0.19
Eosin	0.1 N NaOH	0.23*
Rhodamine B	Ethylene glycol	0.89
Rhodamine B	Ethanol	0.97
Rhodamine B	Ethanol	0.69*
1-Amino-naphthalene-3:6:8-sulfonate	Water	0.15
1-Dimethylamino-naphthalene-4-sulfonate	Water	0.48
1-Dimethylamino-naphthalene-5-sulfonate	Water	0.53
1-Dimethylamino-naphthalene-7-sulfonate	Water	0.75
1-Dimethylamino-naphthalene-8-sulfonate	Water	0.03
9-Amino acridine	Water	0.98
9-Amino acridine	Ethanol	0.99
2-Methoxy-6-chloro-9-N-glycyl acridine	Water	0.74
Acriflavin	Water	0.54
Anthracene	Benzene	0.29
Riboflavin	Water, pH 7	0.26
Uranyl acetate	Water	0.04
Fluorene	Ethanol	0.53
Fluorene	Hexane	0.54
Anthracene	Ethanol	0.30
Anthracene	Hexane	0.31
Phenanthrene	Alcohol	0.10
Naphthalene	Alcohol	0.12
Naphthalene	Hexane	0.10
Sodium salicylate	Water	0.28
Sodium p-toluensulfonate	Water	0.05
Sodium sulfanilate	Water	0.07
Phenol	Water	0.22
Indole	Water	0.45
Skatole	Water	0.42
Chlorophyll a	Benzene	0.32
Chlorophyll a	Ether	0.32
Chlorophyll a	Acetone	0.30
Chlorophyll a	Ethanol	0.23
Chlorophyll a	Methanol	0.23
Chlorophyll a	Cyclohexanol	0.30
Chlorophyll a	Dioxan	0.32
Pheophytin a	Benzene	0.18
Chlorophyll b	Benzene	0.11
Chlorophyll b	Ether	0.12
Chlorophyll b	Acetone	0.09
Chlorophyll b	Ethanol	0.095
Chlorophyll b	Methanol	0.10

a Most of the values presented are from Weber and Teale.[12] A few values obtained by Parker and Rees[9] are indicated by an asterisk.

the wavelength of the exciting light.[16–18a] In all spectral regions where molecules have some absorption, fluorescence can be brought on even by radiation which is at the extreme ends of an absorption band, emission having the same relationship to absorption as it does at the point of maximum absorption. In those instances where there is found to be a change in fluorescence efficiency with wavelength, additional chemical species are to be considered.[18b, 18c] Lavorel[19] has demonstrated that at high concentrations of fluorescein, thionine, and chlorophyll dimerization occurs. The absorption spectra of these dimers differ from the parent compounds and, in these instances, the dimers have different spectral properties. It is apparent that when more than one molecular absorbing species is present in a solution fluorescence efficiency will not be constant over the entire spectrum.

VI. EXCITATION AND FLUORESCENCE SPECTRA

From the previous discussions it is apparent that fluorescence is characterized by specific excitation spectra* and fluorescence spectra. The excitation spectrum is, in theory, identical with the absorption spectrum. It will only differ from it as a result of instrumental artifacts. The fluorescence spectrum is, as we have seen, displaced towards higher wavelengths. Its measurement also involves the correction of many instrumental artifacts. As an approximation, the fluorescence spectrum has been considered by some as the mirror image of the absorption spectrum. There are a sufficient number of examples of such mirror-image relationships between the two spectra as to make this quite plausible. An example of absorption and fluorescence spectra showing this mirror-image relationship is shown in Fig. 5, for anthracene. However, most polyatomic molecules, in aqueous solution, have only one fluorescent band which is associated with the absorption band of longest wavelength. When such molecules are excited by light coincident with an absorption band at shorter wavelength, the excited molecules undergo internal conversion with loss of sufficient energy so as to pass over into the state corresponding to the longest-wave absorption band. As shown in Fig. 5, quinine in dilute acid solution exhibits more than one absorption band, yet shows only one fluorescence band which is independent of the exciting wavelength. Although present-day molecular theory rests on a relatively firm base, the prediction of fluorescent spectra of polyatomic molecules requires a more intimate knowledge of structure and probabilities than is generally available. On

---

* Excitation and fluorescence spectra will be defined in operational terms in the section on instrumentation.

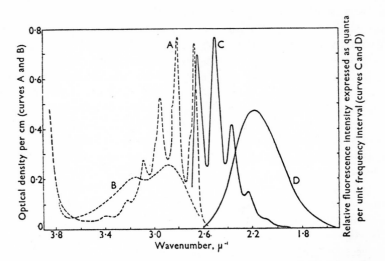

FIG. 5. Absorption and fluorescence spectra of anthracene (in ethanol) and quinine (in 0.1 N sulfuric acid). Curve A, anthracene absorption; curve B, quinine absorption; curve C, anthracene fluorescence; curve D, quinine fluorescence (from Parker and Rees[9]).

the other hand, it is possible, as we shall see below, to make some generalizations concerning structure and fluorescence based almost entirely on experimental observations.

VII. CHEMICAL STRUCTURE AND FLUORESCENCE

If fluorescence results from specific electronic changes within the molecule, then information concerning fluorescence characteristics, spectrum, efficiency, decay time, temperature coefficient, etc., can be used to determine molecular structure. Theoretical organic chemists are making more and more use of fluorescence and phosphorescence in their attempts to elucidate molecular structure and mechanisms of molecular interaction. As for the analyst, a most practical reason for studying fluorescence-structure relationships is to enable him to predict which molecules will fluoresce, under what conditions they will fluoresce, and what their fluorescence characteristics will be.

From the preceding discussion of the mechanism of the fluorescence process it would appear that all molecules which absorb light energy should fluoresce. We know, however, that under ordinary conditions, even with sensitive and appropriate detecting devices, this is not so. Actually, what this means is that the fluorescence efficiency of most absorbing molecules is very low. If fluorescence is the converse of the absorption of light energy, then we can explain the nonfluorescence of the large number of strongly absorbing substances only by alternative processes for the utilization of the electronic excitation energies of the molecules. In other words, fluorescence is not as widespread as light absorption because of competing "quenching" processes tending to decrease the quantum yield of fluorescence. The latter result from any interferences with the persistence of the excited electronic state. The type of interference which is responsible for most instances of nonfluorescence results from competitive electronic transitions within the excited molecule leading to phenomena other than emission of radiation. A general term for this most common type of quenching is "internal conversion." There must be many types of internal conversion. One such process can be the radiationless transition (electron spin reversal) from the excited singlet electronic level to a triplet level with somewhat lower energy. This leads to phosphorescence. However, in many instances the latter is not an efficient process at normal temperatures so that the singlet-triplet conversion may yield no emission. Molecules containing bromo, iodo, nitro, and azo groups show little fluorescence for this reason. The marked quenching effects of halide atoms on fluorescence becomes evident on comparing the fluorescence efficiency of the uranin dyes in aqueous solution (Table IV). It is also apparent that iodide is much more effective than bromide in this respect.

Another type of radiationless transition occurs when the excited singlet electronic state leads to an electronic state in which a chemical bond is broken. These are termed *predissociative transitions*. Molecules containing linkages with a bond strength lower than the energy required for electronic excitation will have a great tendency for predissociation. Some absorbing molecules which do not fluoresce as a result of predissociation are shown in Table V. It is apparent that when a molecule absorbs in a region where

TABLE IV

THE EFFECT OF HALOGEN SUBSTITUTION ON THE QUANTUM EFFICIENCY
OF FLUORESCENCE[a]

Compound	Quantum efficiency of fluorescence
Uranine (sodium fluorescein)_____	0.70
Tetrabromouranine (eosine)_____	0.15
Tetraiodouranine (erythrosine)_____	0.02

[a] Data from Bowen and Wokes.[10]

TABLE V

NONFLUORESCENCE OF SOME ABSORBING MOLECULES DUE TO PREDISSOCIATION

Compound	Absorption maximum		Weakest linkage		Fluorescence[a]
	mμ	kcal./mole	bond strength kcal./mole	bond	
Butadiene	210	140–160	125	C=C	−
Decapentaene	300	95	125	C=C	+
Nitrobenzene	300–320	90	55–60	C—N	−
m-Nitrodimethyl-aniline	400–450	60–70	55–60	C—N	+
Iodobenzene	300–320	90	60	C—I	−
Erythrosine (tetra-iodouranine)	560	50	60	C—I	+

[a] +, Fluorescent; −, nonfluorescent.

the energy is greater than the bond strength of one of the linkages, then excitation energy is lost by predissociation and fluorescence does not take place. These same bonds in molecules absorbing at higher wavelengths (less energy) do not bring about quenching. When dealing with solutions the solvent may influence predissociation, either inhibiting it or enhancing it, and thereby increase or decrease the fluorescence yield. Interaction with solvents, other solutes, or with other molecules of the same species can also drain off the electronic excitation energy of a molecule and cause its dissipation as heat. In the latter instance fluorescence efficiency increases as the concentration is lowered. Other solutes can quench fluorescence by interacting with the fluorophore to form an addition product with different fluorescence characteristics. They can also quench fluorescence due to de-activating collisions. This may be represented in the following manner (after Bowen and Wokes[10]):

$$M + h\nu \rightarrow M^* \text{ (light absorption)}$$

$$M^* \rightarrow M + h\nu^1 \text{ (fluorescence emission)}$$

$$M^* + Q \rightarrow \text{quenching}$$

where M is the fluorescent and Q the quenching molecule. Such quenching effects are usually diminished at low temperature and high viscosities.

Many quenching reactions represent fairly specific interactions between the fluorophore and the quenching agent quite frequently the alteration in fluorescence being the only means of detecting the combination. Several examples of such specific quenching reactions will be presented in subsequent chapters. The mechanisms of most quenching interactions still require much study before they are understood. One may merely summarize by saying that the electronic excitation energy of most absorbing molecules is dissipated more readily in other ways than by fluorescence emission. However, it is the great dependence of the radiationless transitions upon the environment that makes fluorescence intensity a sensitive indicator for chemical interactions, i.e., dissociation, complex formation, etc.

Having discussed, in general terms, those factors which limit fluorescence emission of absorbing molecules, let us now turn to an examination of the types of chemical structures which do fluoresce. Much of what has been published previously on this subject pertains to the solid and gaseous state. Furthermore, until very recently, instruments were not available for detecting fluorescence in the ultraviolet and infrared regions. Obviously, one of the most important limitations of fluorescence is the availability of appropriate detecting devices. For this reason the following discussions of fluorescence and chemical structure may well have been presented after the section on instruments. The present format has been used for the sake of continuity. However, reference to instrumentation will be made in this section wherever necessary.

As was stated previously, there is a vast literature relating to fluorescent molecules and the characteristics of the radiation which they emit. Summaries and discussions have appeared in a number of books and review articles* and the reader is referred to these for many excellent presentations concerning the fluorescence of dyestuffs and other aromatic and polyaromatic compounds in the solid state and in organic solvents. A particularly good discussion of structure and fluorescence was presented by Förster.[20]

There has been a more recent study of fluorescence-structure relationship of aromatic compounds in aqueous solution and the influence of pH thereon (Williams[21]). The paper provides data which are closer to the interests of the biologist and for this reason it will be discussed at some length (Table VI). Williams' studies were carried out with a commercial fluorescence spectrometer, using a xenon arc and a 1P28 photomultiplier detector. Wavelength data are presented as uncorrected instrumental readings. However, in almost all cases excitation maxima were found to agree fairly well with absorption maxima (± 5 mμ). The paper starts with a discussion of those substituents which bring about detectable fluorescence when attached to the benzene ring. The fluorescence characteristics of some monosubstituted benzenes are shown in Table VI. It is apparent

---

* See general references at the end of this chapter.

TABLE VI

FLUORESCENCE CHARACTERISTICS OF MONOSUBSTITUTED BENZENES

C_6H_5R	Excitation maxima $m\mu$	Fluorescence maximum $m\mu$	pH at which fluorescence is	
			Maximal	Zero[b]
OH	270 (215)[a]	310	1	13
OCH₃	270 (220)	300	1–14	—
NH₂	280 (240)	340	7–10	1
NHCH₃	290 (240)	360	7–10	1
NO₂				
COOH				
CH₂COOH	Nonfluorescent at all pH values.			
N(CH₃)₃⁺				
NHCOCH₃				

[a] Minor excitation peaks are given in parentheses.
[b] In some instances fluorescence may be minimal rather than zero. Measurements were made with an Aminco-Bowman spectrophotofluorometer (after Williams[21]).

that the ortho and para directing groups favor fluorescence whereas the meta directing groups do not. This, however, requires further explanation. Unsubstituted benzene is itself fluorescent in organic solvents, although this is not shown in Table VI. However, even if benzene did fluoresce in aqueous solution, its excitation would be at lower wavelengths than any of the compounds possessing bathochromic† substituents. It is at these lower wavelengths (200–250 mμ) that the energy output of the xenon arc falls off sharply and where instrumental data is in particular need of correction. However, from a practical standpoint, it is apparent that substitution of benzene by ortho and para directing bathochromic groups results in fluorescence which is detectable with available instrumentation. Although the nitro group is also bathochromic, it readily leads to various types of internal conversion of the absorbed energy so that little fluorescence is emitted.

Most compounds of biological interest are ionizable and can therefore exist in aqueous solution in various stages of dissociation, depending upon the hydrogen ion concentration. Williams' paper[21] continues with observations and a discussion on the relationships between molecular dissociation and fluorescence. Thus, phenol fluoresces maximally at pH 1; fluorescence

† As in absorptiometry, groups which shift the excitation maximum to longer wavelengths are termed bathochromes. Those which cause a shift to shorter wavelengths are termed hypsochromes.

diminishes as the pH is raised and is essentially zero at pH 13. This suggests that the fluorescent species is the unionized phenol molecule. The phenolate ion, in aqueous solution, does not fluoresce when excited at any wavelength, including its absorption maximum, 287 mμ.* In agreement with this, anisole, the methyl ether of phenol, and therefore unable to ionize, fluoresces maximally at all pH values. Catechol, resorcinol, and quinol also fluoresce in the un-ionized forms only. On the other hand, certain other phenols fluoresce only in the ionized forms. Monohydroxy and dihydroxybenzoic acids fluoresce in alkaline solution where they are maximally ionized. In the case of the polycyclic and heterocyclic phenols, such as α- and β-naphthol and ortho- and para-hydroxydiphenyl, the ionized form is also the fluorescent species. In hydroxyphenylacetic acid and related compounds fluorescence again occurs only when the phenol is undissociated. However, it is maximal when the carboxyl group is dissociated so that the maximal fluorescence occurs at neutral pH. Aniline and methyl substituted anilines fluoresce only in their undissociated forms. The relationships of fluorescence to structure and dissociation are summarized in Table VII.[22] Such findings suggest that one may make use of fluorescent compounds as indicators of pH. This is actually so and a discussion of fluorescent pH indicators and their applications is presented in Appendix I.

Williams points out that in many cases where the undissociated molecule is apparently the fluorophor, rearrangements to quinone structures can take place and these quinones are most likely the actual fluorescent species. Resonance between quinone structures is shown below for naphthol and hydroxydiphenyl:

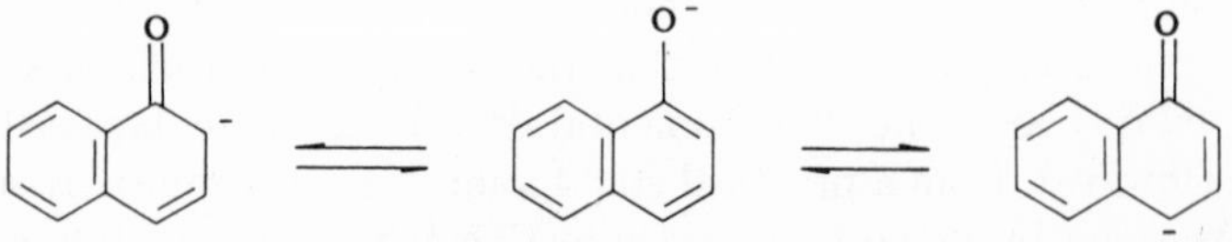

Ionized α-naphthol

Ionized p-hydroxydiphenyl

---

* The reverse of this was reported by W. West, *in* "Chemical Applications of Spectroscopy" (W. West, ed.), Par. 715. Interscience, New York, 1956. However, the report of "a long wavelength absorption band of the phenolate ions with a corresponding long wavelength fluorescence extending into the visible" is based on work carried out 50 years ago [H. Ley and K. V. Engelhardt, *Z. physik. Chem.* **74**, 1 (1910)]. The solvent in that study was not water but alcohol.

Williams also notes that in many instances fluorescence spectra are influenced markedly, both in intensity and in position, by changes in hydrogen ion concentration which have little effect on the absorption spectrum. Such findings have also been reported for the indoles,[23] where changes in pH produce large changes in fluorescence intensity without producing comparable changes in absorbency. In the case of 5-hydroxyindoles, strong acid produces a marked shift in fluorescence spectrum, from 330 to 540 mμ, without altering the absorption (excitation) spectrum (Chapter 5). Fluorescence is, without a doubt, a more sensitive index of molecular structure than is absorption.

It is apparent from all the foregoing discussions that it should be possible to predict, with some degree of assurance, whether or not a given compound will fluoresce and even the pH at which maximal fluorescence will occur. In Table VII are summarized various fluorescent and nonfluorescent structures. The examples presented are certainly not sufficient to make sweeping generalizations concerning the relationships between structure and fluorescence. Although a perusal of the literature indicates that many thousands of compounds have been investigated for fluorescence, much of the earlier work cannot be utilized for structural considerations. This is so because of questionable techniques, failure to distinguish between fluorescence in solution and in the solid state, and lack of information as to the purity of the compounds themselves. To be certain of all of these points discussions concerning structure and fluorescence are based mainly on data which has been obtained recently. Although the data in Table VII are of modest proportions, they represent a summary of some fluorophores which the biologist should be able to recognize and use for comparison with similar molecules in which he may be interested. It is hoped that the recent interest in fluorescence will provide additional data which will make predictions concerning fluorescence more certain.

Certainly the first requirement for fluorescence is an absorbing structure, the greater the absorbancy of a molecule the more intense will its fluorescence be. Thus, phenylalanine and tyrosine are both absorbing structures, the former at 260 mμ and the latter at 275 mμ. However, the molecular extinction of phenylalanine and its fluorescence efficiency are both much lower than the corresponding values for tyrosine.[15] Under ideal conditions one would expect phenylalanine to exhibit no more than 1–2% of the fluorescence of tyrosine. In practice, the value is less than 0.1–0.2%. This is so because phenylalanine absorbs and fluoresces lower in the ultraviolet region, where available lamps and detectors are much less efficient. For all practical purposes, then, the fluorescence of phenylalanine and similar benzenoid compounds are too low to be of use for microanalytical purposes. Absorption and fluorescence are enhanced in aromatic structures

TABLE VII

Fʟᴜᴏʀᴇsᴄᴇɴᴄᴇ ᴀɴᴅ Mᴏʟᴇᴄᴜʟᴀʀ Sᴛʀᴜᴄᴛᴜʀᴇ ᴀɴᴅ Dɪssᴏᴄɪᴀᴛɪᴏɴ[a]

The purpose of this table is to summarize, with examples, in what ways molecular structure and dissociation influence fluorescence. The term non-fluorescent, as used in the table, indicates that the particular structure does not yield useful fluorescence in solution, with available instrumentation.

Fluorescent structures	Nonfluorescent structures

R = OH
  = OR
  = NH_2
  = F

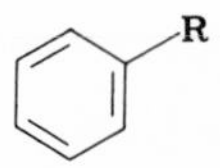

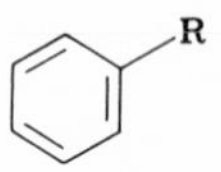

R = H, NO_2, COOH
  = Alkyl group
  = Cl, Br, I

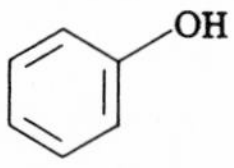

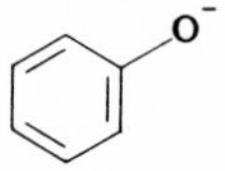

Phenol

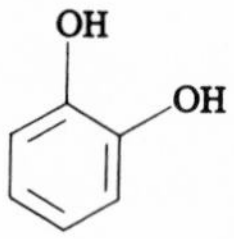

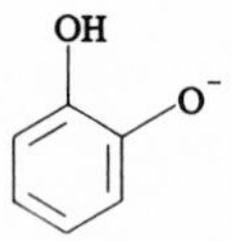

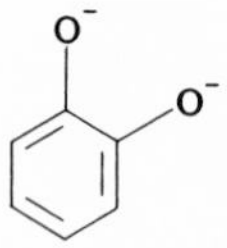

Catechol

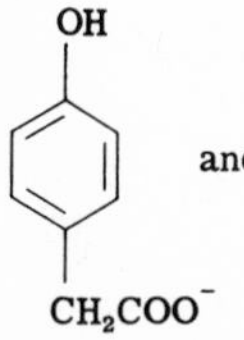

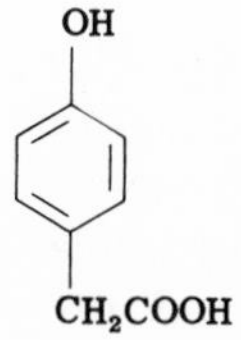

(maximal fluorescence)

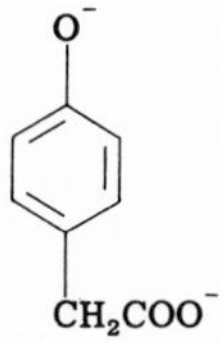

p-Hydroxyphenylacetic acid

[a] Measurements made at room temperature.

TABLE VII *(continued)*

Fluorescent structures	Nonfluorescent structures

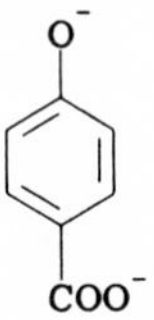

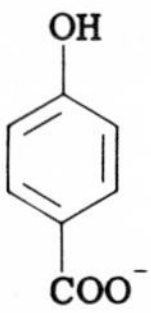

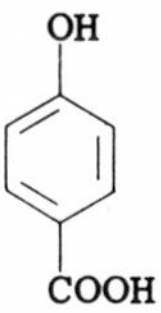

p-Hydroxybenzoic acid

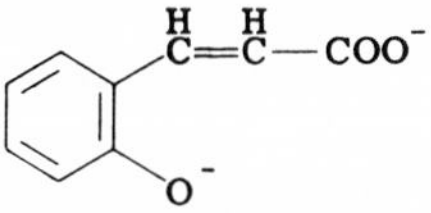

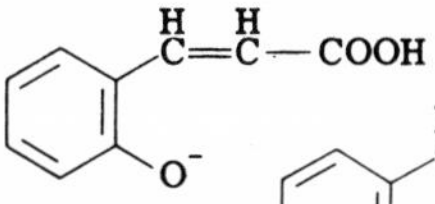

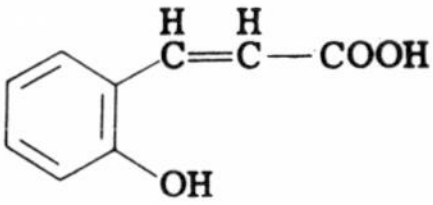

and

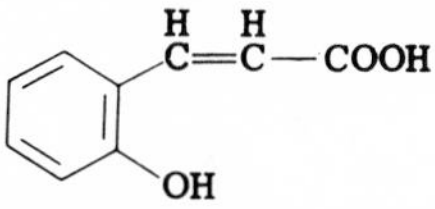

o-Coumaric acid

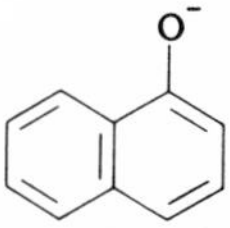

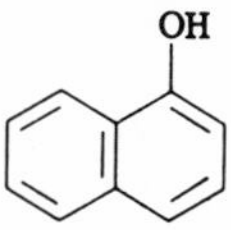

α-Naphthol

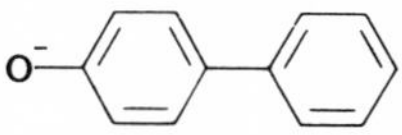

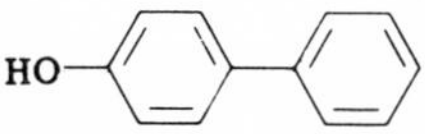

p-Hydroxydiphenyl

TABLE VII *(continued)*

Fluorescent structures	Nonfluorescent structures

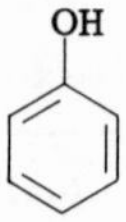

Phenol   Phenolic ester

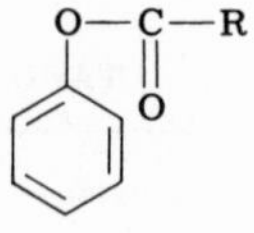

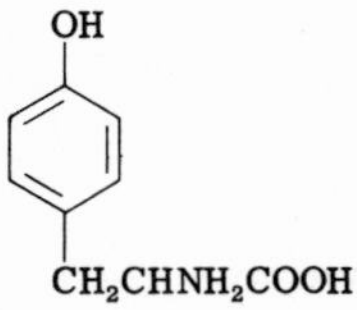

Tyrosine   Diiodotyrosine

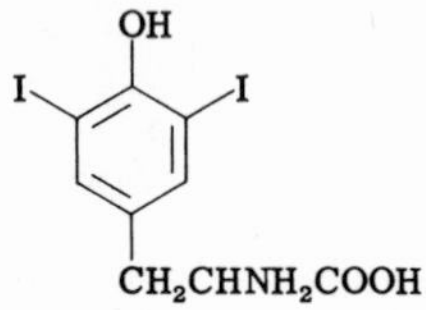

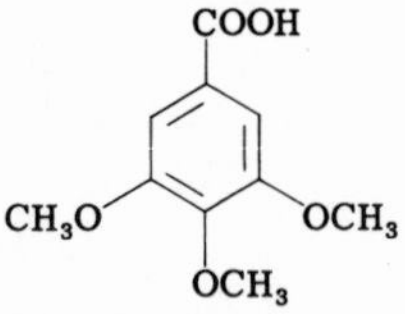

Trimethylgallic acid   Trihydroxybenzoic acid

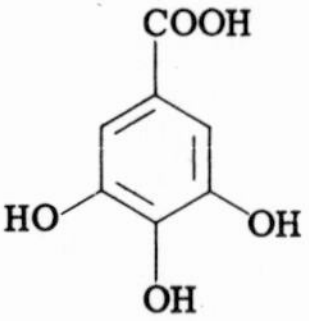

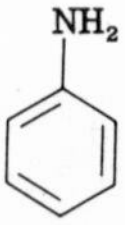

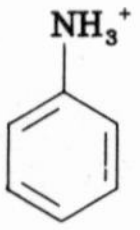

Aniline

TABLE VII *(continued)*

Fluorescent structures	Nonfluorescent structures
NH$_2$ Aniline	NHCOCH$_3$ Acetanilide
Indole	Quinoline
3-Hydroxyquinoline	Quinoline
Dihydroxynaphthalene	Naphthoquinone
$H_2C=\overset{H}{C}-\overset{H}{C}=\overset{H}{C}-\overset{H}{C}=\overset{H}{C}-\overset{H}{C}=\overset{H}{C}-\overset{H}{C}=CH_2$ Decapentaene	$H_2C=\overset{H}{C}-\overset{H}{C}=CH_2$ Butadiene

that are substituted by electron donating groups in conjugated unsaturated systems capable of a high degree of resonance. Dissociations which lead to increased resonance energy enhance fluorescence, those which decrease resonance energy decrease fluorescence. Accordingly, fluorescent compounds may be classified on the basis of several general fluorophoric structures.

A. Phenols

Many of the compounds dealt with in biology are phenolic and therefore fluorescent. These include estrogens, tocopherol, pyridoxine and its analogs, salicylate, and the various phenols and catechols related to tyrosine and epinephrine. However, the substitution of iodine, an electron accepting group, into the tyrosine molecule (mono and diiodotyrosine) abolishes fluorescence.

B. Aromatic Amines

p-Aminobenzoic acid, anthranilic acid, and kynurenine are some of the fluorescent aromatic amines of interest to the biologist. The fluorescence of aromatic amines and of phenols is lost upon acetylation, indicating again that the substituent requirement is not for nitrogen or oxygen, as such, but for a group capable of donating electrons to the aromatic ring.

C. Polycyclic Compounds

1. HOMOCYCLIC

A large proportion of the carcinogenic hydrocarbons fall into this category. Synthetic derivatives of vitamin K are also examples of highly fluorescent homocyclic structures.

2. HETEROCYCLIC

By far the largest number of fluorescing structures, of biological interest, is found in this category. It includes all the indole precursors and metabolites of tryptophan, the purines and their nucleosides and nucleotides, riboflavin and its various derivatives and conjugated forms, quinolines such as kynurenic acid, xanthurenic acid, and the cinchona alkaloids. Other heterocyclic compounds that are highly fluorescent are the acridines, substituted coumarins, porphyrins, and a host of dyes.

D. Conjugated Polyenes

The best example in this category is vitamin A. Although few other conjugated polyenes of biological interest are known to be fluorescent, it may well be that with the development of newer instruments and techniques, fluorophors wil be found among the unsaturated fatty acids and among steroid precursors or their derivatives.

One may generalize, therefore, that a given structure will be fluorescent if it falls into one of the preceding categories. Does that mean then that fluorescence assay is limited to compounds having structures similar to those shown in Table VII? Certainly not. For many years chemists have used colorimetric methods to determine colorless compounds. They do this by converting the colorless compound to a colored derivative, or *chromophore*, which can then be measured with a photometer. Applications of this type of colorimetry are too numerous and too well known to require listing. The reactions used in colorimetry are, in general, condensations with reagents which lead to resonant structures which absorb visible light. Just as chemical reactions can be used to form chromophores, so can they be used to form fluorophores. In some instances chromophores formed in long-used colorimetric reaction have been found to be fluorescent thereby increasing the sensitivity of the procedure manyfold. This was the case with the nitrosonaphthol procedure for tyrosine estimation.[24] The guiding principle in seeking fluorescent derivatives is to form those which will meet the requirements for fluorescence which may be deduced from Table VII. This can be done by condensing with a compound which is already fluorescent and separating the fluorescent derivative from the excess of fluorescent reagent. Fluorescent antibody techniques make use of such a procedure, utilizing reagents such as fluorescein isocyanate to react with the free amino groups of the proteins. Nonfluorescent steroids are converted to fluorophores by dehydration in concentrated sulfuric acids. This converts the cyclic alcohols to phenols. Epinephrine and norepinephrine are phenolic and therefore fluoresce in their native forms. However, this fluorescence is not sufficiently specific nor sensitive for their detection in the presence of the huge number of biologically occurring phenols. Specific and sensitive fluorescence procedures have been developed for these important compounds. One involves oxidation and internal condensation, the other condensation with a diamine. In both cases, highly fluorescent polycyclic compounds are formed. Many of the procedures described in the subsequent sections involve such chemical condensations to convert a nonfluorescent compound to a fluorophor and thereby make it amenable to fluorescence assay.

VIII. FLUORESCENCE POLARIZATION

Light may be considered to be a propagated train of electromagnetic waves consisting of alternating electric and magnetic fields mutually at right angles to the direction of propagation. When all the radiation has its electric vector in the same plane, the wave is said to be plane polarized and the direction of the electric vector is customarily used to describe the plane of polarization. Light ordinarily consists of a bundle of plane-polarized waves with their electric vectors randomly oriented. There are two common procedures for bringing about polarization. One utilizes the fact that oblique reflection from a polished surface is maximal for waves with their electric vector perpendicular to the surface. Polarization is most frequently accomplished by the use of doubly refracting crystals (Iceland spar) which by their optical properties convert unpolarized light into two plane-polarized components with their planes at right angles. The separation of the two components is accomplished by providing a reflecting surface inside the crystal that eliminates one of the components. Polarizers made in this fashion (Nicol, Wallaston) are generally marked to indicate the direction of the electric vector. These crystals can also be used to analyze radiation for polarization since the transmission of light through them will not change with orientation unless the light is partially polarized. If one crystal is used as a polarizer to provide plane-polarized light and a second crystal is used to analyze the light from the first crystal, the intensity of the transmitted light varies with the relative angle between the crystals as $\cos^2$ of the angle. A substance placed between the polarizer and analyzer can therefore be examined for its ability to depolarize (randomize the electric vector) or to rotate the electric vector (compounds with asymmetric centers such as sugars and amino acids). Without the polarizer, the system (light source and analyzer) can be used to determine the polarizing properties of a given substance or the analyzer alone can be used to examine any source of radiation for the presence of partial polarization.

It is recognized that the absorption and emission of light by a molecule requires interaction of the electromagnetic radiation with the electric field produced by oscillating electrons rigidly bound into the molecular structure.[25] This interaction is maximal when the electric vectors of the oscillating structure and the incident radiation are parallel. When polarized light is used as a source, it preferentially excites those molecules whose oscillators of absorption have an appreciable component in the plane of the electrical vector of the exciting beam. Therefore, although molecules in solution are randomly dispersed, the excited molecules represent a selected group with a preferred orientation. If the lifetime of the excited state is long compared with the time required for rotational diffusion (Brownian motion), the oscillators will become randomly orientated before appreciable fluorescence is emitted. For this reason, molecules of low molecular volume, in an environment of low viscosity, exhibit little fluorescence polarization. When rotational diffusion is small compared to the lifetime of the excited state, as with large molecules or in viscous media, fluorescence will result before randomization is complete. Under these conditions oscillators of fluorescence emission will still be oriented in a preferred plane so that the fluorescent light will be partially polarized. In highly viscous solvents, even small molecules will maintain detectable polarization during the lifetime of the fluorescence emission. Large molecules, such as proteins, preserve their orientation for periods of time which are relatively long compared to the decay time of the fluorescence emission (10^{-8} second). As a result, the fluorescence will show partial polarization. The degree of fluorescence polarization of a protein solution depends on many environmental and structural factors. The fraction of fluorescent light which is polarized, P, is obtained from the equation

$$ P = \frac{F_{||} - F_{\perp}}{F_{||} + F_{\perp}} $$

where $F_{||}$ is the intensity of fluorescence measured with the two prisms oriented with their electric vectors parallel and $F_{\perp}$ is the fluorescence intensity measured with the prisms crossed at right angles. A detailed discussion concerning fluorescence polarization and its application to studies on protein structure and enzyme interaction is presented in Chapter 6.

General References

Bartholomew, R. J., Spectrofluorimetry and its application to chemical analysis, *Rev. Pure and Appl. Chem.* **8**, 265 (1958).

Bladergroen, W., "Problems in Photosynthesis," Chapter II. C. C Thomas, Springfield, Illinois, 1960.

Bowen, E. J., "The Chemical Aspects of Light." Oxford Univ. Press, London and New York, 1946.

Bowen, E. J., and Wokes, F., "Fluorescence of Solutions." Longmans, Green, London, 1953.

DeMent, J., "Fluorochemistry." Chemical Publ. (Tudor), New York, 1945.

Förster, T., "Fluoreszenz Organischer Verbindungen." Vandenhoeck and Ruprecht, Göttingen, 1951.

Harvey, E. N., "A History of Luminescence." American Philosophical Society, Philadelphia, Pennsylvania, 1957.

Kauzmann, W., "Quantum Chemistry." Academic Press, New York, 1957.

Lewis, G. N. and Kasha, M., Phosphorescence and the triplet state, *J. Am. Chem. Soc.* **66**, 2100 (1944).

Pringsheim, P., "Fluorescence and Phosphorescence." Interscience, New York, 1949.

Pringsheim, P., and Vogel, M., "Luminescence of Liquids and Solids." Interscience, New York, 1946.

Radley, J. A., and Grant, J., "Fluorescence Analysis in Ultraviolet Light," 5th ed. Van Nostrand, Princeton, New Jersey, 1954.

Reid, C., "Excited States in Chemistry and Biology." Academic Press, New York, 1957.

Teale, F. W. J., and Weber, G., Ultraviolet fluorescence of the aromatic amino acids, *Biochem. J.* **65**, 476 (1957).

West, W. (ed.), "Chemical Applications of Spectroscopy." Interscience, New York, 1956.

REFERENCES

1. Brice, B. A., Nutting, G. C., and Halwer, M., *J. Am. Chem. Soc.* **75**, 824 (1953).
2. Jablonski, A., *Z. Physik* **94**, 38 (1935).
3. Lewis, G. N., and Kasha, M., *J. Am. Chem. Soc.* **66**, 2100 (1944).
4. Weber, G., *Nature* **180**, 1409 (1957).
5. Szymanowski, W., *Z. Physik* **95**, 460 (1935).
6. Pringsheim, P., "Fluorescence and Phosphorescence." Interscience, New York, 1949.
7. Kasha, M., and Nauman, R. V., *J. Chem. Phys.* **17**, 516 (1949).
8. Mehler, A. H., Bloom, B., Ahrendt, M. E., and Stetten, DeW., Jr., *Science* **126**, 1285 (1957).
9. Parker, C. A., and Rees, W. T., *Analyst* **85**, 587 (1960).
10. Bowen, E. J., and Wokes, F., "Fluorescence of Solutions." Longmans, Green, London, 1953.
11. "Instruction and Service Manual No. 768A," p. 4. American Instrument Co., Silver Spring, Maryland, 1959.
12. Weber, G., and Teale, F. W. J., *Trans. Faraday Soc.* **53**, 646 (1957).
13. Melhuish, W. H., *New Zealand J. Sci. Technol.* **B37**, 142 (1955).
14. Shore, V. G., and Pardee, A. B., *Arch. Biochem. Biophys.* **60**, 100 (1956).
15. Weber, G., and Teale, F. W. J., *Biochem. J.* **65**, 476 (1957).
16. Wavilov, S. I., *Z. Physik* **42**, 311 (1927).
17. Neporent, B. S., *Zhur. Fiz. Khim.* **21**, 1111 (1947).
18a. Teale, F. W. J., and Weber, G., *Biochem. J.* **65**, 476 (1957).
18b. Jablonski, A., *Acta. Phys. Polon.* **13**, 240 (1954).
18c. Weber, G., and Teale, F. W. J., *Trans. Faraday Soc.* **54**, 640 (1958).
19. Lavorel, J., *J. Phys. Chem.* **61**, 1600 (1957).
20. Förster, T., "Fluoreszenz Organischer Verbindungen." Vandenhoeck and Ruprecht, Göttingen, 1951.
21. Williams, R. T., *J. Roy. Inst. Chem.* **83**, 611 (1959).
22. White, A., *Biochem. J.* **71**, 217 (1959).
23. Udenfriend, S., Bogdanski, D. F., and Weissbach, H., *Science* **122**, 972 (1955).
24. Waalkes, T. P., and Udenfriend, S., *J. Lab. Clin. Med.* **50**, 733 (1957).
25. Weber, G., *Advances in Protein Chem.* **8**, 415 (1953).

3

INSTRUMENTATION

I. INTRODUCTORY REMARKS

THE fluorescent phenomena which have been known for the longest time
are those which are excited by sunlight and occur in the visible region, no
instrumentation being required. Natural pigments, bacterial cultures
(decaying animal and vegetable matter), and minerals were known to
emit characteristic radiation, when irradiated by visible light, long before
the advent of modern science. Exactly when fluorescence was first put to
practical use is hard to say. However, it is certainly true that geologists
have been using fluorescence to characterize minerals for a long time.

II. FLUORESCENCE DETECTION WITH PORTABLE
ULTRAVIOLET LAMPS

Modern applications of fluorescence detection began with the development
of sources of ultraviolet light. These have included various arcs and other
devices leading up to modern high-pressure gaseous discharge lamps.
Ultraviolet lamps, emitting "black light," have been used widely to detect
and measure the visible fluorescence of solids, solutions, paper chromatograms, and under the microscope. Several types of portable ultraviolet
lamps are now available from commercial sources. These are, in general,
mercury vapor lamps mounted in such a manner as to provide a handle
and appropriate shielding. Some of these are battery operated and can be
used in the field. In the United States, Ultra-Violet Products Inc., and
Hanovia Lamp Division produce a number of such mercury vapor lamps.
Lamps have been designed for far short-wave ultraviolet (253 mμ) and
long-wave ultraviolet (366 mμ) by varying the internal pressure of the
lamps and by using filters to limit the emitted light. Portable ultraviolet
lamps have received wide use in the field of geology and many identifica-
tions of ores and minerals are now based on their fluorescence characteristics. At the present time, portable lamps are actually used on field trips
and even provide final identification of radioactive ores along with Geiger
counters. The application of ultraviolet light along with fluorescent dyes
has, of course, received wide use in criminology and in the field of enter-

tainment. In biology, portable ultraviolet lamps are now standard equipment for laboratories engaged in analysis by paper chromatography. They are also used as light sources for fluorescence microscopy and for inspection of food.

III. FLUORESCENCE MEASURING DEVICES, GENERAL

The use of portable ultraviolet lamps, as described in the preceding section, is excellent for detection purposes since it provides a means of identification of a substance by the characteristic color of the emitted fluorescence. There have been many important biologic applications with such simple devices many of which will be discussed in the appropriate sections below. However, the more exact applications of fluorescence are as procedures for quantitative assay in solution at the microgram and submicrogram level. To make use of fluorescence in this way it is necessary to use some type of measuring device. Many such devices have been reported and a large number of them are commercially available. These include relatively simple visible instruments, more complex and precise photoelectric fluorometers, and the more sophisticated fluorescence spectrometers. However, all instruments designed to utilize fluorescence as a tool for quantitative assay are based on the same general principles and present similar problems. Before turning to a discussion of specific instruments, let us consider some of the general problems involved in measuring the fluorescence emitted from a solution by means of a typical fluorescence measuring device. The diagram shown in Fig. 1 indicates the various components of

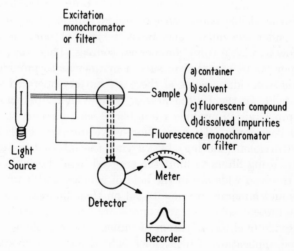

FIG. 1. Schematic diagram of a typical fluorescence measuring device.

such an assay (instrumental and otherwise) which, in other sections, are considered from both theoretical as well as practical standpoints.

The nature of the light source and its stability characteristics are important points to consider in an instrument. The methods used to isolate specific regions of the spectrum for excitation of fluorescence include filters and various types of monochromators. The cells used as containers for the solutions to be assayed are important considerations, too, since they can either limit the exciting or emitted light. They can also scatter light or emit fluorescence of their own. The solvent employed for most biological studies is water. However, many assays are carried out in other solvents. Some solvents fluoresce, some absorb light, and others may quench the fluorescence. Even water may have to be specially purified for certain types of assays. Of course, each molecular species has its own requirements as to solvent, pH, stability, excitation wavelength, and fluorescence wavelength, all of which must be ascertained experimentally. Relationships between structure and fluorescence have already been discussed. In biological studies the fluorescent substance must first be isolated from some animal or vegetable source before fluorescence assay can be carried out. A useful isolation procedure must remove all substances having fluorescences of their own, similar to the compound of interest, and still permit quantitative recovery from the biological material. If the compound does not fluoresce or if its native fluorescence is not sufficiently intense or specific, then it is necessary to convert it to a suitable fluorophore by some chemical reactions. The combination of extraction procedures with chemical reactions can lead to fairly specific analytical methods.

The fluorescent light emitted by the sample comes to the detector along with scattered light and light emitted by the fluorescence of the container, the solvent, and dissolved impurities. It is necessary to provide some means for removing as much of this interference as possible. This is carried out with suitable filter arrangements or by monochromators or by a combination of both. Finally, a suitable detecting device must be used which, in turn, activates an appropriate meter or recording device. Of course, the eye and the camera have long been used for this purpose. However, modern instruments make use of photocells and photomultiplier tubes. The latter are, of course, the most sensitive devices available for measuring light emission. Here again the nature of the emitted light, whether in the ultraviolet, visible, or infrared, determines the most desirable detector.

It is obvious that there are many important considerations which must be taken into account in order to carry out a successful fluorescence assay. The reason for discussing each of these points in some detail is not to make experts out of those who wish to use this procedure. The elementary discussions of optics and electronics are merely presented to acquaint the reader with the principles of the various commercially available instruments so that he can decide for himself which is most desirable for his particular problem and available facilities.

A. Visual Comparators

The earliest instruments for quantifying fluorescence made use of a source of ultraviolet light for excitation and used the eye for detection. The addition of filters to remove the exciting light and to cut down scattered light results in an instrument which though simple and inexpensive can be quite useful in many biological studies. Visual fluorometers can also be used for fairly precise studies. Many such instruments have been reported in the literature and patents have been granted on a number of them. Today they are seldom used in the research laboratory but still find some use in field studies and in some diagnostic laboratories. Detection of fluorescence in the urine with a visual comparator has long been used as a simple procedure for checking on antimalarial therapy in troops at out of the way stations.

Fluorescence assay is the bases for a simple test for gastric hydrochloric acid, the principle being displacement of quinine from a quinine resin by the acid in the stomach. The displaced quinine appears in the urine and can be assayed visually with a fluorescence comparator. Such instruments are so inexpensive and simple that they should be within reach of any diagnostic laboratory. One such instrument which was once commercially available in the United States is shown in Fig. 2. It is a viewing box con-

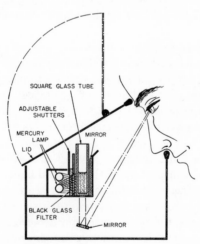

Fig. 2. Comparator for the measurement of fluorescence by visual means (Photovolt Corporation).

taining a mercury lamp, a filter, cells, and a mirror. Unknown samples are placed alongside standards and comparison is made by eye. Although there have been a number of commercial comparator instruments, most of these are no longer available or cannot be considered as standard instruments.

Although visual comparator methods are simple and inexpensive, they do not offer the precision and range of applicability which can be obtained with more sophisticated electronic measuring devices. This was realized very early and resulted in the development of many electronic fluorescence meters, which we may generally classify as filter fluorometers. Reports on photoelectric instruments first appeared in the literature in the late 1930's. Since then there have been hundreds of publications concerning designs of fluorometers. A summary of commercially available fluorometers through 1942 was presented by Loofbourow and Harris.[1] Radley and Grant,[2] Pringsheim,[3] and Bowen and Wokes[4] have presented some information about instruments available through 1950. Photofluorometers which are presently available from commercial sources will be considered in the following section.

The commercial instruments have a variety of designs and electronic circuits, but with few exceptions they have the same general plan (see Fig. 1). Excitation of fluorescences is most usually produced by a high pressure mercury arc and is therefore limited to the 313-, 365-, and 405-mμ mercury lines. A variety of arrangements are used for condensing and focusing the light, including glass and quartz lenses and mirrors. Colored-glass and interference filters are used to limit the light from the source and to remove those portions which overlap the emitted fluorescence (primary filter). Solutions are usually assayed in glass cells. Some instruments have quartz optics and require quartz cells when measurements are made below 320 mμ. Filters are also placed between the sample and the detector to absorb stray exciting radiation (secondary filters). The latter are frequently a composite of two or more filters. Photoelectric detectors of one sort or another are used in all the instruments.

Because of the similarities in instrumental design, it seems most reasonable to discuss the component parts before discussing the individual instruments. There are many excellent references on the subject. However, one which was written particularly for the biologist is a review entitled, "Generation, Control, and Measurement of Visible and Near-Visible Radiant Energy" by Withrow and Withrow.[5]

B. Excitation Sources

A variety of light sources have been used for fluorescence assay. Most photofluorometers make use of mercury-arc discharge lamps because of

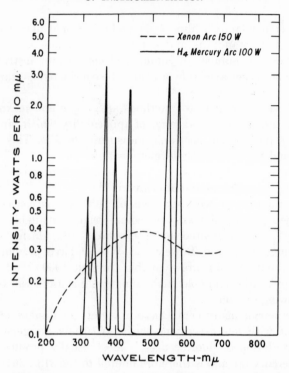

FIG. 3. Spectral characteristics of the 150-watt xenon arc and of the Pyrex-jacketed 100-watt H–4 mercury lamp.

their high light intensity. Hydrogen discharge lamps have also been used. More recently, the xenon arc has been introduced and has proven to be especially useful in fluorescence spectrometry. Characteristics of the 100-watt H-4 mercury lamp and the 150-watt xenon arc, two of the light sources which are more commonly used in fluorescence assay today, are shown in Fig. 3. It is apparent that there is great variation in the spectral distribution of radiation from the different sources. Thus, the ultraviolet emission of the Pyrex jacketed H-4 lamp is mainly at 365 mμ with a smaller radiation peak at 313 mμ. The H-3 lamp (85-watt) has similar characteristics. Removal of the Pyrex jacket makes available mainly the 254-mμ mercury line. High-pressure mercury lamps, such as the H-6 (1000-watt) are several times richer in energy throughout the entire spectral range, extending from the near ultraviolet through the near infrared. However, the lamp is not too well suited for measurement of spectra since the energy is not evenly distributed but is mainly concentrated at the several characteristic bands of the mercury arc. Because of the great heat of the 1000-watt lamp, an elaborate water cooling system is required. Hydrogen discharge lamps offer the advantage of a continuum through the ultraviolet and visible regions. However, the intensity of the smaller lamps which are used in spectrophotometry is quite low, particularly above 300 mμ, and the radiation is diffuse. The 1000-watt hydrogen lamp has greater intensity and could be useful for fluorescence studies; however, it also requires water cooling. The xenon high-pressure arcs have contributed considerably to the field of fluorescence assay because the high-intensity continuum which they produce makes it possible to devise relatively simple fluorescence spectrometer instruments. Of the available xenon arc lamps, the Osram XB165 (D.C. operated) has been found to be the most stable and to possess the longest life (Fig. 4). The Hanovia 150-

Fig. 4. The Osram XB165, D.C. operated, xenon arc lamp. (70 per cent of actual size shown.)

watt xenon arc (A.C. operated) has similar characteristics, but its instability limits its use for precise fluorescence assay. Xenon-mercury arcs may also be useful to provide radiation throughout the ultraviolet and visible range and, in addition, to yield extra high intensity at the mercury lines. The additional line intensities may prove useful for certain assays where extremely high sensitivity is required. However, they yield distorted spectra. The problem of lamps will be discussed again in the section on fluorescence spectrometers.

Since the intensity of emitted fluorescence is dependent upon the intensity of the exciting source, it would appear that the sensitivity of fluorescence assay can be increased without limit by using the most intense sources. However, increased light output also increases scattered light and causes heating. All the light sources require additional transformers or ballast circuits to maintain appropriate voltage. The more powerful the light source the larger, more complicated, and more expensive the ballast circuit. For these reasons the more intense sources are not as useful as they might otherwise be. Furthermore, the sensitivity of modern photomultiplier detectors removes much of the need for intense light sources for fluorescence assay.

C. Sample Cuvettes

With filter fluorometers, standard test tubes can be used once they have been matched by selecting those which give the same response with a standard solution of some fluorescent material. Dilute solutions of quinine in 0.1 N H_2SO_4 are frequently used for this purpose. Following a great deal of usage, the tubes may become scratched and scatter light excessively. For excitation below 320 mμ, quartz cells are required. However, different varieties and grades of quartz exhibit different amounts of fluorescence in the ultraviolet. According to Price et al.[6] Corning quartz possesses a lower native fluorescence than do many other varieties. However, even this quartz exhibits sufficient fluorescence to be detected on the Aminco Bowman spectrophotofluorometer, at highest sensitivity, showing excitation at 265 and 330 mu and fluorescence at 500 mμ (uncorrected). Quite obviously the more elaborate the measurements the more care must be given to selection of sample cuvettes.

D. Optical Filters and Monochromators

The most precise methods for obtaining spectral resolution make use of monochromators or spectral dispersing systems. These are used in spectrophotometers and in the more elegant and precise fluorescence spectrometers (described below). In the more simple colorimeters (photometers) and the fluorometers, monochromators are replaced by optical filters for isolating desired spectral regions. Filters are also used to remove stray radiation and unwanted spectral orders. In the usual filter fluorometer, filters are used for both purposes. A *primary filter* is used to isolate a specific spectral region of the light source. The most desirable combination of filter and light source is one which selectively passes a high intensity of light corresponding to the absorption maximum of the compound to be assayed; as much of the light above and below the absorption maximum should be removed. As shown in Fig. 1, the detector is most frequently placed at right angles to the exciting light. Theoretically, with an ideal primary filter the detector should "see" only the emitted fluorescence. However, aside from the leakage of the primary filter, the exciting light also produces Tyndall scattering, Rayleigh scattering, Raman scattering, and nonspecific emission from the solvent, the container, and from lenses and other optical accessories. These may be much more intense than the emitted fluorescence and must be removed. For this purpose a *secondary filter* is employed.

Ideally, the secondary filter "cuts off" all the exciting energy and passes all the emitted fluorescence. When the absorption and fluorescence maxima are widely separated this is readily accomplished. A comparison of the

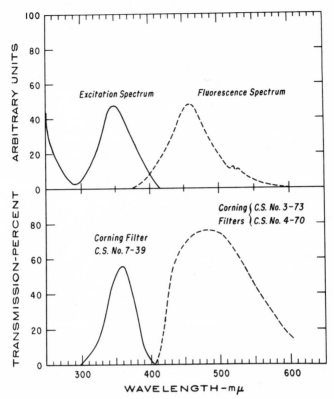

FIG. 5. Excitation and fluorescence spectra of quinine and the spectral characteristics of the filter combination which has been generally used for quinine assay.

filters generally used for quinine assay along with the spectral properties of quinine is shown in Fig. 5. A similar comparison for riboflavin is shown in Fig. 6. It is apparent that the riboflavin primary filter does not correspond too well to the excitation maxima of the compound. It should also be evident that the filter fluorometers do not lend themselves as readily to the assay of compounds with excitation and fluorescence maxima which are fairly close to each other. Secondary filters are frequently composed of a combination of optical filters. One reason for this is that the "cutoff" filter may itself fluoresce when excited by stray light from the exciting source. If this fluorescence differs from that of the sample a second filter can be used to eliminate it.

Fluorometers make use of two types of filters. Of the two, colored glasses with selective absorption characteristics, have been most widely used. Tinted glass filters of the Corning Glass Company have been most

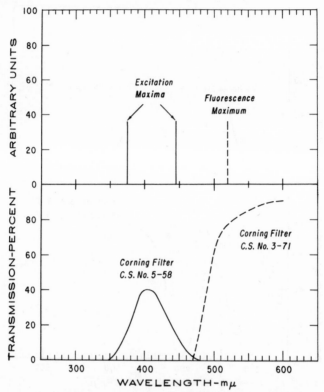

FIG. 6. Excitation and fluorescence maxima of riboflavin and the spectral character-istics of a filter combination which has been widely used for riboflavin assay.

generally used in filter instruments in the United States. Transmission characteristics of the Corning optical filters which have been most widely used in fluorescence assay are shown in Fig. 7. Wratten filters supplied by the Eastman Kodak Company, have also been widely used. These are composed of gelatin dyed with organic dyes which are sandwiched between two sheets of glass. Since Wratten filters deteriorate with time, particularly when exposed to intense light, their use should be limited to that of sec-ondary filters. The fluorometers abroad have made use of glass filters from the Jena Glass Works (Germany) and the Chance-Pilkington Optical Works (England).

Interference filters are now being used in fluorometers to a greater extent. These are made by depositing two or more metal films on a glass plate, each film being separated from the next by a spacer of nonabsorbing material. A second glass plate is then cemented on for protection. The dis-

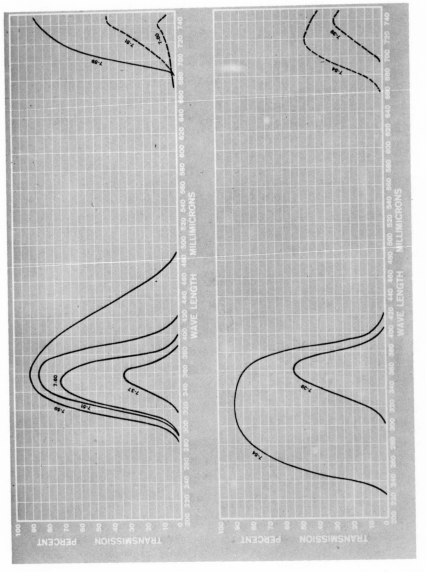

FIG. 7 (A)

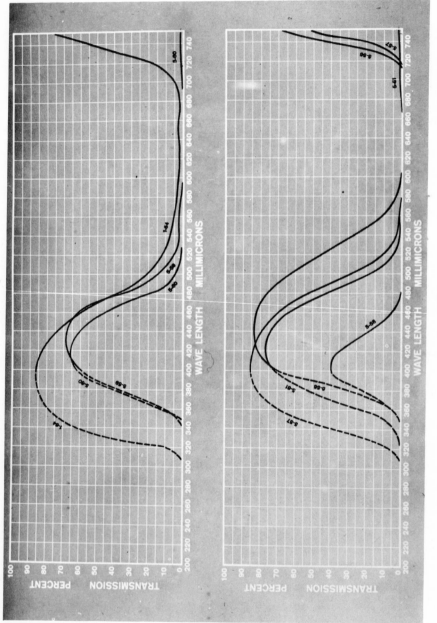

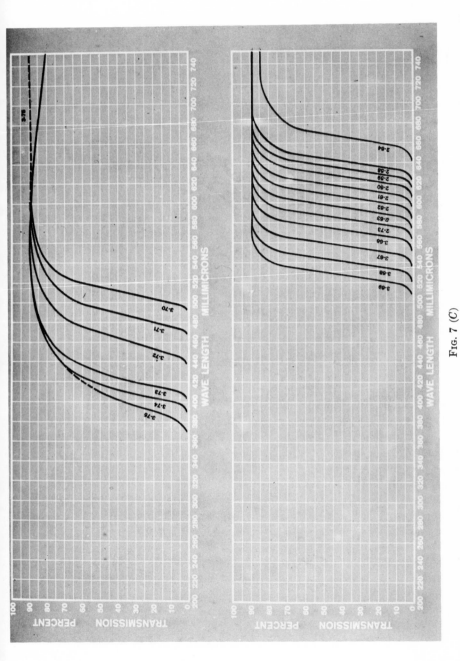

FIG. 7 (C)

FIG. 7. Transmission characteristics of Corning optical filters which have been widely used in fluorescence assay. (Corning Glass Works, "Bulletin CF–1.") A. Ultraviolet transmitting visible absorbing filters. B. Blue filters. C. Sharp-cut yellow and red filters.

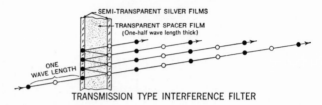

TRANSMISSION TYPE INTERFERENCE FILTER

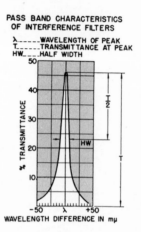

FIG. 8. Characteristics of transmission type interference filters. (Bausch & Lomb Inc., Bull. D–248, November, 1952.)

tance between the metal films determines the light which the filter will pass. Light of wavelength or half-wavelength equal to the distance between the filters will be passed, other light will be reflected. This is shown in Fig. 8. It is apparent that the angle at which light hits the filter determines the distance between layers and therefore the spectral characteristics. The spectral properties of typical interference filters are shown in Fig. 9. Narrow-band interference filters from 340 to 800 mμ are available from a variety of sources including Baush & Lomb Inc., Farrand Optical Company, Photovolt Corporation, Jena Glass Works, Schott and Gen. Gesachton Anstalt, Balzers, Barr and Stroud, and Axler Associates. Many firms will prepare filters to the specifications of an investigator. Interference filters have many advantages to offer. In addition to providing narrow bands with relatively high transmission, they can be used with high-intensity light sources with little fear of deterioration. The stability to heat is due to their removal of undesired radiation by transmission and reflection and not by absorption.

The most precise instruments for isolating a narrow spectral band are monochromators. These are spectroscopic devices utilizing slits, lenses, or

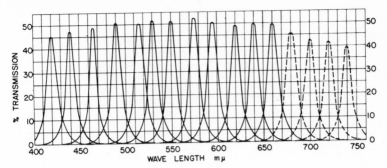

Fig. 9. Transmission characteristics of interference filters available from Photovolt Corporation.

mirrors and either prisms or gratings for dispersing the radiation. It is apparent that monochromators are fairly complex and therefore too bulky and too expensive to be used in place of filters except in much more sophisticated instruments such as fluorescence spectrometers. Different types of monochromator design have been utilized by various instrument manufacturers and will be presented in subsequent sections. For simplified and critical discussions of the various types of optics used in monochromators, the reader is referred to Harrison et al.[7] and Withrow and Withrow.[5] Here will be found comparisons of the relative merits of lenses versus mirrors for focusing light and of prisms versus gratings for dispersing the radiation.

E. Photodetectors

Most of the fluorometers developed over 5 years ago used barrier layer photocells for detection in the visible region of the spectrum. Such photocells depend almost entirely on outside electrical circuitry for amplification in order to convert the relatively weak fluorescent light to a measureable signal. However, recent advances in electronics have made it possible to use photomultiplier tubes as the detecting elements in fluorometers. The latter are much more sensitive. The photomultiplier is a vacuum tube in which many photosensitive electrodes are arranged in series, in such a manner that, when radiation impinges on the first electrode, the resulting electrons produced by the photoelectric effect are made to impinge on the second electrode; electrons produced here then impinge on the third electrode, and so on through the entire series. As electrons pass from electrode to electrode the number is multiplied so that large electrical pulses are obtained at the anode in response to relatively weak radiation striking the photosensitive cathode. The photomultiplier is thus an amplifier as well as a photodetector. The spectral characteristics of some de-

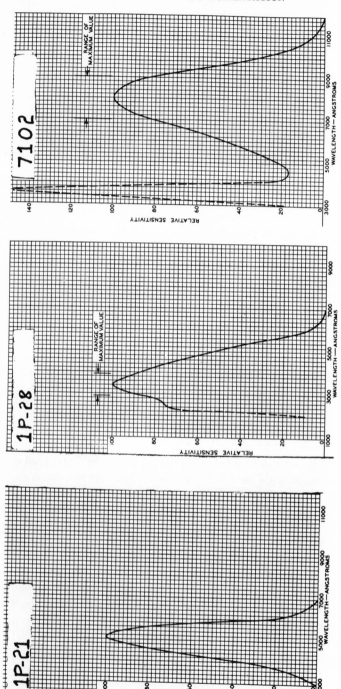

Fig. 10. Relative spectral characteristics of some RCA photodetector tubes.[8] The curves shown for the 1P-28 and 7102 tubes are for equal values of radiant flux at all wavelengths. The curve shown for the 1P-21 tube is the one obtained by irradiation with a tungsten source and does not truly represent the spectral response of this tube. The latter can be found in the "RCA Tube Handbook."[8]

tector tubes commonly used in present day filter fluorometers are shown in Fig. 10. An excellent discussion of the fundamental principles and practical applications of photoelectric methods of detection was presented by Harrison et al.[7] and by Withrow and Withrow.[5] A more complete discussion of photomultiplier tubes may be found in the "RCA Tube Handbook."[8]

IV. FILTER FLUOROMETERS

Filter photofluorometers differ from each other in the various components which are utilized and in the manner in which these are assembled. The following instruments are presently available mainly in the United States.

A. Coleman Model 12C Photofluorometer

The Coleman Model 12C photofluorometer (Fig. 11) has been used in a large number of laboratories for almost 20 years. In its present form it comprises a single unit with an H-3 mercury vapor lamp as a light source.

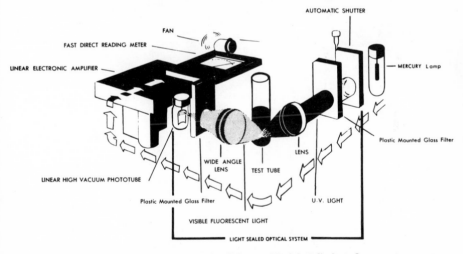

FIG. 11. Simplified diagram of the Coleman Model 12C photofluorometer.

Colored-glass primary and secondary filters are provided in spectral adaptors and glass lenses are used for focusing the exciting light and the emitted fluorescence. Matched 19-mm. test tubes may be used as cuvettes. Eight milliliters of sample is required in the standard cuvettes, but microadapters have been designed for smaller volumes. A blue-sensitive phototube is used as the detector with a linear amplifier and a fast direct reading meter. The instrument is applicable to many standard analyses of vitamins and drugs. The use of a simple phototube limits its sensitivity so that it is not applicable to the measurement of normal amounts of norepinephrine in the blood. As with most filter fluorometers the glass optics limit application to those compounds whose absorption coincides with one of the mercury lines above 300 mμ.

B. Photovolt Fluorescence Meter Model 540

The Photovolt fluorescence meter Model 540 (Figs. 12 and 13) is a multicomponent instrument. It contains only 1 lens made of quartz for

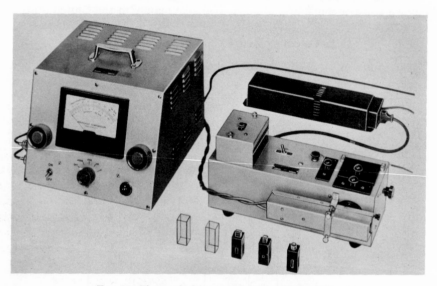

FIG. 12. Photovolt fluorescence meter Model 540.

focusing the exciting light. A variety of light sources can therefore be used, including quartz-jacketed H-3 mercury vapor lamps for the entire mercury range, low-pressure mercury lamps (253 mμ), and phosphor coated mercury lamps (320–390 mμ). The filter holders and sample compartments are made to permit all types of adaptation. The use of a photomultiplier tube (1P21) as a detector also permits high sensitivity in the ultraviolet and visible. Detectors for fluorescence in the infrared are also offered as attachments. This instrument was designed for fairly broad applications of fluorescence assay. It has been utilized for measurement of the various vitamins and to assay norepinephrine in tissues.

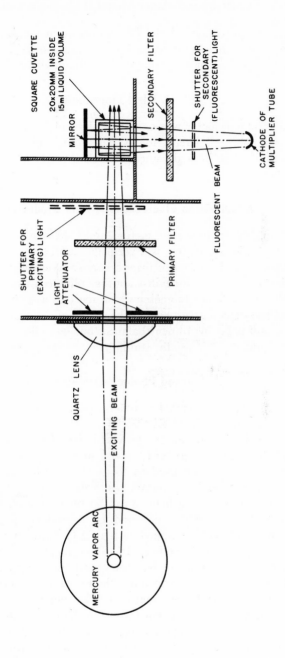

Fig. 13. Diagrammatic sketch of the Photovolt fluorescence meter Model 540.

C. Pfaltz and Bauer Model B Fluorophotometer

The Pfaltz and Bauer Model B Fluorophotometer was one of the first commercial instruments introduced for vitamin assays. It is a filter instrument utilizing a barrier layer photocell as the detector.

D. Klett Fluorimeter

The Klett Fluorimeter was one of the first filter instruments to include a balanced photocell circuit for stability. An additional feature of the instrument is that it is designed to measure samples from very small volumes up to 25 ml.

E. Lumetron Fluorescence Meter Model 402-EF

The Lumetron fluorescence meter Model 402-EF is also produced by Photovolt Company. It, too, is a versatile instrument, but is not as sensitive as the former because it utilizes a barrier layer cell. However, the output of the cell is balanced against another photocell using a bridge circuit with slide wire and galvanometer which receives a small amount of radiation from the light source (Fig. 14). The use of a split beam arrangement in such a null point instrument balances out lamp fluctuations and makes for an instrument with greater stability.

F. Farrand Photoelectric Fluorometer

The Farrand photoelectric fluorometer (Fig. 15) is another versatile and sensitive instrument. It makes use of an H-3 mercury lamp as a light source. A variety of filters are provided with the instrument, including Corning colored-glass (standard 2-inch squares) and the interference type filters. Quartz lenses are used to focus the exciting radiation onto the sample cuvette (standard 10-mm. test tubes) and to focus the emitted fluorescent beam on the cathode of a 1P21 photomultiplier tube. In earlier instruments the latter was powered by a battery-pack power supply. More recent models are equipped with an A.C. power supply. The use of a photomultiplier detector endows the instrument with great sensitivity and the Farrand fluorometer has been used for many analyses requiring the highest sensitivity. This includes the measurement of epinephrine and norepinephrine in biological material. The quartz optics also make it possible to adapt the instrument for use in the ultraviolet region. The H-3 lamp with its envelope removed emits light below 300 mμ. Interference filters are now available to isolate bands in the far-ultraviolet region. Quartz test tubes and adapters for optically flat cuvettes are also available.

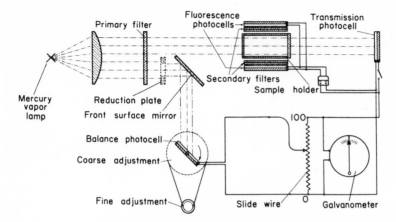

FIG. 14. Diagram of the Lumetron fluorescence meter Model 402–EF.

G. Turner Model 110 Fluorometer

The Turner Model 110 fluorometer is an instrument which has been introduced most recently (Figs. 16 and 17). All the components are housed in a single unit. The light source which is routinely supplied is a low-pressure 4-watt mercury lamp, exciting mainly at 360 mμ. Corresponding 2 $\times$ 2-inch colored-glass primary and secondary filters are supplied. Lamps and filters for excitation at 254 mμ are also available. The use of low-intensity light sources may be useful when measurements are made of photosensitive

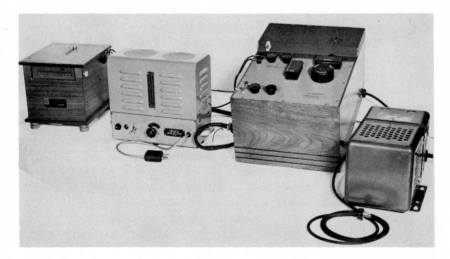

FIG. 15. Farrand photoelectric fluorometer.

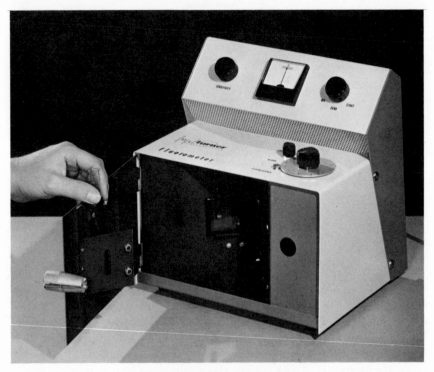

FIG. 16. Turner Model 110 fluorometer.

substances. The instrument uses no lenses, the fluorescent light passing through a slit to the photomultiplier tube. Instead of measuring the fluorescent light directly, it is balanced against a small amount of the light from the source, the balancing being carried out by a rapidly revolving light interrupter which alternately presents to the photomultiplier tube the fluorescent light and the source light. The two are balanced by appropriate circuitry and thus provide a stable null-point instrument. The absence of focusing lenses has the advantage of permitting fluorescent applications which require excitation below 300 mμ. The photomultiplier detector endows the instrument with sufficient sensitivity to permit its application to most of the presently known assays. The combination of sensitivity with the stability of a null-point instrument make the Turner instrument adaptable as a monitor of continuous flow devices such as chromatographic columns. For such applications, the sample holder, which is in the door of the instrument, can be replaced by another door with a

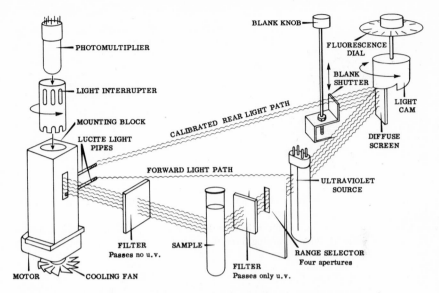

Fig. 17. Diagram of the Turner Model 110 fluorometer optical system.

continuous flow cuvette. The Turner Model 111 fluorometer is a self-balancing modification designed for use with recording equipment.

H. Jarrell Ash G-M Fluorometer

The Jarrell Ash G-M fluorometer is a relatively simple instrument which was designed originally for measurement of fluorescence of various metal ores. Recently it has been adapted to permit the measurement of fluorescence in solution.

I. Hilger Spekker Photoelectric Fluorometer

The Hilger Spekker photoelectric fluorometer has been widely used in Great Britain and in Europe. It utilizes a standard mercury lamp and colored-glass filters. A photomultiplier tube is used as fluorescence detector and is connected with a galvonometer in a null indicating circuit with a compensating photocell which is arranged to "see" a small amount of the light from the mercury lamp. The optical and electrical systems are shown diagrammatically in Fig. 18.

J. Hitochi Fluorophotometer (Type FPL-2)

The Hitochi fluorophotometer (Type FPL-2), produced in Japan,

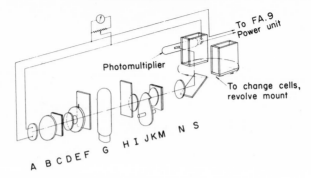

Fig. 18. Simplified diagram of the optical system and electrical circuit of the Hilger Spekker photoelectric fluorometer.

A, photocell; B, window; C, filter; D, lens; E, iris diaphragm; F, heat absorbing filter; G, mercury lamp; H, heat absorbing filter; I, window; J, variable shutter; K, lens; M, filter (U.V.); N, lens; S, reflector.

utilizes a high-pressure mercury lamp for excitation. Interference filters are used for isolating the fluorescent light which is detected by a photomultiplier tube. The detection limit for vitamin B_2 is given as 0.001 p.p.m.

K. Fluorescence Accessory for the Beckman Model DU Spectrophotometer

Recently, Beckman Instruments Inc. made available a fluorescence accessory for converting the DU instrument for use as a photofluorometer in the visible region of the spectrum.[9a] The accessory provides mainly the exciting lamp and utilizes the sample compartment and phototube of the instrument (Fig. 19). Light from a tungsten lamp is directed through a diaphragm mask to a quartz lens and isolating filter. The filters used determine the radiation transmitted to the diagonal mirror. The mirror deflects the beam upward through the bottom of the cuvette (corex cells

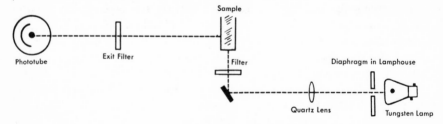

Fig. 19. Optical diagram of the fluorescence accessory for the Beckman Model DU spectrophotometer.

with bases and two sides optically polished). The emitted fluorescent light passes through an aperture plate in the sample compartment then through another filter which absorbs scattered light. The standard phototube detector of the DU instrument can be used. However, for maximal sensitivity the photomultiplier assembly is recommended.

L. Fluorometer for Ultraviolet Fluorescence

Konev and Kozunin[9b] have developed a relatively simple filter fluorometer for measurement of the fluorescence which has its excitation and fluorescence in the ultraviolet region and have applied it to the measurement of proteins in milk (Chapter 6). A quartz-jacketed mercury lamp is used for excitation and the spectral region 240–290 mμ is isolated by means of a quartz cell filled with chlorine gas along with an appropriate optical filter. A piece of uranyl glass is placed in front of the photocell so that the ultraviolet light (340–350 mμ) emitted by the protein solution excites it and causes it to emit visible fluorescence. The uranyl fluorescence is detected with a sensitive photocell which has its maximum spectral sensitivity at 510 mμ. The use of a fluorescent glass to convert ultraviolet to visible fluorescence permits the construction of relatively simple and sensitive fluorometers for this region of the spectrum.

It is apparent that although the various instruments contain many of the same components they do differ in a number of characteristics; the greatest differences are in sensitivity and applicability to different spectral regions. With the photomultiplier instruments sensitivity can be extremely great, a limiting factor being the amounts of stray light which reach the detectors. Instrumental design is therefore of great importance in achieving maximal sensitivity. In determining the sensitivity of a fluorometer for a particular assay, one should not merely set the reagent blank to zero and measure the meter deflections of a given concentration of the fluorophore. The absolute meter deflection produced by a reagent blank should be measured first. Then, without zeroing the reagent blank, measurement of the meter deflection of the sample, in the same reagent, should be made. Thus, two instruments, A and B, may each claim that they can detect 0.02 μg. of quinine, but if one examines the data in Table I, it is evident that instrument A is at the limit of detectability giving only a small deflection above the blank. On the other hand, instrument B shows severalfold deflection over the blank. Such comparisons should be made in selecting an instrument for a given assay.

The commercially available filter fluorometers should be able to answer many of the routine requirements for fluorescence assay in the laboratory.

TABLE I

COMPARISON OF INSTRUMENTAL SENSITIVITIES

	Instrument A		Instrument B	
	With blank set to zero	Direct meter reading	With blank set to zero	Direct meter reading
Blank_____	0	170	0	5
Sample_____	20	190	40	45

Of course, individual laboratories will continue to devise their own instruments for specific needs and, as in the past, reports of "home-made" fluorometers will continue to appear in the journals.

V. FLUORESCENCE SPECTROMETRY

Although the fluorometers are an obvious advance over visual comparator instruments, they are merely more exact devices for measuring fluorescence at fixed excitation and emission wavelengths. We have seen that emission of fluorescence is as characteristic of molecules as is light absorption. Molecules which absorb light have a characteristic absorption spectrum. Correspondingly, they must have a characteristic excitation spectrum which should be identical with the absorption spectrum. The emitted fluorescence also exhibits a characteristic spectrum. Furthermore, molecules may be excited to fluorescence throughout the range of radiation corresponding to electronic excitation (200–800 mμ). The process is not limited to a few of the mercury lines. In addition, emission of fluorescence can also occur throughout this range, including the ultraviolet and infrared. Some of the filter fluorometers have utilized quartz optics, quartz cuvettes, and appropriate detectors to permit measurements outside the visible region. These are certainly excellent extensions of the instruments. However, it is quite obvious that the same instrumental sophistication as in the field of light absorption, which led to the development of modern spectrophotometers, should be applicable to fluorescence.

Physicists have long been aware of this and measurements of fluorescence spectra have been carried out in many laboratories. At first these were made with photographic methods. A description of such a method for the measurement of fluorescence spectra by Jacobson and Simpson[10] is presented below. It is typical of the techniques used before the advent of the phototube and modern fluorescence spectrometers.

The fluorescence spectra were photographed by means of a Hilger wide-aperture spectrograph, with a quartz mercury lamp as light source. Visible light was removed with ultraviolet glass filters and with a 0.5-cm. quartz cell containing a 20% solution of $CuSO_4$. Solutions were contained in a 2-cm. quartz cell, the light being focused on to it by means of a quartz rod. The light incident on and emerging from the specimen was focused with quartz lenses. If the slit was sufficiently narrow to separate clearly the components of the mercury yellow doublet, an exposure from 30 seconds to 10 minutes was generally adequate for the solutions but was increased to 30 minutes or longer in some cases. Solids and tissue sections needed at least 6 to 9 hours for good photographs to be obtained. Photomicrometer tracing of the plates was then carried out.

It is apparent that the lengthy exposures required for the photographic method would not lend themselves well to many labile compounds. However, in spite of the limitations and tedious nature of the method, Jacobson and Simpson[10] were able to carry out excellent studies on the pterins in tissues and extracts. These will be discussed in the section on vitamins.

More recently, individually designed photoelectric spectrometers have been used and many such instruments may be found in the literature. The development of photoelectric instruments for measuring fluorescence spectra is relatively recent, less than 10 years old. This was done by adapting commercial spectrophotometers, usually by arranging to pass the emitted fluorescent light through the spectrophotometer for analysis, excitation usually being carried out by a mercury arc lamp with appropriate filters.

A. Attachment for Beckman Spectrophotometers

Fluorescence attachments were described some time ago for both the Beckman DU spectrophotometer and the recording Model DK instrument.[11] Both attachments make use of a standard mercury vapor lamp which uses the same power supply unit as is normally provided to operate the hydrogen lamp. However, other light sources can be used with the fluorescent attachment and an arrangement is provided for coupling with a second monochromator, as described by Gemmill,[12] to permit more exact control of the excitation energy. As shown in Fig. 20, the fluorescence attachment housing contains the light source, a filter, sample holder, and the optics needed to reflect the emitted fluorescence into the DU or DK monochromator. The latter converts this into a fluorescence spectrum. The photomultiplier attachment must be used to obtain the highest sensitivity. With appropriate light sources the modified Beckman instruments can be made to measure fluorescence spectra of compounds which are excited in the ultraviolet region.[12] The Beckman Model B has been

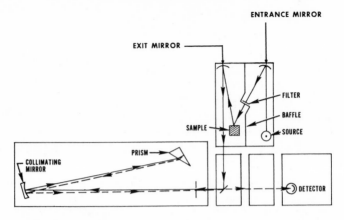

Fig. 20. Attachment for measuring fluorescence spectra with the Beckman Model DU and DK spectrophotometers.

modified in a similar manner to measure fluorescence spectra in the visible range.[13]

B. Accessory for Cary Spectrophotometers

A fluorescence accessory is also available for the Cary spectrophotometers Models 10 and 11 which permit them to be used for recording fluorescence spectra. This attachment (Fig. 21), which is interchangeable with the standard cell-compartment and lamp housing, also makes use of a mercury arc and filters. An attempt has been made to correct fluorescence for variations in the output of the exciting lamp to increase the stability of the instrument. This is accomplished by balancing the fluorescence against a portion of the output of the exciting lamp.

C. Hilger Uvispek Spectrophotometer—Fluorescence Attachment

The Hilger Uvispek spectrophotometer is also provided with a fluorescence attachment. In this case, a 12-volt 36-watt tungsten filament lamp is used as a light source and colored-glass filters are used for limiting the exciting and fluorescent light. This attachment does not make use of the monochromator but converts the spectrophotometer into a photofluorometer. In other words, it is not designed to measure fluorescence spectra. The use of a tungsten lamp for excitation and photocells for detection may limit the sensitivity of the instrument. It is applicable to assay of compounds which absorb in the region of the spectrum transmitted by glass.

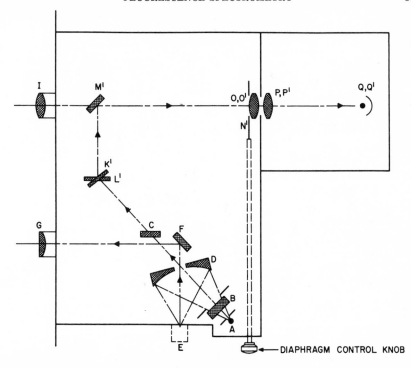

Fig. 21. Optical diagram of the fluorescence accessory for the Cary Models 10 and 11 spectrophotometers.

A, mercury source; B, filter; C-L'-K', monitoring beam; D, mirror; E, sample cuvette; F, mirror; G, focusing lens for monochromator entrance; I, focusing lens for monochromator exit; N, diaphragm; O, O' and P, P' are lenses; Q, Q', photodetector.

D. Fluorescence Attachment ZFM4 for the Zeiss Spectrophotometer PMQ11

The fluorescence attachment ZFM4, for the Zeiss spectrophotometer PMQ11 is shown in Fig. 22. As shown in the optical diagram, Fig. 23, the exciting light is provided by a mercury lamp; condensing lenses and glass filters select and focus the light on the sample container. Lenses and mirrors are then used to pass the emitted fluorescence through the monochromator for dispersion. Measurement of fluorescence spectra can be made by rotating the monochromator by hand. A provision is made for the measurement of surface fluorescence from concentrated solutions or from solids. The monochromator can also be used to disperse the exciting light. However, the instrument is not really suited to the measurement of excitation spectra. This is so because the only continuum light source provided is the standard hydrogen lamp, thereby limiting such measurements

Fig. 22. Fluorescence attachment ZFM4 for the Zeiss spectrophotometer Model PMQ 11.

to relatively concentrated solutions. The use of glass filters further limits the measurement of excitation spectra to compounds which fluoresce in the visible range.

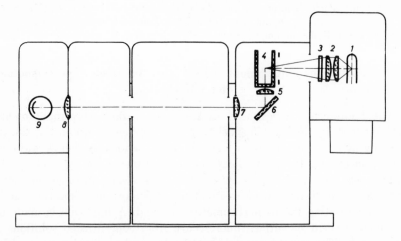

Fig. 23. Optical diagram of the Zeiss spectrophotometer with the fluorescence attachment.

1, mercury lamp; 2, lenses; 3, filter; 4, cuvette; 5, lens; 6, mirror; 7, lens leading into monochrometer; 8, lens at monochromator exit; 9, photodetector.

E. Fluorescence Spectrometer Designed by Bowman *et al.*

It is apparent that attachments for spectrophotometers can be utilized for measuring fluorescence spectra. However, most of the commercially available attachments generally permit such measurements to be made only when absorption maxima coincide with one of the mercury lines. Furthermore, they are not well suited to measuring excitation spectra. The need for more sophisticated instrumentation was long apparent to many investigators. What was needed was an instrument which could selectively irradiate with monochromatic light throughout the ultraviolet and visible regions (200–600 mμ), detect the emitted fluorescence throughout this same region, and, furthermore, measure both the excitation and fluorescence spectra. Such an instrument would be to fluorescence assay what the spectrophotometer is to absorptiometry. The term fluorescence spectrometer seems most appropriate. It was obvious that such an instrument would require two ultraviolet monochromators, a continuous high-energy light source, and sensitive detectors. Reports of fluorescence spectrometers of various sorts have appeared from time to time in the

Fig. 24. Experimental fluorescence spectrometer designed by Bowman.

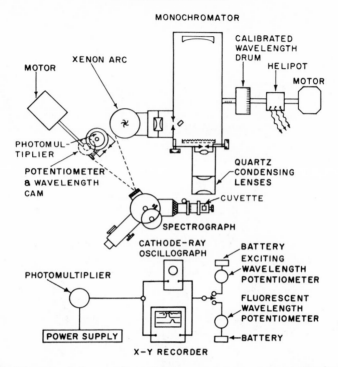

FIG. 25. Diagram of the experimental fluorescence spectrometer designed by Bowman (Bowman et al.[14]).

literature. However, the development of a recording fluorescence spectrometer covering the complete range from 230–600 mμ and sensitive to the submicrogram level was first reported by Bowman et al.[14] in 1955. This apparatus, shown in Fig. 24, was in effect a conglomeration of available optical equipment which Dr. Robert L. Bowman was able to put together into a functional laboratory instrument. As shown in the schematic diagram (Fig. 25), it consisted of a 125-watt xenon arc (Hanovia) to provide uniform light output from the ultraviolet through the visible with a Bausch & Lomb grating monochromator to select the excitation wavelength. The sample, about 1 ml., was placed in a 1 cm.[2] silica cuvette. The fluorescent light was analyzed by passing it through a modified quartz prism microspectrograph equipped with a mechanical scanning device containing a 1P28 photomultiplier tube. The photomultiplier output was coupled either to the vertical axis of a cathode-ray oscilloscope or to a pen and ink recorder.

In operation, the phototube scanning the emitted fluorescent light plotted a wavelength versus intensity diagram, *fluorescence spectrum*, on the oscilloscope or recorder; with the fluorescence analyzer set to the peak output, the incident light could then be varied through the visible and ultraviolet to yield an *excitation spectrum*

It was not intended to obtain "true" spectra with this experimental instrument. However, although the spectra were skewed due to various optical defects, the observed maxima were usually accurate to within ±5 mμ of reported values. The instrument was initially designed to determine whether a significant number of compounds emitted fluorescence in the ultraviolet region and whether it was feasible to develop simpler instrumentation with which to carry out routine measurements of ultraviolet fluorescence. It was, of course, found that a host of biologically important compounds fluoresced in the ultraviolet region.[15,16] However, by the time the instrument was completed it was fairly obvious that filter instruments that would cover both the ultraviolet and visible regions of the spectrum would require a multitude of attachments. The instrument really revealed itself as a powerful research tool in the studies on 5-hydroxyindole metabolism which were being carried out in the author's laboratory at that time. With it, methods were developed for the assay of 5-hydroxytryptamine in tissues at levels of 0.01–1 μg./g. of tissue.[17] Not only that, but identification of the amine could be made at the microgram level by comparing excitation and fluorescent spectra with those of the authentic compound. Many additional applications of spectrophotofluometry were made in the author's laboratory with the experimental instrument. Although these studies aroused wide interest, it was apparent that not many biologists would apply this technique if each one had to make his own instrument as did Goldzieher *et al.*,[18] Goldstein *et al.*,[19] Jokl,[20] and others. Fortunately, the interest in fluorescence spectrometry did not go unnoticed by the instrument manufacturers. Following along the suggestions of Dr. Robert L. Bowman, two commercial recording fluorescence spectrometers were developed and became generally available, the Aminco-Bowman spectrophotofluorometer and the Farrand spectrofluorometer. It should be pointed out that both were originally designed primarily for quantitative assay where one of the most important requirements was sensitivity. Simplicity, size, and cost were also factors in determining instrument design.

F. Aminco-Bowman Spectrophotofluorometer

The Aminco-Bowman spectrophotofluorometer is an adaptation of the laboratory model developed by Dr. Robert L. Bowman. The instrument is made up of several components all of which are shown in Fig. 26 except for the recording equipment. A schematic diagram of the optical part of the instrument is shown in Fig. 27. In operation, light from an Osram

FIG. 26. Aminco-Bowman spectrophotofluorometer.

XB165 xenon arc* is dispersed by the exciting monochromator into radiation which shines on the sample. Fluorescent light from the sample is dispersed by a similar monochromator into monochromatic radiation which shines on a photomultiplier detector. There the radiation is transformed to a weak electrical signal and fed to a photometer, where it is amplified. The photometer output is coupled to the vertical (Y) axis of an X-Y recorder (or to an oscillograph). For quantitative analysis the signal can be indicated directly on the meter. The two gratings are oscillated by motor-driven cams to which are coupled graduated disks for visual observation and adjustment of wavelength. Potentiometers, coupled to the gratings, supply the wavelength information in the form of a D.C. signal to the recorder X-axis. When the recorder X-input is connected to the scanning fluorescence monochromator (solid line circuit, Fig. 27) and the excitation monochromator is set at a wavelength for maximal excitation, a wavelength vs. intensity diagram is plotted. This is an instrumental, uncorrected, fluorescence spectrum. Similarly, when the recorder X-input is connected to the scanning excitation monochromator (dashed circuit, Fig. 27) and the fluorescence monochromator is set at a wavelength for maximal fluorescence, an instrumental, uncorrected, excitation spectrum is plotted. The coupling between the drive motors and wavelength disks are equipped with a clutch to permit manual variation of wavelength when the motors are stopped.

The amplified electron current from the photomultiplier tube is collected at its anode and is passed through a precision resistor connected to

---

* Earlier models were equipped with Hanovia xenon arcs and an A.C. ballast. The light output from these lamps was not sufficiently stable and they were replaced by the Osram XBo–150 W xenon arc, with a D.C. ballast.

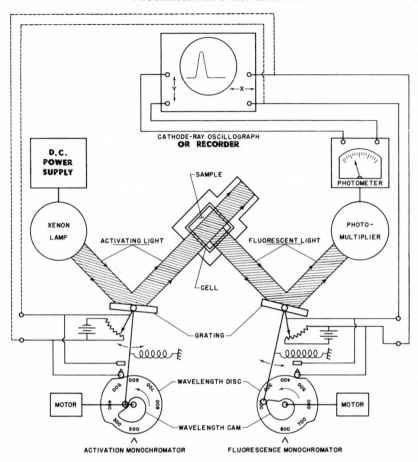

FIG. 27. Schematic diagram of the Aminco-Bowman spectrophotofluorometer.

a meter multiplier switch where it develops a voltage which is applied to the grid of an input-stage amplifier tube. This voltage is amplified by a two-stage resistance-coupled amplifier which drives a 0–1 milliammeter. A portion of the amplifier output voltage is applied to two pairs of output terminals for operating X-Y recorders or oscillographs. The sensitivity is controlled by a meter multiplier switch and a potentiometer. Gross sensitivity adjustments are made with the meter multiplier switch which reduces recorder output signals together with meter readings, in steps of 1, 1/3, 1/10, 1/30, 1/100, 1/300, and 1/1000. Fine sensitivity adjustments are made with a sensitivity potentiometer which continuously adjusts the recorder output signal and meter readings over a range of 3.5 to 1. A dark-

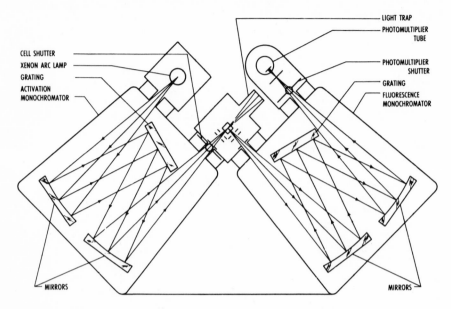

LIGHT TRAP
PHOTOMULTIPLIER TUBE
CELL SHUTTER
XENON ARC LAMP
GRATING
ACTIVATION MONOCHROMATOR
PHOTOMULTIPLIER SHUTTER
GRATING
FLUORESCENCE MONOCHROMATOR
MIRRORS
MIRRORS

FIG. 28. Diagram of the optical system of the Aminco-Bowman spectrophotofluor-ometer.

current control cancels photomultiplier-tube dark current by the applica-tion of an equal current of opposite polarity to the anode current.

A schematic diagram of the optical unit of the Aminco-Bowman instru-ment is shown in Fig. 28. The monochromators are of the Czerny-Turner type with plane gratings having a ruled area of 50 × 50 mm. and 600 grooves/mm. The two gratings are identical except for the wavelength at which they are blazed. The excitation grating is blazed in the 300-mμ region and the fluorescence grating in the 500-mμ region. Light from the xenon arc is made parallel by the first mirror which directs it toward the grating Dispersed light from the grating is directed to the second mirror which focuses a monochromatic image of the arc in the center of the cell compartment. The magnification of the system is unity; therefore, the image in the cell, with slits removed, is equal to the size of the xenon arc. Since the latter is 1.8 by 3.5 mm., this arrangement permits the use of sample volumes less than 1 ml., depending upon the dimensions of the cell which is used. The equivalent aperture is $F/4.3$. Slits of various sizes can be inserted which serve as baffles to define the light in the vicinity of the cell and the photomultiplier tube. The fluorescent light from the sample is dispersed in the same manner as is the light from the xenon lamp and is focused on a slit in front of the photomultiplier tube. The latter may be either a 1P21 or 1P28 tube, both of which are adapted to the same housing.

An infrared detector attachment is also provided. The latter utilizes an RCA Type 7102 photomultiplier tube. Because of the high dark-current of the red sensitive tube, the attachment contains a dry-ice chamber to cool the tube and housing to a temperature which permits balancing of the dark current. Use of the infrared detector also requires changing the grating in the fluorescence monochromator to one blazed at 1000 mμ.

A polarizer attachment is also provided. It consists of two polarizers, each containing a Glan-Thompson prism rotatable through 90°, mounted on the cell holder to permit both horizontal and vertical polarization of the exciting and fluorescent light beams.

G. Farrand Spectrofluorometer (Catalog No. 104242)

The Farrand spectrofluorometer is also based on a design suggested by the experimental instrument developed by Bowman et al.[14] It, too, is a multicomponent system, as shown in Fig. 29. A schematic diagram of the Farrand spectrofluorometer, with a recorder, is shown in Fig. 30. The instrument utilizes as a light source, an Osram XB165 xenon lamp, with an appropriate transformer. Two Farrand grating monochromators, range 220–650 mμ, are used for dispersing the exciting and fluorescent light. Excitation and fluorescence spectra (uncorrected) can be scanned, either with a recorder or an oscilloscope. Fluorescence can also be measured manually. The optical unit, consisting of the xenon lamp, the two monochromators, and the sample chamber, is mounted on a metal base plate. A photomultiplier photometer power supply and control unit comprise the detector system. The detector head is attached directly to the exit slit of the analyzing monochromator and positioned so that the cathode of the photomultiplier tube is at the focal point of the monochromator. The instrument is normally supplied with a 1P21 photomultiplier tube. However, photomultiplier tubes 1P28, 931A, and 1P22 are interchangeable with the 1P21 tube. An attachment which utilizes a lead-sulphide phototube is also available for investigations of fluorescence in the infrared region. The fluorescence monochromator must then be replaced by an infrared monochromator. The latter is available as an additional accessory.

The current from the detector tube is measured by a sensitive microammeter. For automatic operation two types of recorders are available. The simpler of the two is a Varian strip chart recorder which is provided with a connecting cable and plug to the ammeter. In operation, chart paper is fed through at the rate of 2 inches/minute. Instrumental excitation and fluorescence spectra are plotted against time and the time scale is converted to wavelength units by appropriate calibration. For studies requiring greater precision a Brown Elektronik recorder is also available, ap-

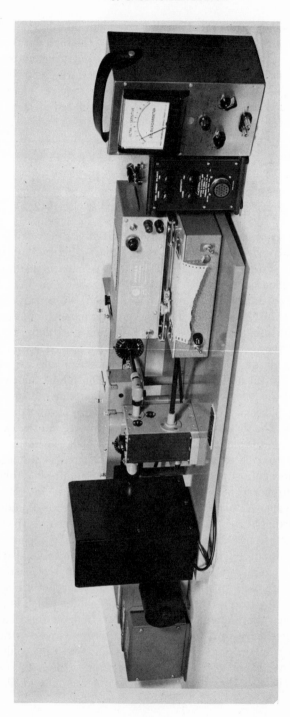

FIG. 29. Farrand spectrofluorometer.

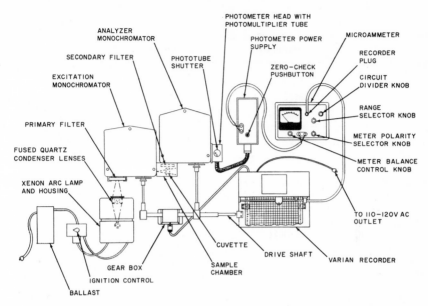

FIG. 30. Diagram of the Farrand spectrofluorometer with a Varian recorder.

propriately modified for use with the Farrand microammeter. This recorder is larger and faster. Spectral output is also plotted against time.

The Aminco-Bowman spectrophotofluorometer and the Farrand spectrofluorometer were designed to provide the biologist with a sensitive instrument which, while capable of yielding as precise spectral data as possible, was still sufficiently simple and inexpensive as to be widely used. For this reason they do not contain additional circuitry to convert the instrumental spectra to "true" spectra. When the true excitation spectrum is desired, it may be calculated from the observed spectrum by measuring and correcting for the spectral intensity reaching the sample cell. For true fluorescence spectra, many instrumental calibrations must be made.[21] These calibrations are certainly tedious, but they are not difficult to set up. Details of such calibration are presented in Chapter 4. Although the over-all spectral curves actually obtained with these two commercial instruments are skewed, they do yield excitation maxima which agree quite well with the observed absorption maxima, once the relatively simple "setting-up" calibrations suggested by the manufacturers have been carried out. A comparison of absorption maxima with excitation maxima obtained with the Aminco-Bowman spectrophotofluorometer is shown in Table II. This is really not too bad considering that less than 5 years ago not only were most of these fluorescent data not available, but

TABLE II

EXCITATION MAXIMA MEASURED WITH A COMMERCIAL FLUORESCENCE SPECTROMETER
AND THE CORRESPONDING ABSORPTION MAXIMA[a]

Compound	Excitation maxima mμ	Absorption maxima mμ
Phenol	270, 215	270, 210.5
Phenolate ion	Not fluorescent	287, 235
Anisole	270, 220	269, 217
Salicylic acid, singly ionized	295	296
Doubly ionized salicylate	310	306
Doubly ionized m-hydroxybenzoate	315	312
2,4-Dihydroxybenzoic acid (7)[b]	295, 250	292, 248
3,4-Dihydroxybenzoic acid (10–11)	300	303, 276
α-Naphthol (14)	330, 250	335, 245
β-Naphthol (14)	350, 280	348, 280
2-Hydroxyquinoline	325, 280	325, 270
3-Hydroxyquinoline (1)	350, 250	340, 245
3-Hydroxyquinoline (14)	350, 250	355, 245
7-Hydroxyquinoline (14)	370, 280	360, 275
DPNH	340	340

[a] The data are from Williams[22] and were obtained with an Aminco-Bowman spectrophotofluorometer.
[b] Figures in parenthesis represent pH values.

only a handful of individuals possessed the knowledge or the equipment with which to carry out such studies. Today, hundreds of laboratories throughout the world possess these commercial recording fluorescence spectrometers. The instruments are sufficiently sensitive and accurate for the biologist to conduct a large proportion of his studies which are quantitative in nature, with no need of corrections. Of course, where instrumental wavelength data for either excitation or fluorescence are reported, it should be stated whether they have been corrected or not (see Chapter 4). Obviously, large discrepancies between absorption data and excitation data should be investigated, since they indicate the necessity for further calibration of the instrument.

H. Zeiss Spectrofluorometer

More recently, Carl Zeiss has introduced a spectrofluorometer. This is made up of a series of prism monochromators and other components on an optical bench, as shown in Fig. 31. A compensating automatic recorder is

FIG. 31. Zeiss spectrofluorometer.

in preparation as well. The optical arrangement shown in Fig. 32 indicates that the exciting radiation passes into the side of the cuvette (K) and that the fluorescent radiation passes out of the cell at right angles through the polished bottom and to the second monochromator (M2). The arrangement is such that normally 2 ml. of solution is used. However, with special

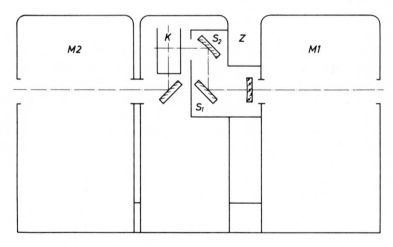

FIG. 32. Optical diagram of the Zeiss spectrofluorometer.
M1, excitation monochromator; S_1, mirror; S_2, mirror; K, cuvette; M2, fluorescence monochromator.

cuvettes, as little as 0.5 ml. can be used for measurement without sacrificing sensitivity. To ensure stability of the arc an Osram XBO1 xenon arc, D.C. operated, is used. A 500-watt lamp is used in this instrument as compared to the 150-watt lamps used in the Aminco-Bowman and Farrand instruments. The more powerful lamp is used to make up for the light losses which are inherent in a prism instrument. Grating instruments will, with equal intensity of light source, always be more sensitive. It remains to be seen whether the 500-watt lamp can endow the Zeiss instrument with sensitivity comparable to the two grating instruments. In any event the prism instrument should provide for greater resolution. This property may make it useful for distinguishing closely related substances. It remains to be seen how much more useful the greater resolution will prove to be. The Zeiss spectrofluorometer, like the Aminco-Bowman and the Farrand instruments, makes no attempt to correct the instrumental spectra automatically. Here, too, calibrations can be made as described in Chapter 4.

VI. AUTOMATICALLY CORRECTED FLUORESCENCE SPECTROMETERS

It is obvious that an instrument which would present true spectra, without the need of calculated corrections, would be the ideal fluorescence spectrometer. Recognizing this, several investigators have designed instruments which can automatically correct for instrumental artifacts.

A. Instrument of Parker[23]

One of the most ingenious procedures for obtaining corrected fluorescence excitation spectra was devised by Parker.[23] A schematic diagram of this recording fluorescence spectrometer is shown in Fig. 33. The exciting light from the monochromator M_1 is focused on the sample in cell C, but a small proportion of the beam is reflected by the clear silica plate B on to the front surface of the fluorescent screen F. This consists of a silica cell, 5-mm. optical depth, containing a suitable fluorescence solution the fluorescence of which is viewed through the back face of the cell by the monitoring photomultiplier P_2. A suitable band of fluorescent light from the sample in C is selected by the monochromator M_2 and is received by the photomultiplier P_1. After amplification the outputs of P_1 and P_2 are passed to the ratio recorder R. The frequency drum on M_1 is motor driven, and as the frequency is varied, the slits are adjusted so as to maintain the output of P_2 approximately constant (within a factor of 2). If the contents of F are suitably chosen, the output of P_2 can be made proportional

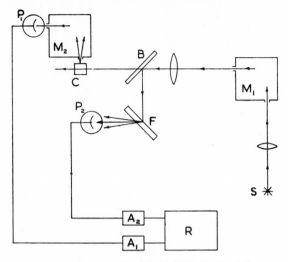

FIG. 33. Diagram of the spectrometer designed by Parker[23] for recording true excitation spectra.

S, xenon arc; M_1, excitation monochromator; B, clear silica plate for reflecting small portion of beam to F; C, cuvette; M_2, fluorescence monochromator; F, fluorescent screen; silica cell containing fluorescein solution; P_1, photomultiplier for fluorescence detection; P_2, monitoring photomultiplier; R, ratio recorder; A_1 and A_2 are amplifying circuits.

to the quantum output of M_1 over a wide frequency range. R then records the true fluorescence excitation spectrum of the sample in C.

Fluorescent solutions chosen for the monitoring cell F should ideally have constant fluorescence yield and absorb substantially the whole of the beam within a depth of about 1 mm. or less, over the whole spectrum range concerned. Thus, the fluorescent light has to penetrate a further 4 mm. of solution before reaching P_2, and the solution therefore acts as a filter for the exciting light and also ensures that substantially the same band of fluorescent light reaches P_2, irrespective of the depth of penetration of the incident beam. In these experiments a solution of 4.4×10^{-3} M fluorescein in a mixture of sodium carbonate and bicarbonate (both 0.1 N) was used. Although the fluorescence yield of this solution is not precisely constant, it was considered to be sufficiently so for the preliminary trials. The apparent relative efficiency of the monitoring screen was found to be accurate to about $\pm 5\%$. The over-all deviation from linearity, $\pm 10\%$, was considered adequate for many purposes without applying the correction. The linearity can no doubt be improved by choice of a more suitable solution for F.

In Fig. 34 the excitation spectrum of 1:2-benzanthracene recorded with the Parker instrument is compared with the absorption spectrum as meas-

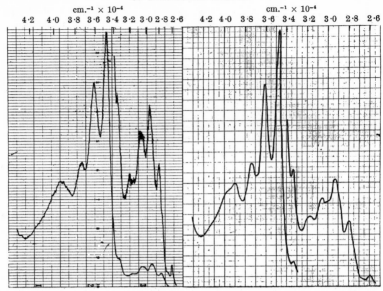

FIG. 34. Comparison of excitation spectrum of 1:2 benzanthracene (in ethanol) obtained with the Parker fluorescence spectrometer (left) with the absorption spectrum obtained with a Unicam Model SP 500 spectrophotometer (right) (after Parker[23]).

ured with a Unicam SP 500 spectrophotometer. The agreement between the two is excellent and probably the best that has yet been achieved. The two are not identical because of the differing bandwidths which were used. The experimentally determined excitation spectrum of a substance of biological interest, 5-hydroxytryptamine, is shown in Fig. 35. The fluorescence of this substance occurs entirely in the ultraviolet region. With the corrected instrument the excitation spectrum was found to coincide fairly well with the absorption spectrum. This is clearly a much more useful record than the second curve which was recorded at constant slit width without the use of the fluorescent screen monitor.

The recording of corrected spectra requires the use of comparatively narrow apertures and one would expect that the instrument would therefore not be very sensitive. However, in a recording of the excitation spectrum of a 10^{-7} M 5-hydroxytryptamine solution which was presented in the same paper the maximal response was more than 10 times the reagent blank. This instrumental design is certainly worth considering by those who require the most exact spectral information.

B. Instrument of Bartholemew and Moss

Bartholomew[24,25] utilized two quartz-prism monochromators and an arrangement of detector circuitry so as to correct, at least in part, for

cm.$^{-1}$ × 10^{-4}
4·2   4·0   3·8   3·6   3·4   3·2   3·0

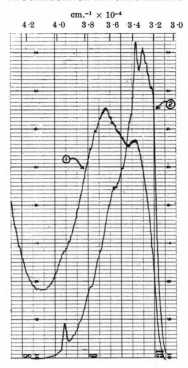

Fig. 35. Excitation spectra of serotonin (5-hydroxytryptamine); ① spectrum obtained with correcting monitor, ② uncorrected spectrum (after Parker[23]).

variations in exciting light. Details of this instrument were made available in a later publication by Moss.[26] The instrument described by Moss has a 375-watt xenon arc lamp with a quartz envelope (Siemens). A quartz-prism monochromator (Unicam SP 500) is used for isolating the wavelength for excitation. A similar monochromator is used for spectral assay of the fluorescence emission. The detector is a quartz-window photomultiplier tube (EMI Type 6256). The output of the detector is measured by counting photons arriving at the cathode. This is done with a scaler or a recording ratemeter. Photon counting is certainly sensitive but quite slow compared to the more conventional electronic procedure for amplification and measurement of photomultiplier output. The most significant aspect of this instrument is the use of a portion of the exciting light as a monitoring beam to excite a fluorescent screen which is used as a standard of reference to correct for instrumental variations. The screen is a piece of solid plastic scintillator, p-terphenyl and 2,5-di(4-biphenylyl)-oxazole in solid solution in polyvinyl toluene.* A shutter permits either monitor

* Obtained from Nash and Thompson Ltd., Chessington, Surrey, England.

fluorescence or sample fluorescence to be selected for measurement. The monochromator scales are motor driven and can scan an excitation spectrum from 200 to 450 mμ in 20 minutes. Being a prism instrument, the scanning involves the coupling of a cam to the slit in order to maintain constant band width.

The efficiency of the scintillator monitor as evaluated by comparing its output with the quantum intensity of the source (determined by ferrioxalate actinometry) was found to be within 10% between from about 280 to 400 mμ. Fluorescence excitation spectra of quinine obtained with this instrument, when corrected by the scintillator monitor and by chemical actinometry, were found to be almost identical. However, they did not coincide exactly with the absorption spectrum of quinine.

C. Instrument of Lipsett

Lipsett[27a] reported the design of another automatic recording instrument to measure absolute excitation and fluorescent spectra. The spectral sensitivity characteristics of the detecting photomultiplier, together with associated monochromators and mirrors, were corrected by means of a cam driven in synchronism with the wavelength drive of the detecting monochromator. In order to cancel fluctuations in the exciting light source, a fraction of the light coming from the source monochromator was focused onto a second photomultiplier, whose output was used to operate a servo system. The recorded spectrum is therefore the ratio of fluorescent light to exciting light. As yet no spectral data obtained with this instrument has been published.

D. Unicam Spectrofluorimeter

As for commercial instruments, Unicam has designed a spectrofluorimeter which corrects for fluctuation of the exciting lamp. A fraction of the exciting radiation, controlled by a calibrated attenuator, is made to alternate at the detector with the fluorescent light issuing from the second monochromator. Measurement is made by adjusting the beam attenuator until the detected outputs are equal for both the reference beam of the exciting light and the sample beam of fluorescence. A null-indicating meter is used and fluorescence intensity is measured as a fraction of the intensity of the exciting radiation. A diagrammatic presentation of the Unicam spectrofluorimeter is shown in Fig. 36. This instrument has not yet been distributed so that no data have been published concerning its application. However, the instrument was developed in collaboration with Bartholomew[24] and probably yields data comparable to that reported by him.

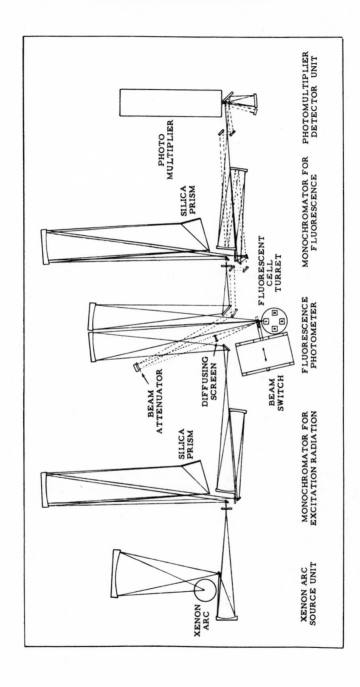

FIG. 36. Diagram of Unicam spectrofluorimeter.

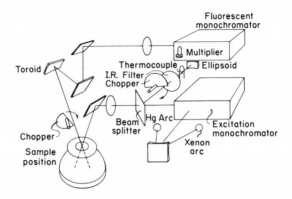

FIG. 37. Schematic diagram of the Perkin-Elmer linear energy spectrophotofluorimeter.

E. Perkin-Elmer Linear Energy Spectrophotofluorimeter

The Perkin-Elmer Corporation has developed a multicomponent linear energy spectrophotofluorimeter[27b] which they report compensates for variations in component characteristics. A schematic presentation of the instrument is shown in Fig. 37. A portion of the dispersed radiation from the excitation monochromator is directed to a thermocouple by a beam splitter. A servo device adjusts the slits of the monochromator to keep a constant signal on the thermocouple, the signal output of which is linearly related to the incident energy and is independent of wavelength. As a result, the radiation reaching the sample is also of constant energy. A second servo device, attached to the slits of the fluorescence monochromator, is programmed so that the detector will produce a signal which is linearly related to the energy incident upon it. The instrument uses either an H-3 mercury arc or a Hanovia 150-watt xenon lamp for excitation and two Perkin-Elmer Model 83 single-pass high-aperture monochromators with fused silica prisms for dispersing exciting light and fluorescence. A 1P28 photomultiplier tube is the detector and the output signal is recorded on an automatic recorder.

The type of spectral data which the Perkin-Elmer linear energy spectrophotofluorometer is capable of recording is shown in Fig. 38. The excitation spectrum of quinine was recorded first with the instrument utilizing a self-regulatory mechanism to excite with constant energy. The second spectrum is that of the same solution of quinine; however, the self-regulating mechanism was not used and the slits remained fixed during the recording. The constant energy spectrum is certainly more closely related to the actual absorption spectrum of quinine than is the one obtained with fixed slits;

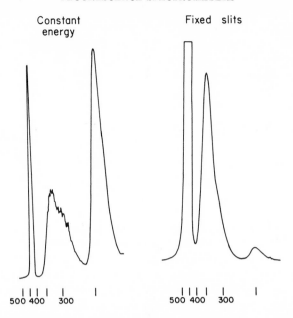

FIG. 38. Excitation spectra of quinine sulfate obtained with the Perkin-Elmer spectro-photofluorimeter; left, using constant energy of excitation; right, with fixed slits. Ordinates are in arbitrary units, abscissa are mμ (after Slavin et al.[27b]).

the 250-mμ absorption peak is known to be larger than the one at 350 mμ. Undoubtedly, such an instrument can obtain directly fluorometric data having greater physical significance than data obtained with other commercially produced instruments.

It should be pointed out that the instrument, as presently set up, can compensate only somewhere above 230 mμ since at this wavelength the slits are wide open. Regulation below this wavelength will require some modifications. However, the region below 250 mμ is not used very frequently. Again, it would appear that the use of attenuated slits to maintain constant energy would lower the sensitivity. However, the instrument is claimed to be able to detect quinine at a concentration of 0.001 μg./ml. although it was not stated how this compared with the level of stray light.

F. C.G.A. Spectrophotofluorometer, Model DC/3000

The C.G.A. spectrophotofluorometer, C.G.A. Model DC/3000,[28] another partially corrected spectrofluorometer, has been developed recently in Italy.* It comprises two prism monochromators and an arrangement for a

---

* Dr. Alberto Ciampolini, Florence, Italy.

reference phototube to operate an automatic servo regulatory slit designed to produce constant excitation energy. This yields partially corrected excitation spectra. Fluorescence spectra are not corrected. It will be interesting to see how this instrument performs when it is introduced into the laboratory.

VII. MEASUREMENT OF FLUORESCENCE IN LIVING TISSUES

Since fluorescence is such a sensitive and specific indicator of the state of certain molecules (oxidized vs. reduced; ionized vs. unionized; free vs. bound), it can be applied as a means of measuring metabolic changes within living cells. There are, of course, many ways in which this can be accomplished. One method is to use suspensions of large numbers of cells or cell particles. The laboratory of Duysens[29, 30] has carried out pioneer work on the development of instrumentation and techniques for applying fluorometry to the study of metabolic processes associated with photosynthesis. Application of this instrumentation to the measurement of reduced pyridine nucleotides and chlorophyll in photosynthetic bacteria and within isolated chloroplasts are described in subsequent sections of this book (Chapter 9).

In the presence of large numbers of macro particles, the problem of light scattering and light absorption become so great that it is not possible to measure the emission at 90° as is most frequently done for fluorescence in dilute solution. Instead, the fluorescent light is emitted at a glancing angle. With such an optical arrangement, measurements of absolute fluorescence spectra on complex mixtures may be subject to some error. However, difference spectra (before and after irradiation or enzymatic reduction) can be fairly precise.

A. Instrument of French

An automatically corrected fluorescence spectrometer for measuring fluorescence of intact cells and protoplasts in the visible and infrared regions of the spectrum has been reported by French.[31] This instrument (Fig. 39) has been applied to many problems relating to chlorophyll and other plant pigments. The exciting source in this instrument is a Bol type high-pressure mercury lamp, similar to the GE H-6 but dissipating twice the power in a narrower capillary.* The lamp is run on alternating current so that it yields 120 light pulses per second and the image is focused on the slit of the exciting monochromator.† The incident light may be measured

---

* Obtained from the Huggins Laboratories, Menlo Park, California.
† The monochromators contain 10 × 10 cm. gratings.

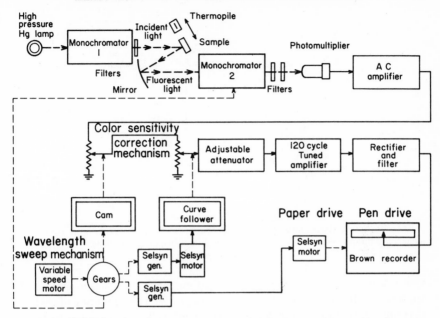

FIG. 39. Diagram of an automatically corrected fluorescence spectrometer for measuring fluorescence of intact cells and chloroplasts in the visible and infrared (French[31]).

(for calibration purposes) by means of a thermopile made available in a sliding mount. Fluorescence emitted by the sample is sent into the second monochromator by means of a mirror. Before reaching the photomultiplier tube the light is passed through filters to remove reflected incident light and to partially correct for the spectral characteristics of the tube. The output of the photomultiplier tube is amplified and passed through an attenuator linked to the wavelength drive of the fluorescence monochromator by a cam which is cut to correct for variations in photomultiplier sensitivity and monochromator transmission with wavelength. Residual errors are removed by a photoelectric curve follower which regulates another attenuator drive. The final electrical signal is proportional to the intensity of the emitted fluorescence and is separated from random noise by means of a 120-cycle tuned amplifier, rectified to direct current and used to drive a recorder. According to French, the instrument once calibrated maintains the calibration for months. This is obviously an elegant instrument and many of its features may be worth incorporating into commercial instruments.

B. Instrument of Duysens and Amesz

Duysens and Amesz[29] have described an apparatus for using the mercury lines 313 and 366 mμ as exciting radiation in intact cells and particles and

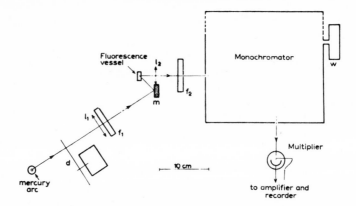

FIG. 40. Schematic diagram of the apparatus used by Duysens and Amesz[29] for measuring fluorescence spectra of intact cells and subcellular particles.

measuring the spectra of the resulting fluorescence from 350 to 600 mμ. Its application to the measurement of changes in the amounts of reduced pyridine nucleotide within photosynthesizing cells is considered to be one of the most important advances in the field of photosynthesis.

The apparatus devised by Duysens and Amesz[29] for measurement of fluorescence spectra is shown in Fig. 40. The mercury arc is imaged in a vessel containing the fluorescing solution or cell suspension. The fluorescing image, emitted from the same surface is focused by lens l_2 on the entrance slit of a Bausch & Lomb 600-mm. 1200 lines/mm. grating monochromator. Filter set f_1 (various combinations of glass and interference filters) is used to isolate the 313 and 366 mμ lines of the mercury arc for excitation; filter set f_2 absorbs the incident radiation and scattered light. A wavelength drum is driven by a synchronous motor and a sectioned disk, d, mounted on another synchronous motor, is used to modulate the incident light. The alternating current component of the current caused by the light falling on an RCA 1P21 photomultiplier is amplified and fed into a recorder. The instrument is calibrated with mercury lines and with a calibrated tungsten lamp. The precision of the calibration of the energy emitted by the instrument can be calculated to be within 3% throughout the spectral regions studied.

An example of the data obtained with this instrument is shown in Fig. 12, Chapter 9. It is apparent that addition of the substrate, ethanol, to yeast cells causes a large increase in the fluorescence emitted at 443 mμ when excited at 366 mμ. This is characteristic of reduced pyridine nucleotides. Other applications of this instrument to metabolic studies are presented in Chapter 9.

Olson[32] modified the apparatus of Duysens and Amesz[29] to permit the determination of excitation spectra. A 1000-watt high-pressure xenon lamp is used in conjunction with a Bausch & Lomb grating monochromator to furnish monochromatic excitation light at wavelengths between 260 and 400 mμ. The excitation beam is modulated as previously described and its quantum intensity is monitored by a calibrated photocell which receives a small and constant fraction of the total light coming from the monochromator. The image of the exit slit of the monochromator is focused on the sample cuvette (thickness, 1 mm.). The fluorescent area of the sample is imaged onto a photomultiplier by means of a high-aperture lens system. Suitable filters permit only light in the region 420–480 mμ to reach the photomultiplier. Unmodulated light from a tungsten source, with appropriate filters, is used to furnish photosynthetically active light for studies on reduced pyridine nucleotide changes in photosynthesizing cells. Excitation spectra of light induced fluorescence changes in several intact microorganisms are shown in Fig. 15, Chapter 9. The fluorescence of the intact cells compares with that of enzyme bound DPNH. More detailed discussions of the applications of these instruments are presented in subsequent sections of this book.

C. Differential Microfluorometer

Having demonstrated fluorometric changes within living cells, the next step is to localize quantitatively these changes within specific cell structures. A differential microfluorometer for the localization of reduced pyridine nucleotides within living cells was recently developed by Chance and Legallais.[33] The instrument was devised so as to measure with the greatest sensitivity and with a minimum absorption of excitation energy in the biological specimen the difference in fluorescence intensity between two spots that are approximately 15 μ apart, each spot being 5 μ in diameter. The operation of the apparatus depends on the synchronous action of three components: (1) a light flash, (2) the position of a diaphragm selecting a portion of the magnified image, and (3) a switch for measuring the photocurrent at the time of the flash. This operation occurs 120 times per second (at a line frequency of 60 c.p.s.). The diaphragm and the switch circuits are operated by the reed on a Brown Instrument "converter." A differential detecting circuit is used so that the output represents the difference of the peak amplitudes of the two pulses. The circuit, therefore, discriminates against common fluctuations of the measurement and reference intensities. A 1000-watt AH-6 water-cooled mercury arc lamp usually is used for excitation with a constant voltage regulator that does not introduce waveform distortion. The radiation from the mercury arc is isolated by appropriate filters. For pyridine nucleotide studies a Wratten A2 filter which transmits wavelengths greater than 410 mμ is placed in front of the phototube. The combination of the transmission characteristics of this filter and the response characteristics of the S-4 type phototube which is used gives a peak at 440 mμ and a half-power response at 420 and 540 mμ. This may be compared with a 443 mμ fluorescence peak and a half-power emission at 400 and 500 mμ for intracellular reduced pyridine nucleotide.

An inverted or metallurgical type of microscope is used with an objective that has an adjustable aperture stop. The specimen is illuminated by means of a cardioid dark-field condenser which is critically adjusted for optimal response. For scanning of the image pneumatic bending of quartz plates is used, as suggested by Caspersson.[34] The scale factor for this movement is about 20 mm. = 1 μ and the reproducibility is better than 1 μ. The microscope is equipped so that it can be changed from visual observation to photoelectric recording, the photocell being located at the aperture which is normally used for a camera. A vibrating diaphragm with an amplitude of vibration of 4.5 mm. is used. With a 1.5-mm. hole and 300 $\times$ magnification the optical aperture is 5 μ. A 1-inch diameter end-on phototube, Type 9524A is used, and is operated from dry batteries.

The electrical circuits used in conjunction with the differential microfluorometer are such that a signal-to-noise ratio of 22 may be attained for a 3 $\times$ 10^{-16}-amp. primary photocurrent and a 2-second time constant. The signals observed from a variety of biological material are in the range of 10^{-15} amp. For example, the Nebenkern of the grasshopper spermatid (an aggregation of the mitochondria of the cell) gives a signal of about 5 $\times$ 10^{-16} amp. A pentaploid yeast cell, on the other hand, gives a signal which may exceed 10^{-15} amp. The response of the circuit represents a recording of the difference in fluorescence from two different portions of a cell.

The microfluorometer has been applied to a number of studies (see Chapter 9) and offers great promise for following changes in the amount of reduced pyridine nucleotide within mitochondria and other subcellular particles within individual cells. This makes possible *in vivo* kinetic studies. With other filter combinations it should be applicable to intracellular studies on flavine and chlorophyll metabolism.

D. Microspectrofluorometer

A continuously scanning microspectrofluorometer has been developed by Olson[35] to obtain fluorescence spectra of material mounted on a microscope slide. It has been possible to obtain fluorescence spectra of chlorophyll and its photooxidation product in a small number of chloroplasts on a microscope slide. The operation of the microspectrofluorometer depends upon the

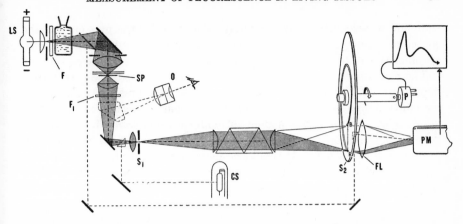

FIG. 41. Recording microspectrofluorometer of Olson[35].

LS, light source; F, filter for isolating exciting light; SP, specimen; F₁, filter to remove exciting light; O, observation ocular; S₁, entrance slit; S₂, spiral scanning slit; RB, spectrum; FL, field lens; PM, photomultiplier; CS, calibration source; P, potentiometer.

mechanical scanning of emission spectra by a rotating linear spiral exit slit. Energy admitted by the spiral slit at each wavelength position along the spectral image of cell structures is collected by a photomultiplier tube and the resulting signal is displayed along the vertical axis of an oscilloscope. A continuous rotation potentiometer attached to the shaft which turns the scanning slit controls the position of the oscilloscope trace along the horizontal axis and provides a positive linkage to the wavelength position resulting in the graphic display of the emission spectrum for each full-slit rotation or scan. Details of the complete apparatus are shown in Fig. 41. Additional features are included which provide (1) wavelength calibration, (2) monitoring of source or photomultiplier variability, and (3) alternate visual observation of the microscope field and cell structures showing fluorescence.

An alternate modification of the apparatus developed for the continuous simultaneous recording of the time course of spectral changes at several wavelengths was also reported. Scanning slit and dispersion optics are replaced by a system in which the total emission from the fluorescent cell structures is collimated and scanned by a sequence of appropriate interference filters. The energy transmitted by each filter is displayed along the Y axis, the oscilloscope time base amplifier providing the X axis. The latter, when adjusted to a slow sweep (several seconds per centimeter), permits a repetitive display of successive filter sequences such that the peaks provided by each filter appear along more intense continuous lines These lines represent the time course of emission changes for the corresponding wavelength region. As many as four such lines plus one for the source monitor can thus be displayed simultaneously and recorded photographically without serious interference caused by trace interlace.

The microspectrofluorometer of Olson[35] is not as sensitive as the microfluorometer of Chance and Legallais[33] since it measures only portions of the emitted fluorescence light rather than the entire spectral output of the fluorescence emission. This is inherent in all spectrometers. The instrument developed by Olson is not a differential one but depends upon spectral resolution for specificity. It should be possible to utilize it to confirm and expand the changes in intracellular fluorescence emission with metabolism, which have been observed by other means. It should also be possible to utilize a continuous light source for excitation and a second monochromator to permit the measurement of excitation spectra with the microspectrofluorometer.

VIII. FLUORESCENCE ON PAPER CHROMATOGRAMS

Chromatography and electrophoresis on paper are now so widely used that it is not surprising that fluorescence is one means used to detect and measure chemical compounds on developed paper strips. In some instances the compounds appear as absorbing substances which quench the background fluorescence of the paper. However, in most cases the compounds are themselves fluorescent or are converted to fluorescent derivatives by appropriate treatment of the developed chromatograms.

Detection of the fluorescent spots is carried out by exploring the chromatograms under ultraviolet light in a darkened room. Portable lamps which emit either the 254 or 360 mμ lines of the mercury vapor spectrum are available for this purpose. For those who do not have a room which can be readily darkened or where examination in ultraviolet light is a frequent procedure, there are now available portable laboratory cabinets for this purpose. The Chromato-vue, made by Ultra-Violet Products Inc., San Gabriel, California, is arranged to be operated even in brilliantly lighted rooms. It contains appropriate lamps, filters and reflectors for both 254- and 365-mμ excitation. In addition, the operator can switch on white light for outlining the spots on the chromatogram once they are detected by fluorescence.

As with light absorption it is also possible to make quantitative measurement of the fluorescence emitted by paper chromatograms. Some commercial instruments are specifically designed to do this. However, many available instruments or components can be utilized for this purpose. Van Horst et al.[36] carried out a quantitative study of the development of amino acid fluorescence on paper using a Model 525 Photovolt densitometer with a phototube detector appropriately connected to a Varian automatic graphic recorder. A mercury vapor lamp was used for excitation.

There is now available an accessory for the Turner Model 110 or 111 fluorometer for quantitation of fluorescence on paper. A paper chromatogram holder is provided which can be used to scan paper strips up to 5 × 35 cm. with several selected exciting wavelengths in the ultraviolet and visible region. Emitted light may be quantitated in the region from 360 to 650 mμ.

Various types of spectrophotometers have also been modified to measure the fluorescence emitted from paper chromatograms. Wadman et al.[37] utilized the Beckman spectrophotometer for measuring the fluorescence of reducing oligosaccharides on paper chromatograms following condensation with an amine reagent. Each spot was cut out and fastened with a rubber band in front of a hole in a wooden block. The block was then placed in the cell holder compartment with the paper as near the phototube as possible to expose it to the maximum fluorescent light. A filter, to absorb the exciting light and to transmit the fluorescent light, was placed between the paper and the phototube. The slit was opened wide and the sensitivity set near maximum. Fluorescence was read on the per cent transmission scale. Transmittance without paper was checked frequently to correct for variations in sensitivity of the instrument. The selector switch was set at 1.0 or 0.1, depending upon the intensity of fluorescence.

Use of a spectrophotometer in this manner offers many advantages. Instruments of this type are available in most laboratories and most people know how to use them. With sufficient fluorescent material even the standard lamps can be used. However, it may be more desirable to use a more intense light source and operate at lower electronic amplification where the instrument is more stable. More intense light sources are available as attachments to most spectrophotometers. Of course, any one of the commercially available fluorometer or fluorescence spectrometers can be used (with appropriate modification) to measure fluorescence emitted from paper chromatograms. Fluorescence can be excited by shining the exciting light through the paper, as did Wadman et al.[37] Semm and Fried[38] used a similar arrangement although complete details were not given. It can also be done by placing the paper at a 45° angle across the cell compartment and measuring the fluorescence emitted from the surface.

If one sets up to measure fluorescence from paper in a spectrophotometer or in a fluorescence spectrometer, it should be possible to determine the entire fluorescence spectrum of a given compound. Korte and Weitkamp[39] report procedures for measuring fluorescence spectra of berberin, fluorescein, xanthopterin, and thiochrome on paper chromatograms. Some of these data will be reported in later sections of this book.

The different grades of filter paper vary in their content of fluorescent materials which may interfere with attempts to measure fluorescence emission from the paper. It would appear that problems concerning purity of filter paper would depend upon the compound under analysis. Van Horst et al.[36] report that Whatman No. 1 paper is not suitable for measuring fluorescence of amino acids on paper chromatograms and advise Whatman No. 4 or 54 for this purpose. Shore and Pardee[40] suggest Whatman No. 52 or Schleicher and Schuell No. 57 for amino acid analysis. This same group reported previously[37] that Whatman No. 1, Whatman No. 4, and Schlucher and Schull No. 597 were best suited for fluorometric assay of reducing oligosaccharides on paper chromatograms.

GENERAL REFERENCES

Bowen, E. J., Fluorimeter design, *Analyst* **72**, 377 (1947).
Willard, H. H., Merritt, L. L., and Dean, J. A., "Instrumental Methods of Analysis," Chapter III. Van Nostrand, Princeton, New Jersey, 1958.

REFERENCES

1. Loofbourow, J. R., and Harris, R. S., *Cereal Chem.* **19**, 151 (1942).
2. Radley, J. A., and Grant, J., "Fluorescence Analysis in Ultraviolet Light." Van Nostrand, Princeton, New Jersey, 1954.
3. Pringsheim, P., "Fluorescence and Phosphorescence," p. 10. Interscience, New York, 1949.
4. Bowen, E. J., and Wokes, F., "Fluorescence of Solutions," Chapter 8. Longmans, Green, London, 1953.
5. Withrow, R. B., and Withrow, A. P., *in* "Radiation Biology" (A. Hollaender, ed.), Vol. III: Visible and Near Visible Light. McGraw-Hill, New York, 1956.
6. Price, J. M., Kaihara, M., and Howerton, H. K., *Appl. Optics* **1**, 521 (1962).
7. Harrison, G. R., Lord, R. C., and Loofbourow, J. R., "Practical Spectroscopy," Chapter 12. Prentice-Hall, Englewood Cliffs, New Jersey, 1948.
8. "RCA Tube Handbook," HB–3. Radio Corporation of America, Electron Tube Division, Harrison, New Jersey.
9a. Beckman Instruments, Inc., Fullerton, California, Instruction Manual 305A.
9b. Konev, S. V., and Kozunin, I. I., *Dairy Sci. Abst.* **23**, 103 (1961).
10. Jacobson, W., and Simpson, D. M., *Biochem. J.* **40**, 3 (1946).
11. Beckman Instruments, Inc., Fullerton, California, Bull. 19 (1957).
12. Gemmill, C. L., *Anal. Chem.* **28**, 1061 (1956).
13. McCarter, J. A., *Anal. Chem.* **30**, 158 (1958).
14. Bowman, R. L., Caulfield, P. A., and Udenfriend, S., *Science* **122**, 32 (1955).
15. Duggan, D. E., Udenfriend, S., Bowman, R. L., and Brodie, B. B., *Arch. Biochem. Biophys.* **68**, 1 (1957).
16. Udenfriend, S., Duggan, D. E., Vasta, B. M., and Brodie, B. B., *J. Pharmacol. Exptl. Therap.* **120**, 26 (1957).
17. Udenfriend, S., Weissbach, H., and Clark, C. T., *J. Biol. Chem.* **215**, 337 (1955).
18. Goldzieher, J. W., Bauld, W. S., and Givner, M. L., *Can. J. Biochem. Physiol.* **38**, 233 (1960).

19. Goldstein, J. M., McNabb, W. M., and Hazel, J. F., *J. Chem. Educ.* **34**, 604 (1957).
20. Jokl, J., *Chem. listy* **52**, 1370 (1958).
21. White, C. E., Ho, M., and Weimer, E. Q., *Anal. Chem.* **32**, 438 (1960).
22. Williams, R. T., *J. Roy. Inst. Chem.* **83**, 611 (1959).
23. Parker, C. A., *Nature*, **182**, 1002 (1958).
24. Bartholomew, R. J., *Revs. Pure and Appl. Chem.* (*Australia*) **8**, 265 (1958).
25. Bartholomew, R. J., Ph.D. thesis, University of London, London, 1958.
26. Moss, D. W., *Clin. Chim. Acta* **5**, 283 (1960).
27a. Lipsett, F. R., *J. Opt. Soc. Am.* **49**, 673 (1959).
27b. Slavin, W., Mooney, R. W., and Palumbo, R. W., *J. Opt. Soc. Am.* **51**, 93 (1961).
28. Valzelli, L., *Symposium on Spectrophotofluorimetry, Milan, April 25, 1960*.
29. Duysens, L. N. M., and Amesz, J., *Biochim. et Biophys. Acta* **24**, 19 (1957).
30. Duysens, L. N. M., *in* The Photochemical Apparatus, Its Structure and Function, *Brookhaven Symposia in Biol. No.* **11**, 10 (1959).
31. French, C. S., *in* "The Luminescence of Biological Systems" (F. H. Johnson, ed.), p. 51. Am. Assoc. Advance. Sci., Washington, D. C., 1955.
32. Olson, J. M., *in* The Photochemical Apparatus, Its Structure and Function, *Brookhaven Symposia in Biol. No.* **11**, 316 (1959).
33. Chance, B., and Legallais, V., *Rev. Sci. Instr.* **30**, 732 (1959).
34. Caspersson, T., *J. Roy. Microscop. Soc.* **60**, 8 (1940).
35. Olson, R. A. *Rev. Sci. Instr.* **31**, 844 (1960).
36. Van Horst, S. H., Tang, H., and Jurkovich, V., *Anal. Chem.* **31**, 135 (1959).
37. Wadman, W. H., Thomas, G. J., and Pardee, A. B., *Anal. Chem.* **26**, 1192 (1954).
38. Semm, K., and Fried, R., *Naturwissenschaften* **38**, 326 (1952).
39. Korte, F., and Weitkamp, H., *Angew. Chem.* **70**, 434 (1958).
40. Shore, V. G., and Pardee, A. B., *Anal. Chem.* **28**, 1479 (1956).

4

PRACTICAL
CONSIDERATIONS

I. GENERAL REMARKS

AN UNDERSTANDING of the theoretical principles of fluorescence is, of course, important in the application of the method. However, as with other procedures, there are many practical considerations which are derived through trial and error. Unfortunately, many aspects of fluorescence assay are still in the trial and error stage. However, there has accumulated a body of practical information which can be most helpful to the analyst using fluorescence for assay purposes. As with other analytical methods, fluorescence has certain requirements as to purity of solvents, practical isolation procedures, calibration of instruments, etc. In addition to these, fluorescence assay has certain unique problems inherent in the physical nature of the fluorescence phenomenon. These include interference by scattered light, quenching by other light absorbing species or by solutes which interact in such a manner as to alter the fluorescence characteristics or to decrease the fluorescence yield. Because biologists frequently employ fluorescence assay near the limits of sensitivity, problems of solvent purity and of isolation are more serious considerations than for other analytical techniques.

II. SOLVENTS

It was pointed out in the previous chapter that even the highest grades of quartz and glass have appreciable fluorescence. The fluorescence of a glass or quartz cuvette may be increased or decreased when it contains a solvent, depending upon the fluorescence, light scatter, or absorption of the solvent. Solvents which inherently absorb light are, of course, not suitable for fluorescence in the regions of their absorption. Benzene, acetone, and phenol absorb in some regions of the ultraviolet and cannot be used in these regions. Even traces of these solvents, such as may remain as residues from earlier stages in an isolation procedure, may interfere with a fluorescence assay.

Spectral grade solvents, which have been made available for spectrophotometric studies, may be better for fluorescence assay than are ordinary reagent grade materials. However, apparently even these high-grade solvents are not always directly suitable for fluorescence assay. Recognizing this, the American Society for Testing Materials has set up a committee* to investigate the possibility of encouraging the production of "fluorescence grade" solvents by industry.[1]

For many purposes, however, solvent purification is not needed. In other instances relatively simple procedures can be used. n-Butanol, ether, ethylene dichloride, benzene, and heptane have been suitably purified for fluorescence assay by shaking with 0.1 N HCl, 0.1 N NaOH, and then with water, as described by Brodie et al.[2] Where necessary, further purification can be achieved by passing the solvents over silica gel which removes many fluorescent impurities from the solvents.

The best grades of absolute ethanol require purification when used in fluorescence studies where the highest sensitivity is needed. Distillation from potassium hydroxide has been found suitable for this purpose. In the procedure for tocopherol assay, Duggan[3] employed "hexane." The material obtained from Matheson was reported to be the best available yet required considerable purification for the assay. The hexane was repeatedly washed with fresh portions of concentrated sulfuric acid until the acid layer remained colorless. It was then washed, first with dilute alkali, then with water. It was then dried with anhydrous Na_2SO_4 and passed over a column of silica gel which removed some dark yellow material. Following this, it was distilled. A procedure somewhat comparable to this was employed by Hess et al.[4] to purify heptane for the determination of reserpine in tissues. In the same study it was found necessary to purify reagent grade isoamyl alcohol. Five hundred milliliters of reagent grade isoamyl alcohol were shaken twice with 100 ml. of 25% sulfuric acid, twice with 100 ml. of water, and finally with a dilute solution of $NaHCO_3$. Aldehyde materials were removed by shaking the alcohol layer with 100 ml. of 40% sodium bisulfite for 30 minutes and esters were removed by boiling the alcohol for 2 hours with 100 ml. of 20% acqueous KOH. The alcohol was then dried over solid NaOH, then over anhydrous Na_2SO_4, and finally distilled. The fraction boiling between 129° and 131° was collected. The material prepared in this manner showed extremely low fluorescence and light absorption. Of course, for many other purposes much simpler purification would suffice.

The procedures just presented are but examples of the many types of purification which have been used for the preparation of solvents for

---

* *Subcommittee VI on Fluorescence Spectroscopy* of Committee E–13 on Absorption Spectroscopy, American Society for Testing Matherials (A.S.T.M.).

fluorometry. Although many solvents can be used without purification in certain assays it is always necessary to check a given solvent for reagent blanks in a procedure. If the solvent blank is appreciable it can usually be reduced to reasonable values by distillation or by washing with acid, alkali, and water, as just described. It is a good idea, in any extraction procedure, to leave the organic solvent saturated with water so as to avoid volume changes during subsequent extractions. In the following sections of this book, unless it is stated otherwise, all solvents used for extraction should be considered as having been purified so as to minimize their contribution to the reagent blank. The term water signifies distilled water. Additional solvent purification procedures will be presented under the appropriate analytical methods.

Recently the Hartman-Leddon Company of Philadelphia has made available a large number of "fluorescence grade" solvents. These include many of the alcohols, chlorinated solvents, hydrocarbons, and water. Howerton* has compared the fluorescence characteristics of many of these fluorescence grade solvents with other reagent and spectral grade solvents and has found them to have, in most cases, less than 10% of the fluorescence of the others. As an example, whereas a spectral grade of chloroform gave 10 divisions of fluorescence at the highest sensitivity of a fluorescence spectrometer (excitation, 250–350 mμ; fluorescence, 390 mμ) the fluorescence grade solvent showed no fluorescence. At these same instrumental settings of sensitivity a sample of deionized and distilled water yielded 5 divisions of fluorescence with excitation at 320 mμ and fluorescence at 410 mμ. By contrast the purified water available from the Hartman-Leddon Company yielded less than one-half a division of fluorescence. Such purified solvents should be helpful in many ways. A most important use to which they can be put is in the checking of instrumental performance by providing "low-fluorescent" solvents in which to dissolve chemical standards for calibration. The fluorescence grade solvents, even if they prove to be too expensive for large-scale routine analytical procedures, provide the investigator with a highly purified reference standard to guide him in the development of laboratory procedures of solvent purification.

III. OTHER SOURCES OF CONTAMINATION

A. Stopcock Grease

The most highly purified solvent can, of course, be readily contaminated by fluorescent or absorbing materials "picked up" in the analytical procedures. Obviously, the biological materials themselves contain many

---

* H. K. Howerton, American Instrument Co., personal communication.

interfering substances which require separation before analysis. However, there are other sources of contamination which, once recognized, can then be avoided. When employing solvent extraction procedures it is customary in many laboratories to use separatory funnels. These have stopcocks and, of course, may require some lubrication so as not to freeze. Most stopcock greases should be avoided since even traces can yield sufficient fluorescence to completely invalidate an analytical procedure. In the author's laboratory solvent extractions are usually carried out in bottles and transfers are made by pipet. Where separatory funnels must be used they are frequently used without any lubrication. Where it is absolutely necessary to use some form of stopcock lubrication, this should be checked carefully as to interference in the particular assay procedure. Rubber and cork stoppers are additional sources of contamination, particularly when organic solvents are employed.

B. Cleansing Agents

Another source of contamination in fluorescence assay is the glassware itself. It has been found that many synthetic detergents, which are now so popular in laboratory wash rooms, are fluorescent. Excitation and fluorescent spectra of laboratory detergents should be checked. Although careful rinsing can remove most of the fluorescence, it has been found in practice that traces of fluorescence appear, from time to time, on various pieces of glassware washed with these agents. Inorganic cleaning agents such as Calgonite have been found suitable for fluorometry. When chromic acid is used it must be carefully removed by rinsing since even traces of it may absorb sufficient light in the ultraviolet region so as to interfere with an assay. Chromic acid should not be used on cuvettes. These should be cleaned, when necessary, in hot, concentrated nitric acid. Burch[5] had advocated cleaning all glassware which is used for fluorescence assay, first, by boiling in half-concentrated nitric acid, rinsing, boiling in distilled water, and then rinsing in distilled water. Such extreme care becomes more necessary as one approaches the limits of detection of a given fluorescent compound. It becomes necessary, for instance, in the measurement of plasma norepinephrine. In most instances such cleansing procedures are reserved for the glassware used in the terminal stages of an assay, particularly the cuvettes. Of course, those laboratories which have the best laboratory helpers will have the least trouble with their glassware.

C. Chemical Reagents

Purification of chemical agents other than solvents is also frequently required. This is particularly true when the fluorescence assay procedures approach the limits of detection of a given substance. In the procedure for the measurement of epinephrine and norepinephrine in plasma, as developed by Weil-Malherbe and Bone[6] (Chapter 5), it is necessary to remove fluorescence contamination from the alumina adsorbent and to purify a number of the chemical reagents extensively in order to permit their use in the assay. Many commonly employed chemicals, such as ascorbic acid and buffer ingredients, may contain traces of fluorescent contamination which only become apparent when used in the most sensitive procedures. Of course, the growth of microorganisms in a buffer or in any solution will nearly always contribute to fluorescent contamination. When used directly such solutions will also produce appreciable light scattering.

D. Filter Paper

Filter paper can be another source of contamination due to the residues of aromatic compounds, particularly phenols, which are present in the original wood. Concentration of a substance by a procedure involving elution from paper chromatograms may require preliminary purification of the paper. This can be accomplished by washing the paper with the solvent used for elution. The application of paper for filtration purposes should also be checked for interference with a given analytical method. There are now available a large variety of filter papers produced by different manufacturers. Information concerning porosity, strength, and even chromatographic properties is available for many makes and grades of filter papers. However, little is known about the trace fluorescent impurities in the paper.

IV. PROBLEMS RELATING TO ANALYSIS IN DILUTE SOLUTION

It should be kept in mind that in applying fluorescence assay, one is generally dealing with compounds in extremely dilute solutions (0.001–1.0 μg./ml.). Aside from the specific problems of the fluorescent method, one should expect to encounter all the problems pertaining to the handling of molecules in dilute solution. It is generally recognized that very dilute solutions are not as stable as more concentrated ones. For this reason most analytical laboratories keep concentrated (stock) solutions of standards from which they make periodic dilutions for "working" standards. The latter may be one-hundredth or one-thousandth the concentration of the stock standard.

Reasons for the more rapid deterioration of the highly dilute solutions are numerous. However, it should be pointed out that similar deterioration processes go on in the more concentrated solutions, but go undetected because only a "small constant fraction" of molecules may be involved. As the solutions are diluted, this small constant fraction can become a significant part of the total and so become apparent.

A. Adsorption onto Surfaces

Quinine, when dissolved in organic solvents such as chloroform or benzene, is quite stable. Solutions in these solvents at 50 μg./ml. can be recovered quantitatively by extraction into dilute 0.1 N H$_2$SO$_4$. However, in the application of an extraction procedure to the assay of quinine and related alkaloids at the submicrogram level, Brodie et al.[2] found that small amounts of quinine were adsorbed onto the glassware from the organic solvent. This occurred at many stages of the analysis; onto bottles, pipettes, test tubes, and cuvettes. In fact, by intentionally exposing benzene solutions of quinine at 1 μg./ml. to large amounts of glass surface, essentially all the quinine could be removed. The alkaloid could then be recovered by rinsing the glassware with 0.1 M H$_2$SO$_4$. Of course, the loss of 1 μg./ml. from a 50 μg./ml. solution would be just barely detectable.

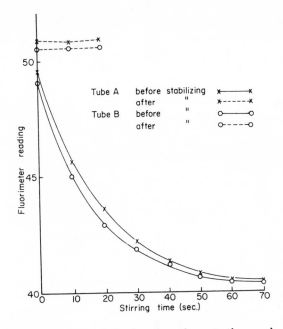

Fig. 1. Loss of quinine from solution by adsorption onto the vessel wall. The quinine, 1 μg./ml., was dissolved in 0.1 N sulfuric acid. Stabilization involved treatment with hot alkali or saturation of the glass surface with quinine (after Bird[7]).

Bird[7] observed that there is great variability in the ability of glassware to adsorb quinine from solution. With occasional batches of glassware large losses due to adsorption can occur even from solutions made up in 0.1 N sulfuric acid (Fig. 1).

The loss of organic substances by adsorption onto surfaces is a general phenomenon and becomes particularly troublesome at or below the microgram level. There are many types of adsorption losses. The adsorption from organic solvents onto glass surfaces is one which is most frequently encountered. This problem is greatest with aromatic substances and becomes worse the less polar the organic solvent. When it is encountered the amount of glass surfaces should be minimized. New glass surfaces may be more adsorptive and the glassware may have to be put through cleaning solution or heated in strong acid or in strong alkali[7] before use. The addition of small amounts of a more polar solvent to the nonpolar solvent frequently reduces adsorptive loss. Thus, for phenylethylamine extractions to be quantitative it is advisable to use mixtures of chloroform and isoamyl alcohol.[8] The alcohol's function is to prevent adsorptive losses. In many instances mixed solvents are not desirable for the extraction. In such cases, after separation of the phases, a few drops of an alcohol can be added to the organic solvent. The presence of the alcohol markedly reduces adsorptive losses during further manipulations. The loss of chloroquine was found to be much greater in heptane than in ethylene-dichloride but was completely prevented by the addition of ethanol[2] (Table I). This procedure

TABLE I

ADSORPTION OF CHLOROQUINE ONTO GLASS FROM HEPTANE AND ETHYLENE DICHLORIDE[a]

Time minutes	Recovery of chloroquine (%)			
	Ethylene dichloride		Heptane	
	Without ethanol	With ethanol	Without ethanol	With ethanol
0	100	100	100	100
5	98	101	94	97
15	99	100	81	104
30	97	98	76	103

[a] Duplicate 10-μg. samples of chloroquine were extracted from alkaline solution into 150 ml. of solvent; final concentration 0.067 μg./ml. After separation of the phases, 2 ml. of ethanol were added to the organic phase of one sample, the other being used as a control. The aqueous phase was discarded and 15 aliquots of solvent were removed at intervals and analyzed for chloroquine. The value at zero time was arbitrarily taken as 100% (after Brodie et al.[2]).

was also utilized with great success in the assays for atebrine[9] and cinchona alkaloids[10] and has been found useful in many other analyses in this and in other laboratories.

Ion exchange resins and other adsorbents are now widely used for analytical purposes, and one can predict fairly well the fate of a given compound on various types of chromatographic columns. However, at the submicrogram level, behavior on chromatograms may also become variable and less predictable. The best conditions for a given isolation is often found by trial and error.

Analytical measurements usually require prior removal of proteins. This is frequently achieved by addition of an agent which precipitates the protein and, at the same time, leaves the compound of interest in solution. A large number of protein precipitants are known, each devised for a given analytical procedure. The precipitating protein represents a large, freshly formed, adsorptive surface, varying in shape and charge according to the type of precipitant used. Each compound for analysis requires careful study as to the best protein precipitating agent. Furthermore, procedures developed for the recovery of 50 μg. of a compound will frequently not serve for 0.05 to 0.5 μg. of the same compound. As an example, quinine, when added to plasma in amounts which are much larger than are usually found during therapy, can be recovered in protein-free filtrates prepared with a variety of protein precipitants, including trichloroacetic acid and tungstic acid. However, at the microgram level, large and variable losses of quinine are encountered with most precipitants except metaphosphoric acid.[11] The latter agent was found suitable for quinine through trial and error. However, adsorptive losses, even with this precipitating agent, are appreciable at the microgram level unless plasma proteins are diluted considerably before addition of the precipitant (see Chapter 5). Each assay procedure may require a specific precipitating agent. Thus, neutral zinc precipitation was found most suitable for serotonin.[12] Tungstic acid was best for tryptophan and tyrosine in plasma.[13] However, trichloracetic acid was adequate for estimation of tyrosine in tissue homogenates.[14] Because such losses vary with the concentrations employed, it is obvious that in checking the recovery in a given procedure it is necessary to use quantities comparable to those being assayed.

B. Oxidation

There are other ways in which losses may occur from very dilute solutions. The presence of traces of oxidants, including oxygen, may destroy small amounts of a compound even under the best conditions. Such losses become quite apparent in extremely dilute solutions. Norepinephrine is relatively stable in 0.01 N HCl at concentrations of 100 to 1000 μg./ml. or higher, and such solutions can be kept for over a year when stored in the refrigerator. However, solutions below 1 μg./ml. must be made up fresh at frequent intervals. Traces of peroxides in ether, when it is used as a solvent, can markedly lower recoveries. Washing the ether with reducing agents (ferrous sulfate) removes such interference.

C. Photodecomposition

Photodecomposition becomes a serious problem in more dilute solutions. In concentrated solutions the substance itself acts as a filter to prevent decomposition. Dilute solutions of labile compounds which absorb appreciably in the regions of the glass ultraviolet and violet must be protected from light. Dilute standards of such compounds can be stored in low actinic glass containers which are available commercially.

Photodecomposition may also become a problem during the initial stages of a fluorometric assay. Aronow and Howard[15] reported that the fluorophor derived through condensation of norepinephrine with ethylenediamine was unstable in ordinary light. They suggested that the chemical steps be carried out in red light. Whereas, the necessity for using red light in this assay is questionable, it does appear that daylight does alter the nature of the fluorescent product(s) formed in this reaction.

Photodecomposition is most prominent during the fluorometric measurement. Since the intensity of emitted fluorescence depends on the intensity of the exciting light, highly intense sources are employed in order to attain the highest sensitivities. However, many fluorescent substances decompose when exposed to an intense beam of absorbing energy so that the fluorescence deteriorates during the measurement. Even quinine, which has long been used as a reference standard, suffers decomposition at very low concentration. Bowen and Wokes[16] reported that quinine solutions containing 1 μg./ml. or greater are not appreciably affected by prolonged irradiation whereas solutions containing 0.02 μg./ml. or less do deteriorate under continued exposure to the ultraviolet light in a standard fluorometer. The rate of photodecomposition of a 0.01 μg./ml. solution of quinine is shown in Table II. It is apparent that continued exposure of such dilute solutions produces appreciable decomposition. The increased sensitivity of present-day instruments may make it necessary to use such dilute solutions as standards. If so, they should be prepared freshly at frequent intervals by dilution of more concentrated stock solutions. It is apparent that decomposition of a reference standard can lead to serious errors. In the author's laboratory, standards of the substance being assayed are carried through the isolation procedure of each analysis and are compared with solutions

TABLE II

EFFECT OF ULTRAVIOLET IRRADIATION OF AN 0.01 μg./ml. QUININE SOLUTION IN A
SPEKKER FLUOROMETER[a]

Irradiation time minutes	Fluorescence % of inital
0	100
2	94.4
21	92.9
38	88.1
86	74.3
97	69.2
100	68.0

[a] After Bowen and Wokes.[16]

of the same substance prepared by direct dilution of a stock standard. Only when the substance to be assayed is highly unstable or is formed through a complex chemical procedure need another chemical substance be used as a reference standard.

Since the optical systems of commercial instruments differ from one another they will lead to different degrees of photodecomposition during fluorometric assay. A shutter is absolutely essential so that the solution is only irradiated during the short period of measurement. Rapidly responding meters are also desirable to permit rapid assay. In general, it is desirable to increase instrument sensitivity by improving the photodetection system rather than by increasing the intensity of the exciting source. When studying labile compounds it may be necessary to use fairly weak exciting sources and depend almost entirely on the detectors for adequate sensitivity. Intense light focused at a point within the cell can frequently cause marked local decomposition. If detection is also from the same focal point, this will be made apparent by a fall in the photodetector response, and a return to original values each time the shutter is closed as diffusion within the cuvette reestablishes the original conditions at the focal point. Some fluorescent substances which are rapidly destroyed during intense exposures at the wavelength of absorption are difficult to assay in instruments having optical systems which focus at a point within the cell. With such labile compounds it may be advisable to use a broad unfocused beam of less intense light for excitation and to have the detector "see" a much larger area of the sample. Some of the commercial photofluorometers presented in Chapter 3 made use of the General Electric 25W phosphor sources, emphasizing the lower incidence of photodecomposition during measurement.

It should be pointed out that photodecomposition may also be put to use in applications of fluorescence assay. In a mixture of two fluorophors with similar spectral characteristics it may be possible to destroy one preferentially by exposure, at an appropriate pH, to its absorbing radiation. Another practical use of photodecomposition is the conversion of a nonfluorescent or poorly fluorescent absorbing molecule to a more highly fluorescent one. Examples of the latter may be seen in the assays of chloroquine[17] (Chapter 5) and riboflavin[18] (Chapter 7).

V. EFFECT OF TEMPERATURE ON FLUORESCENCE

Although it is apparent from general considerations of fluorescence theory that temperature is an important variable, it may not be fully realized

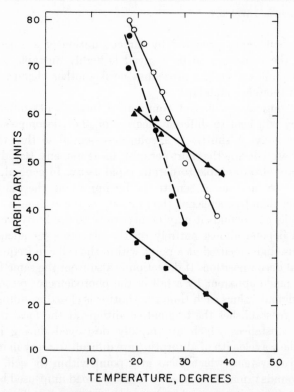

FIG. 2. Variations in fluorescence intensity of several compounds as a function of temperature. All compounds were dissolved in 0.1 M phosphate buffer, pH 7.0, except quinine (J. Green, unpublished observations). ○ — ○, tryptophan or indoleacetic acid; ● — ●, indoleacetic acid in buffer saturated with benzene; ▲ — ▲, tyrosine; ■ — ■, quinine in 0.1 N sulfuric acid.

how necessary it may be to control temperature in carrying out fluorometric assay. Changes in temperature alter the viscosity of the medium and
the number of collisions of the fluorescent species with the solvent, with
others of its kind, or with different molecules in solution. Fluorescence is
exquisitely sensitive to all of these. Just how sensitive fluorescence can be
to changes in temperature is shown in Fig. 2. Indoleacetic acid is of great
interest in biology and is frequently measured by fluorometric procedures.
In aqueous solution at neutral pH, the fluorescence of indoleacetic acid
decreases with increasing temperature with a temperature coefficient of
about 1.5, in the range 20°–30°. This is a much greater temperature sensitivity than is found for quinine or tyrosine (temperature coefficient of
about 1.1 to 1.2 between 20° and 30°). Furthermore, when traces of solvents
such as chloroform or benzene are present, after a typical extraction procedure for indoleacetic acid assay,[19] the lowering of fluorescence with increasing temperature is even more pronounced. It is apparent that with
indoleacetic acid more than usual care must be taken to control temperature. However, this is not only true for indoleacetic acid. Other indoles
such as tryptophan and tryptamine are equally sensitive to temperature
change. Effects such as those produced by traces of organic solvent may
occur with other substances.

The greatest source of temperature variation in fluorometry is the light
source used for excitation. As was pointed out in the previous section on
instrumentation, these are generally arc lamps emitting high energy and
are, therefore, extremely hot. For this reason all instruments make some
arrangements for cooling and insulating the cuvette compartment. However, some increase in temperature occurs in the cuvette compartment of
most instruments. In some instruments the temperature of a sample placed
in the cuvette compartment may rise as much as 10 degrees within a few
minutes. In such an instrument it is almost impossible to measure a series
of samples, such as indoleacetic acid, with any degree of precision. This is
particularly so since in fluorometric assay it is necessary to reread standards frequently and, due to occasional fluctuations in sensitivity, many
samples must also be re-read. Obviously, where the cuvette compartment
is appreciably hotter than room temperature, samples and standards will
be changing in temperature continually. For this reason the cuvette compartment should be kept as close to room temperature as possible. In some
cases it may be necessary to provide special cooling devices. Samples which
may be carried through a procedure involving heating or cooling, such as
use of a cold room or refrigerated centrifuge, should be permitted to come
to room temperature before fluorometric assay.

Obviously, when fluorescence is used in studies with systems which are
themselves sensitive to temperature change, i.e., protein interaction by

polarization methods, enzyme assay, etc., provisions for exact temperature regulation in the cuvette compartment are absolutely essential.

VI. INTERFERENCE DUE TO LIGHT ABSORPTION BY THE SAMPLE

The presence, in solution, of materials which absorb a significant proportion of the excitation or fluorescent radiation will diminish the observed fluorescence by, what is termed, an "inner filter" effect. When the fluorophore itself is present in high concentrations, it should, of course, be diluted so as to yield little absorption. This is seldom a problem. In most extracts of plant and animal tissues or of urine, the compound of interest is present in trace quantities, whereas many other absorbing species are present in much larger amounts. The purpose of extraction and chromatographic procedures in an assay is to remove not only interfering fluorophors, but also other absorbing species, so as to permit maximal excitation and fluorescence to take place.

When the fluorescence of a compound in a tissue extract is lower than it is in purified solvents, a spectrophotometer can be used to corroborate the presence of extraneous material absorbing at the excitation or fluorescence wavelengths. Large amounts of such absorbing material make it impossible to perform the assay. Small amounts, if they remain constant from sample to sample, permit the assay to be carried out provided internal standards are used, i.e., standards prepared in the tissue extract. The magnitude of the filtering effect is seen in Table III which shows the effect of varying amounts of dichromate ion on the observed fluorescence of tryptophan. The dichromate ion has two absorption peaks which overlap the excitation and fluorescence peaks of tryptophan (275 and 348 mμ). It

TABLE III

EFFECT OF DICHROMATE ION ON THE FLUORESCENCE OF TRYPTOPHAN[a]

Sample	Per cent transmission measured at 350 mμ	Fluorescence
1	100	64
2	40	52
3	20	19
4	10	8

[a] All solutions contained 1 μg./ml. of tryptophan in 0.1 M sodium carbonate. Varying amounts of potassium dichromate were added to increase the absorption at both 275 and 350 mμ. Fluorescence is in arbitrary units (Dr. Daniel E. Duggan, unpublished observations).

should be possible to correct for "inner-filter" effects by using adsorption data, but this is not recommended as a routine procedure.

VII. FLUORESCENCE QUENCHING

True quenching is not related to absorption or light scattering, but is due to an interaction of the fluorescent molecule with solvent or with other solutes in such a manner as to lower the efficiency and/or lifetime of the fluorescence process. From a practical standpoint, this means that the compound will exhibit less fluorescence when it is in the presence of the quenching substance. Here again, it requires isolation procedures, in an analysis, to remove such quenching substances. Halide ions, heavy metals, oxygen, and inorganic and organic nitro compounds are known to quench fluorescence. However, the quenching is not random, each instance is indicative of rather specific chemical interactions. Thus, while halides markedly quench the fluorescence of quinine, they have little effect on a host of other substances. The quenching of quinine fluorescence by halide ion indicates a specific interaction between the two. It is of interest though, that this interaction does not alter the ultraviolet absorption spectrum so that quinine exhibits the same molecular extinction in 0.5 M H_2SO_4 and 0.5 M HCl, although the fluorescence in the latter solvent may be only a fraction of what it is in the former. Thus, quenching is actually an indication of the marked sensitivity and specificity of fluorescence to changes in atomic and molecular structure. Where the quencher and fluorophore have a high affinity for each other, even traces of the former can lead to marked diminution of fluorescence. The fluorescence of some dyes in organic solvents is so sensitive to traces of oxygen that they have been used in procedures for oxygen assay at the trace level.[20, 21] However, most compounds, particularly in aqueous solution, are little effected by oxygen.

The presence, in solution, of a second molecule which absorbs radiation at or near the wavelength emitted by the fluorophore will in some instances lead to a diminution in the fluorescence. This type of quenching is due to transfer of the energy from the fluorophor to the second absorbing species. The second molecule may re-emit the energy as heat, it may dissociate, or it may emit its own fluorescence. All of these processes will be at the expense of fluorescence emission by the original fluorophor. Though this process results in interference with an analytical procedure, detection of energy transfer is actually an important aspect of fluorescence which is proving to be the most useful in explaining biochemical mechanisms. It will be discussed in detail in a later section (Chapter 6).

The fact that fluorescence is sensitive to quenching agents and that quenching interactions have a certain degree of specificity offers a practical application in identification and proof of specificity of an assay. Studies with a series of quenching agents may help identify a fluorophor in tissue extracts. Duggan and Udenfriend[13] used such data to establish the identity of plasma tryptophan. Quenching agents may also be used to diminish the fluorescence of one component in a mixture of fluorophors and thereby endow an assay with specificity.

VIII. LIGHT SCATTERING

The theoretical bases for Rayleigh and Raman scattering were discussed in Chapter 2. However, the analyst must learn to contend with these and other forms of scattered and stray light in a practical way since they are most frequently the limiting factors in the sensitivity or reproducibility of an assay procedure. With filter instruments the scattering components comprise part of the over-all "blank" and are controlled by complementary filters which are arrived at through trial and error. In fluorescence spectrometers, scattering is easily distinguished from fluorescence. Rayleigh scattering of the solvent, scattering from colloidal particles (Tyndall effect), and scatter from surfaces of the container occur at the wavelength of excitation. Raman scattering appears at longer wavelengths than the exciting radiation. However, it is an emission which represents a physical property of the pure solvent and always varies from the excitation by a constant frequency or energy difference. The data in Table IV, from a study by Parker,[22] show the Raman peaks of several common solvents

TABLE IV

MOST PROMINENT RAMAN BANDS CORRESPONDING TO VARIOUS MERCURY LINES IN
SEVERAL SOLVENTS[a]

	Wavelength of main Raman band produced by exciting at					Mean frequency shift μ^{-1}
	248 mμ	313 mμ	365 mμ	405 mμ	436 mμ	
	mμ	mμ	mμ	mμ	mμ	
Water	271	350	416	469	511	0.338
Ethanol	267	344	409	459	500	0.292
Cyclohexane	267	344	408	458	499	0.288
Carbon tetra-chloride	—	320	375	418	450	0.070
Chloroform	—	346	410	461	502	0.302

[a] According to Parker.[22]

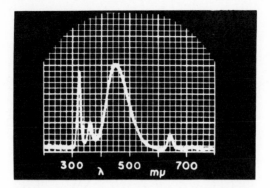

Fig. 3. Observed emission spectrum of a solution of quinine in 0.1 N sulfuric acid, obtained with an Aminco-Bowman spectrophotofluorometer. The various peaks are, from left to right: 320 mμ, Rayleigh and Tyndall scatter coinciding with the wavelength used for excitation; 360 mμ, Raman scatter of water; 450 mμ, quinine fluorescence; 640 mμ, second-order Rayleigh and Tyndall scatter; 720 mμ, second-order Raman scatter (after Price et al.[23]).

observed with a fluorescence spectrometer. At the higher instrumental sensitivities all types of light scattering become evident in fluorescent spectra, even in relatively pure solvents. Rayleigh and Tyndall scatter is usually more intense than the Raman scatter. Smaller second-order scatter and fluorescence peaks are also observed with grating instruments. These are found at twice the wavelength of the corresponding primary peak. An emission spectrum for quinine sulfate, excited at 320 mμ in a commercial fluorescence spectrometer, is shown in Fig. 3.[23] The peak at 320 mμ represents Rayleigh and Tyndall scatter. The smaller peak at 360 mμ is due to Raman scatter in water; the displacement of 40 mμ from the excitation wavelength is almost exactly as reported by Parker,[22] as shown in Table IV. The large peak at 450 mμ is the fluorescence of quinine. The next smaller peak at 640 mμ is due to second-order Rayleigh and Tyndall scatter. At 720 mμ there is a barely discernable peak which is due to the second-order Raman scatter. A second-order fluorescence peak would have been observed at 900 mμ had the instrument been able to detect at this wavelength. All the peaks, except the one due to the fluorescence of quinine, should be observed in a comparable sample of the solvent itself (blank), in this instance 0.1 N sulfuric acid.

When the fluorescence and excitation wavelengths are close together, the distortion due to scattering severely limits sensitivity. Among other things, the slit arrangements in a fluorescence spectrometer or fluorometer limit scatter from the cells and other surfaces. However, the slits also limit fluorescence so that an optimal slit arrangement must be sought in each

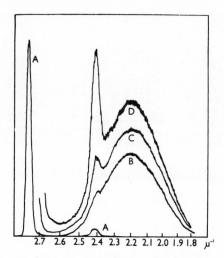

Fig. 4. Increasing magnitude of the Raman scatter with increasing instrumental sensitivity. Curve A, water alone; curve B, 0.1 μg./ml. of quinine at the same sensitivity; curve C, 0.033 μg./ml. of quinine at increased sensitivity; curve D, 0.01 μg./ml. of quinine at still greater sensitivity. The ordinate represents fluorescence intensity in arbitrary units. The Raman peak is the one at about 2.4 μ^{-1} (after Parker[22]).

case. Even in a fluorescence spectrometer slits alone do not provide adequate resolution from scatter at the higher sensitivities. In extending fluorescence assay to the range 0.01–0.001 μg./ml., instruments are operated at such high sensitivities that even Raman scatter can become a serious problem. The increasing magnitude of Raman scatter with increasing instrumental sensitivity is shown in Fig. 4.[22] Crout[24] has demonstrated how the Raman scatter limits the sensitivity of the trihydroxyindole procedure for norepinephrine assay (Chapter 5).

In filter instruments the purpose of the secondary filter is not only to remove the exciting light which may be transmitted by reflection or scatter but also to remove Raman scatter. As pointed out by Parker,[22] it is this last requirement which frequently makes it necessary to use cutoff filters which remove appreciable portions of the fluorescent light, thereby severely limiting sensitivity. Thus, with radiation of wavelength 365 mμ, the most frequently used mercury line in fluorometry, an effective secondary (cutoff) filter would seem to be one which had zero transmission below 410 mμ. However, the Raman band of water excited by radiation of 365 mμ appears at 416 mμ and high blank readings will result from this band unless a secondary filter is used which does not transmit below 430 mμ. The fluorescence of quinine is diminished by the usual filters which cut off radiation below 430 mμ. One of the advantages of fluorescence spectrometers, as compared to filter instruments, is that it is possible to select wavelengths of excitation and of fluorescence which are sufficiently far apart as to minimize Raman scatter. This may, of course, mean that the measurements are made elsewhere than at the peaks of excitation and fluorescence. The information gained with fluorescence spectrometers may also be used in the proper selection of filters for fluorophotometers. Price et al.[23] in their investigations on problems of scattering in a commercial fluorescence spectrometer showed that the scatter components, including Raman scatter, differ from the fluorescence in the degree of polarization and can be removed by the proper use of polarizers. They suggest the use of such polarizers or of optical filters in extending the sensitivity of the commercial fluorescence spectrometers for quantitative analysis. This should also be applicable to filter instruments.

Scattered light does not only distort fluorescence spectra, it also decreases the measured fluorescence since the scattering surfaces deflect the fluorescent light from the detector as well as deflecting the exciting radiation. Figure 5 illustrates an extreme example of scattering due to traces of colloidal silicate in an alkaline protein hydrolysate prepared for tryptophan assay.[25] In this instance the colloidal silicate could be removed by centrifugation in the ultracentrifuge. When this was done the tryptophan fluorescence increased to control levels. Ground-glass stoppers can lead to a similar type of scattering interference. In analytical procedures which

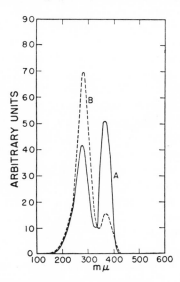

FIG. 5. Excitation spectra of tryptophan in an alkaline hydrolyzate of β-lactoglobulin before (curve A) and after (curve B) centrifugation at high speed. Excitation peaks are at 280 mμ and the scattering peaks at 370 mμ (after Duggan et al.[25]).

end up with extracts having a high degree of scattering, the standards carried through the procedure will, of course, yield less fluorescence than standards prepared in pure water. It is permissable to use such standards provided that the amount of light scatter is constant in all the samples. In a filter instrument variations in the degree of scatter cannot be detected in any simple manner. However, with fluorescence spectrometers the scatter peaks can be checked just as readily as the fluorescence peaks. Monitoring of the scatter peaks in this way can frequently explain erratic analyses.

Interference due to scattering, absorption, or quenching must be minimized and kept constant for the proper functioning of routine analyses. Excessive and variable scattering and absorption can be detected with the fluorescence spectrometer and the spectrophotometer. Quenching materials can be detected only by their effects on the fluorescence intensity. Since interferences of all three types are usually significant in most assays of biological material, internal standards are recommended as the most simple and reliable means of correction. Standards carried through an entire procedure will, of course, also correct for other variables which are not even known to the investigator.

IX. INSTRUMENTAL STANDARDIZATION AND CALIBRATIONS

A. Instrumental Sensitivity

Filter fluorometers are used as comparators for comparing the fluorescence intensities of extracts with that of standards. Once the correct filters are obtained for a given assay, no wavelength calibration is necessary. However, owing to the many factors which can vary the instrumental sensitivity, a solution of a stable fluorophor of known concentration is frequently used to set the instrument to constant sensitivity (comparison standard). For many of the assays which utilize the 365-mμ mercury line, quinine has proved to be an excellent standard. However, even quinine deteriorates on continued exposure (see above) so that fresh standards must be made up from time to time. A standard is particularly important when assaying a light sensitive compound or where the fluorophor is formed through a complex chemical reaction. In the latter instance, the comparison standard also acts as an index of the completeness and reproducibility of the chemical reaction.

As a result of its wide use with filter instruments, which utilize mainly the 365-mμ mercury line for excitation, quinine has come to be regarded as a universal comparison standard. This is a misconception. With the more versatile fluorescence spectrometers, fluorescence can be assayed at wavelengths at which quinine shows no fluorescence. Since the responses of light sources and detectors are not uniform over the entire spectral range, a comparison standard, to be effective, must have fluorescence properties close to those of the compound being assayed. Of course, if the compound being assayed is itself a stable fluorophor, no other comparison standard is needed. Phenols, indoles, quinine, and fluorescein are compounds which have been used as standards from the ultraviolet through the visible region. Appropriate standards for a given assay may be selected from among the many compounds presented in this book.

Parker[22] has suggested the use of the Raman spectrum of the solvent as a means of standardizing instrumental sensitivity from day to day. This is feasible only with fluorescence spectrometers, where, for given wavelength setting and slit openings, the Raman scatter peak is directly proportional to the instrumental sensitivity. When measurements of spectra are carried out on different days and at different instrumental sensitivities, the results can be corrected to a constant sensitivity by correcting for the height of the Raman peak at the time of measurement. Since the Raman peak is always located near the excitation maximum, it offers a more versatile method of standardization than does the use of other fluorescent compounds. When the spectra are recorded the Raman peak appears on the recording as an internal standard of sensitivity. Parker has also suggested that the intensity and location of the Raman peaks of various solvents can be used to check the calibration of the components of a fluorescence spectrometer.

B. Wavelength Calibration

With commercially available fluorescence spectrometers, excitation and fluorescence spectra are reflections of instrumental variables and are not absolute. Variations in light source, detectors, and optical systems markedly influence the shape of the two spectral curves. Of course, for purposes of quantitative analysis or for qualitative comparison of an unknown sample to a standard, instrumental values are entirely adequate. Some calibration is necessary even when a fluorescence spectrometer is used only as a comparator for quantitative assay. Calibration is certainly required for the more sophisticated application of fluorescence spectrometry. Of course, instrument manufacturers carry out calibration at the factory. The emission monochromator may be calibrated with several standard light sources to make certain that it is correct and the fluorescence monochromator is then calibrated against the excitation monochromator.[21] A "Pen-Ray" quartz lamp,* or an equivalent small lamp, is an intense source of monochromatic mercury lines and provides an inexpensive, accurate, and

---

* Ultra-Violet Products Inc., San Gabriel, California.

rapid source for the calibration. The lamp is inserted in the cuvette holder so that its window faces the entrance slit of the fluorescence monochromator. With the narrowest slits available the fluorescence spectrometer is turned on (except for the spectrometer lamp). Calibration is started by rotating the wavelength cam. Starting from the high-wavelength end, the first line observed is the one at 730 mμ, corresponding to the second order of the 365-mμ line. If properly calibrated, the instrument should show a peak meter response each time a wavelength reading on the cam corresponds to one of the mercury lines. If not, then the cam should be adjusted so as to make them correspond. A typical fluorescence monochromator calibration of a commercial fluorescence spectrometer[21] is shown in Table V. Once the fluorescence monochromator is calibrated, the excitation monochromator can be calibrated against it. To do this the spectrometer lamp is turned on and a cuvette containing a highly scattering material (glycogen) or a ground-glass square is placed in the cell compartment. If the two monochromators are in alignment, then, as the cams are rotated, a maximum meter response should occur when the wavelength readings on the two cams coincide. Again, although this is done by the manufacturer, recalibration may be necessary, particularly when components are changed.

TABLE V

CALIBRATION OF THE FLUORESCENCE MONOCHROMATOR OF A COMMERCIAL
FLUORESCENCE SPECTROMETER[a]

Mercury line mμ	Instrument cam reading mμ	Error mμ
365 (730)	730[b]	0
334 (668)	668[b]	0
313 (626)	627[b]	+1
297 (594)	594[b]	0
577	579	+2
546	547	+1
492	491	−1
436	436	0
405	403	−2
365	363	−2
334	332	−2
313	311	−2
297	296	−1
254	253	−1
235	237	+2
225	223	−2

[a] American Instrument Company, Manual No. 768.[26]
[b] Second order.

The data reported by Williams[27] (Chapter 2) was obtained with an Aminco-Bowman spectrophotofluorometer which had been calibrated by the manufacturer with the above two procedures. Such calibration appears to be sufficient to obtain reasonable wavelength values for excitation and fluorescence peaks. On the other hand, Sprince and Rowley[28] reported that the observed excitation maxima of quinine, on another instrument from the same manufacturer, were at 265 and 365 mμ. Since the known absorption maxima for quinine are at 250 and 350 mμ, they applied a correction factor of -15 mμ to their data in order to obtain "corrected" excitation maxima. Such large corrections in this region of the spectrum are unusual. In the author's laboratory, excitation maxima of 250 and 350 mμ had been found previously for quinine.[29] Again, the Aminco-Bowman instrument was used for the measurements. The use of compounds of known absorption characteristics to correct the excitation monochromator is relatively simple and should be carried out where possible. However, where corrections as large as those reported by Sprince and Rowley are observed it may be advisable to adjust the wavelength cam on the excitation monochromator. When using solutions of known fluorophores for calibration purposes they must, of course, be sufficiently dilute so as not to give an inner filter effect which would yield erroneous data. However, even with the best technique this method of calibration is satisfactory only as a rough calibration for a few regions of the spectrum.

Hercules[30] has also discussed the problems and need for calibration of commercial fluorescence spectrometers. With 1-naphthol and other compounds he found deviations in fluorescence maxima of 15 to 20 mμ. Hercules points out that the optical and detecting components are not the only sources of error. Pen and ink recorders can also be responsible for spectral errors unless they are checked at various recording speeds. Hercules[30] did obtain "corrected" spectra for 1-naphthol which were closer to the true values (Fig. 6). He did this by using the response curve provided by the phototube manufacturer. Although in this case the corrected spectra were apparently closer to the true spectra, it must be pointed out that such a correction cannot be accepted as one designed to yield true spectra since it fails to correct for factors other than the phototube response.

Although calibrations of the type suggested by Sprince and Rowley[28] and by Hercules[30] are adequate for most purposes, additional calibration is needed to yield true spectra free of instrumental artifacts. The two additional calibrations are of the exciting radiation through the excitation monochromator and of the phototube response through the fluorescence monochromator. More precise calibration, obtained in the form of correction factors which may be used to correct the experimentally observed data, has been obtained for one of the commercially available fluorescence

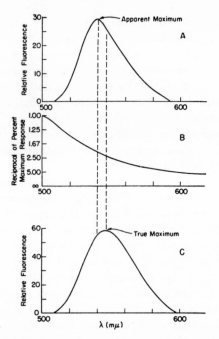

FIG. 6. Correction of the fluorescence spectrum of 1-naphthol for response of the photodetector. A, recorded spectrum; B, photomultiplier response curve provided by the manufacturer; C, corrected fluorescence spectrum (after Hercules[30]).

spectrometers by White et al.[31] However, it can be presented only as an indication of the magnitude of such corrections and the methods by which they may be obtained. They cannot be used for other instruments, even of the same make, because of the many variables involved in each instrument. The corrections shown in Table VI were obtained with a 150-watt xenon arc source and a 1P 28 photomultiplier tube. The corrections for excitation monochromator output were obtained in the following manner: From 240 to 350 mμ the potassium ferrioxalate chemical actinometer,[32] * wavelengths from 250 to 320 mμ were also checked by a photographic method and by the variations in the fluorescence of a fluorescein solution.[33, 34] From 350 to 550 mμ an RCA No. 7200 photomultiplier tube was used. The spectral response data supplied by the tube manufacturer was used to correct the excitation monochromator output at the indicated wavelength intervals.

---

* A chemical actinometer is a device for measuring radiation by determining the photolysis products of a light catalyzed reaction. Potassium ferrioxalate on exposure to light below 490 mμ undergoes decomposition and reduction to ferrous oxalate with a quantum efficiency of the order of unity. The ferrous ion formed as a result of a given exposure to radiation can be readily assayed and is a direct measure of the radiation intensity.

TABLE VI

CORRECTION FACTORS FOR SPECTRAL CALIBRATION OF THE
AMINCO-BOWMAN SPECTROPHOTOFLUOROMETER[a]

Wavelength mμ	Relative intensity (factor)	
	Excitation spectra	Fluorescence spectra
240	0.8	—
245	0.92	—
250	1.02	—
255	1.13	—
260	1.35	—
265	1.51	—
270	1.7	—
275	1.87	—
280	2.0	—
285	2.15	—
290	2.28	—
295	2.45	—
300	2.61	—
305	2.82	—
310	3.0	—
315	3.2	—
320	3.38	—
325	3.52	—
330	3.7	—
335	3.92	—
340	4.15	0.33
345	4.20	0.41
350	4.4	0.42
355	4.48	0.64
360	4.56	0.70
365	4.7	0.83
370	4.9	0.89
375	5.0	0.91
380	5.1	0.98
385	5.25	1.01
390	5.35	1.07
395	5.55	1.12
400	5.8	1.15
405	6.04	1.14
410	5.87	1.19
415	6.0	1.19
420	6.04	1.17
425	6.07	1.16
430	6.0	1.17
435	5.9	1.17
440	5.95	1.17

TABLE VI (*Continued*)

Wavelength mµ	Relative intensity (factor)	
	Excitation spectra	Fluorescence spectra
445	6.1	1.15
450	6.2	1.15
455	6.5	1.11
460	7.2	1.08
465	8.0	1.07
470	7.7	1.05
475	7.0	1.02
480	7.0	1.01
485	6.6	0.99
490	6.34	0.95
495	6.15	0.93
500	6.19	0.91
505	6.19	0.89
510	6.2	0.87
515	6.05	0.84
520	6.03	0.80
525	5.9	0.77
530	5.9	0.74
535	5.7	0.70
540	5.8	0.66
545	—	0.62
550	5.64	0.57
555	—	0.54
560	5.60	0.49
565	—	0.46
570	5.65	0.42
575	—	0.39
580	5.65	0.35
585	—	0.31
590	5.60	0.26
595	—	0.22
600	—	0.17
605	—	0.14
610	—	0.10
615	—	0.09
620	—	0.07
625	—	0.06
630	—	0.05
635	—	0.05
640	—	0.04
645	—	0.04
650	—	0.03
655	—	0.03

TABLE VI (*Continued*)

Wavelength mμ	Relative intensity (factor)	
	Excitation spectra	Fluorescence spectra
660	—	0.03
665	—	0.02
670	—	0.02
675	—	0.02

[a] The correction factors for excitation spectra represent the relative intensity of radiation from the excitation monochromator, at constant slit width. Correction factors for fluorescence spectra were obtained by comparison with a standard lamp. The correction factors are applied by dividing observed readings by the factor (from White et al.[31]).

The region 550–600 mμ was checked by a thermopile. Corrections for the emission monochromator and detector (1P 28 photomultiplier tube) were obtained by comparison with a previously standardized 500-watt lamp† placed so that the light passed directly through the cuvette compartment to the emission monochromator and detector.

With correction factors obtained in this matter, White et al.[31] were able to obtain excitation spectra for quinine, fluorescein, and metal chelates which coincided with the absorption spectra. It is necessary to repeat, however, that owing to variations in different xenon arcs, gratings, and phototubes, and in their positioning, correction factors must be obtained with each instrument. The value of spectrometers which can automatically yield corrected spectra is made apparent by the amount of labor required to properly calibrate an instrument. However, it should be pointed out that fully corrected instruments are complex and if produced commercially will be more expensive than the instruments presently available.

It should also be pointed out that until we know more about all the variables inherent in instrumentation for fluorescence assay that any procedure for correcting spectra, whether by the use of correction factors in a standard instrument or by automatic devices, cannot be regarded as yielding "true" data. At best, we can consider such data as "corrected" and closer to the true physical properties than the direct observations. Perhaps with the development of appropriate instrumentation it will be possible to obtain true excitation and fluorescence spectra. For the present it would be desirable to state, when reporting spectral data, whether it has been corrected or not. If corrected, then some statement should be made

† Standard lamps for calibration may be obtained from the National Bureau of Standards, Washington, D. C.

concerning the methods used and the precision of the measurement (i.e., 325 ± 5 mμ).

X. LIMITS OF SENSITIVITY

With the average fluorescence measuring instrument, quantities as small as 0.1–0.001 μg./ml. of many substances can be readily detected and assayed. With the more sensitive instruments and with compounds having a high absorption and a high quantum yield of fluorescence, 0.001 μg. (a millimicrogram)/ml. can emit sufficient fluorescence so as to permit measurement. Lowry and his colleagues[35] have devised microcuvettes which permit the measurement of fluorescence in as little as 0.05 ml. in an appropriate cuvette holder. Some commercial instruments have adapters which require as little as 0.1 ml. (see Chapter 3). With these small volumes the limits of sensitivity can be extended so that millimicrogram (nanogram) quantities or less of most of the important fluorophors can be measured. The combination of microfluorometry with microchemical techniques, as designed by Lowry et al.,[35, 36] permits the measurement of metabolites and enzymes in a microgram or less of tissue. These microprocedures, some of which are described in subsequent chapters, permit the mapping of metabolic systems within the cell. With some compounds as much as 1 μg./ml. or more are required to yield appreciable fluorescence. Such concentrations are frequently less than are needed for other assay procedures. Even where fluorescence is no more sensitive than absorptiometry or colorimetry, it offers added means for specificity.

XI. HAZARDS

It cannot be overly emphasized that ultraviolet radiation is harmful to the eyes. Care should be taken when calibrating instruments or when examining paper chromatograms that the eyes are shielded from the ultraviolet radiation. With the high-intensity lamps which are presently available, appreciable damage to the eyes can occur before the investigator is aware of it. Two of our colleagues suffered sufficient damage during the calibration of an instrument so as to require medical care.

Some of the lamps which are used for excitation are sealed under considerable pressure and thus present another hazard to the investigator. The xenon arcs which are commonly used in commercial fluorescence spectrometers are under several atmospheres of pressure. Although the lamp manufacturers are careful to point out the dangers inherent in the handling of such high-pressure lamps and instrument manufacturers give warnings in their manuals, some laboratory personnel seem unaware of the explosion hazard. Care should be taken when replacing the lamps or when handling them in any other way. They should also be stored in a safe place when not in use.

GENERAL REFERENCES

Laurence, D. J. R., Fluorescence techniques for the enzymologist, in "Methods in Enzymology" (S. P. Colowick and N. O. Kaplan, eds.), Vol. IV, p. 174. Academic Press, New York, 1957.

Lowry, O. H., Micromethods for the assay of enzymes, in "Methods in Enzymology" (S. P. Colowick and N. O. Kaplan, eds.), Vol. IV, p. 366. Academic Press, New York, 1957.

REFERENCES

1. Robey, R. F., Appl. Spectroscopy 14, 103 (1960).
2. Brodie, B. B., Udenfriend, S., and Baer, J. E., J. Biol. Chem. 168, 299 (1947).
3. Duggan, D. E., Arch. Biochem. Biophys. 84, 116 (1959).
4. Hess, S. M., Shore, P. A., and Brodie, B. B., J. Pharmacol. Exptl. Therap. 118, 84 (1956).
5. Burch, H. B., in "Methods in Enzymology" (S. P. Colowick and N. O. Kaplan, eds.), Vol. III, p. 960. Academic Press, New York, 1957.
6. Weil-Malherbe, H., and Bone, A. D., Biochem. J. 51, 311 (1952).
7. Bird, L. H., New Zealand J. Sci. Technol. 303, 334 (1949).
8. Udenfriend, S., and Cooper, J. R., J. Biol. Chem. 203, 953 (1953).
9. Brodie, B. B., and Udenfriend, S., J. Biol. Chem. 151, 299 (1943).
10. Brodie, B. B., Udenfriend, S., and Dill, W., J. Biol. Chem. 168, 335 (1947).
11. Brodie, B. B., and Udenfriend, S., J. Pharmacol. Exptl. Therap. 78, 154 (1943).
12. Udenfriend, S., Weissbach, H., and Brodie, B. B., Methods of Biochem. Anal. 6, 110 (1958).
13. Duggan, D. E., and Udenfriend, S., J. Biol. Chem. 223, 313 (1956).
14. Chirigos, M. A., Greengard, P., and Udenfriend, S., J. Biol. Chem. 235, 2075 (1960).
15. Aronow, L., and Howard, F. A., Federation Proc. 14, 315 (1955).
16. Bowen, E. J., and Wokes, F., "Fluorescence of Solutions," p. 77. Longmans, Green, London, 1953.
17. Brodie, B. B., Udenfriend, S., Dill, W., and Chenkin, T., J. Biol. Chem. 168, 319 (1947).
18. Yagi, K., J. Biochem. (Japan) 43, 635 (1956).
19. Weissbach, H., King, W., Sjoerdsma, A., and Udenfriend, S., J. Biol. Chem. 234, 81 (1959).
20. Parker, C. A., and Barnes, W. J., Analyst 82, 606 (1957).
21. Kautsky, H., and Hirsch, A., Z. anorg. u allgem. Chem. 222, 126 (1935).
22. Parker, C. A., Analyst 84, 446 (1959).
23. Price, J. M., Kaihara, M., and Howerton, H. K., Appl. Optics 1, 521 (1962).
24. Crout, J. R., Pharmacol. Revs. 11, 296 (1959).
25. Duggan, D. E., Bowman, R. L., Brodie, B. B., and Udenfriend, S. Arch. Biochem. Biophys. 68, 1 (1957).
26. American Instrument Company, Silver Spring, Maryland, Manual No. 768, p. 15, December 1959.

27. Williams, R. T., *J. Roy. Inst. Chem.* **83,** 611 (1959).
28. Sprince, H., and Rowley, G. R., *Science* **125,** 25 (1957).
29. Udenfriend, S., Duggan, D. E., Vasta, B. M., and Brodie, B. B., *J. Pharmacol. Exptl. Therap.* **120,** 26 (1957).
30. Hercules, D. M., *Science* **125,** 1242 (1957).
31. White, C. E., Ho, M., and Weimer, E. Q., *Anal. Chem.* **32,** 438 (1960).
32. Hatchard, C. G., and Parker, C. A., *Proc. Roy. Soc.* **A235,** 518 (1956).
33. Parker, C. A., *Nature* **182,** 1002 (1958).
34. Weber, G., and Teale, F. W. J., *Trans. Faraday Soc.* **54,** 640 (1958).
35. Lowry, O. H., Roberts, N. R., Leiner, K. Y., Wu, M.-L., Farr, A. L., and Albers, R. W., *J. Biol. Chem.* **207,** 1 (1954).
36. Lowry, O. H., *J. Histochem. and Cytochem.* **1,** 420 (1953).

5

AMINO ACIDS, AMINES, AND THEIR METABOLITES

I. GENERAL REMARKS

THE various amino acids give rise to a large number of metabolic products. Some of them are closely related in structure to the parent compound while others differ quite markedly from the original amino acid. For reasons of biochemical continuity, fluorescent methods for the individual amino acids and all their direct metabolites will be presented in one section. This will also include the aromatic amines.

Some amino acids are inherently fluorescent and can therefore be detected fluorometrically without the need for chemical modification. This is true of the aromatic amino acids, tyrosine, 3,4-dihydroxyphenylalanine, tryptophan, kynurenine, and 5-hydroxytryptophan. Thyroxine and triiodotyrosine are not fluorescent. Phenylalanine is fluorescent but weakly so.

Certain amino acids have characteristic structures which permit them to react with specific reagents to yield fluorescent derivatives. Thus, histidine and related imidazoles yield highly fluorescent derivatives upon reaction with o-phthalaldehyde. Arginine and other guanidine compounds are converted to fluorescent derivatives upon treatment with ninhydrin; γ-aminobutyric acid also reacts with ninhydrin to yield a fluorescent product.

Details of specific analytical methods for these amino acids and their congeners will be presented in this section. However, before going on to the specific methods, we will discuss more general reactions for amino acids. Fluorescent amine reagents such as fluorescein isocyanate have been used for labeling intact proteins.[1] They can obviously be made to react with individual amino acids. Separation from excess reagent should then yield fluorescent amino acid derivatives. In a mixture, chromatographic separation would then be required for individual assay. Such an assay by fluorescent derivative methods may be compared to the isotope derivative procedure.[2] Although this has not yet been used for individual amino acids, it may prove of value in certain microanalyses where radioactive procedures are not practical. It should also be applicable to the measurement of amines.

An interesting general procedure for the quantitative assay of amino acids in mixtures is based on their conversion to fluorescent products when heated on paper chromatograms. Patton et al.[3] and Woiwood[4] were able to show that the fluorescence of amino acids on paper is due to their reaction with aldehydic groups derived from modified or short-chain celluloses present in the paper. The reaction generally leads to browning, but under certain conditions can be made to yield mainly highly fluorescent amino acid derivatives. Shore and Pardee[5] were able to develop this into a fairly simple and highly sensitive procedure for assaying amino acids in mixtures. Details of this procedure are presented.

Amino acids are first separated on paper chromatograms. The first dimention is run with n-butyl alcohol-acetic-acid-water as the solvent, and the second with phenol-m-cresol.[6] Before the second dimension is run, the papers are sprayed with pyrophosphate buffer solution, pH 9.3. (Borate buffers interfere with the development of fluorescence). Whatman No. 52 filter paper is most suitable for this method. With purified solvents the amino acid spots are compact, the fluorescence develops well, and the background fluorescence is low. For the development of maximal fluorescence, the phenol and cresol should be purified by distillation.

Following drying of the chromatograms, they are sprayed with a solvent of 4% xylose and 1.5% sodium bisulfite in 0.05 M phosphate buffer at pH 6.5–7. Following spraying, the papers are allowed to dry and then heated in an oven at 80° to 90° for 1.5 to 2 hours. The intensity of fluorescence is dependent upon the heating so that it is desirable to use an oven in which constant temperatures can be maintained.

The fluorescent spots produced by spraying and heating the chromatograms are detected by observing the paper under an ultraviolet lamp. In the work of Shore and Pardee,[5] a Beckman spectrophotometer with the fluorescence accessory was used for quantitative assay. However, similar arrangements can be made with other instruments. The spots are outlined and cut out for fluorometric assay. Each piece of paper with its fluorescent spot is fastened with rubber bands in front of a hole in a wooden block made to fit the cell compartment as described in Chapter 3. Measurement along an entire paper strip is also possible with some instruments. With the Beckman spectrophotometer, as used by Shore and Pardee, a filter (Corning No. 3389) is placed between the paper and the phototube. Its purpose is to absorb the exciting light (365 mμ) and to transmit light above 400 mμ. The slit is set at 2.0 mm., the sensitivity near maximum, and fluorescence is read on the per cent transmittance scale.

The relationship between fluorescence intensity and varying amounts of aspartic acid, arginine, serine, and tyrosine are shown in Fig. 1. Similar results were obtained with other amino acids except for the prolines. The fluorescence procedure is similar in sensitivity to the colorimetric nin-

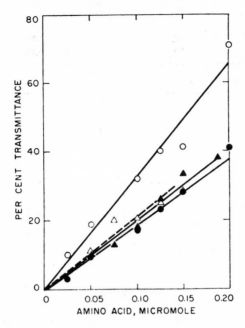

Fig. 1. Fluorescence intensities of the xylose derivatives of several amino acids as a function of the amino acid concentration. O, aspartic acid; ●, arginine; △, serine; ▲, tyrosine (after Shore and Pardee[5]).

hydrin method and can yield data with an accuracy of 10%. It may be possible to extract the fluorescent compounds from the paper and measure fluorescence in solution as described by Wadman et al.[7]

Determination of some amino acids in insulin and β-lactoglobulin were carried out with this procedure. These are shown in Table I along with values obtained by other methods. The procedure has also been applied to the measurement of amino acids in the lipoproteins of human plasma and to determine the amount of azatryptophan incorporated into bacterial proteins. It should also be possible to assay amines by this procedure.

Von Horst et al.[10a] have discussed various problems involved in the application of fluorescence assay to amino acids on paper chromatograms.

II. PHENYLALANINE

The aromatic amino acids are all fluorescent and Teale and Weber[10b] have reported their spectral characteristics in water. Although all are fluorescent, they differ in their fluorescence efficiencies and in molecular extinctions. Furthermore, the light energy emitted by most exciting lamps is not

TABLE I

AMINO ACID COMPOSITION OF INSULIN AND β-LACTOGLOBULIN BY MEASUREMENT OF
FLUORESCENCE ON PAPER CHROMATOGRAMS[a]

Amino acid	Found[b]	Reported values[b]
Insulin		
		Brand (8)[c]
Glutamic acid	137	150
Aspartic acid	55	50
Glycine	90	60
Serine	60	55
Leucine + isoleucine	120	125
Phenylalanine	60	50
Valine	75	67
β-Lactoglobulin		
		Stein and Moore (9)
Alanine	85	80
Lysine	86	77
Aspartic acid	84	86
Leucine + isoleucine	170	163
Glutamic acid	140	130
Valine	60	49

[a] After Shore and Pardee.[5]
[b] Values given in moles per 10^5 g. of protein.
[c] Bibliographic reference in parentheses.

constant throughout the ultraviolet region. As a result the aromatic amino acids vary widely in their ability to yield measurable fluorescence in solution (Table II).

The combination of a low molecular extinction, low quantum efficiency of fluorescence, and excitation at a wavelength where the emission of presently available instrumentation is relatively low make the fluorescence of phenylalanine of little practical value. The observed fluorescence of proteins is due entirely to their tryptophan and tyrosine contents.

It has been reported by Vladimirov[11] that the fluorescence efficiency of phenylalanine is fifty- to sixtyfold higher in the solid state than it is in solution. It may be possible to utilize this phenomenon to permit measurement of phenylalanine fluorescence. It may also be possible to increase the

TABLE II

FLUORESCENCE CHARACTERISTICS OF THE AROMATIC AMINO ACIDS[a]

Amino acid	Excitation maximum	Fluorescence maximum	Mol. extinction at excitation maxima	Fluorescence efficiency	Relative sensitivity[b] (instrumental)
	$m\mu$	$m\mu$		%	
Tryptophan	287	348	6×10^3	20	100.0
Tyrosine	275	303	1.5×10^3	21	9.0
Phenylalanine	260	282	2×10^2	4	<0.5

[a] Data from Teale and Weber,[10b] except for last column.
[b] An Aminco-Bowman spectrophotofluorometer was used. Tryptophan taken as 100.

fluorescence efficiency of phenylalanine by using viscous solvents or very low temperatures. A combination of these procedures with instrumentation designed for the specific wavelengths of phenylalanine excitation and fluorescence may make it possible to assay this amino acid fluorometrically. The same considerations hold for the phenylalanine metabolites, phenylethylamine, phenylpyruvic acid, and phenylacetic acid.

Lowe et al.[12] have reported that phenylalanine yields a highly fluorescent product when treated with alkaline ninhydrin and suggest that this may prove useful as an assay procedure for the amino acid. Phenylethylamine also yields a fluorescent product under the same conditions.

p-Fluorophenylalanine and related fluorine compounds. Recently, Chirigos and Guroff[13] have observed that all commercially available samples of *p*-fluorophenylalanine and its ortho and meta isomers fluoresce in aqueous solution. All of the fluorinated derivatives have an excitation maximum at 260 $m\mu$ and a fluorescence maximum at 315 $m\mu$ (uncorrected). Furthermore, they all emit almost the same intensity of fluorescence per mole, with an estimated quantum yield of the order of 10 to 20%. By contrast, the *p*-chloro and *p*-bromo derivatives are not fluorescent. The fluorescence of the fluorinated amino acids appears to be inherent in the fluorobenzene structure and is not due to a highly fluorescent impurity. This is of great interest from the standpoint of structure-fluorescence relationships. Since *p*-fluorophenylalanine is widely used in biochemical studies, its fluorescence may also be of value for analytical purposes.

III. TYROSINE AND ITS METABOLITES

Tyrosine is utilized in many ways giving rise to a large number of metabolites. Being phenols, many of these products are fluorescent or can be converted to fluorescent derivatives. In addition to tyrosine itself, fluorescent assays have been developed for tyramine, 3,4-dihydroxyphenylalanine (dopa), 3,4-dihydroxyphenylethylamine (dopamine), 3,4-dihydroxyphenylacetic acid, norepinephrine, epinephrine, normetanephrine, metanephrine, and 3-methoxy, 4-hydroxymandelic acid.

A. Tyrosine

HO—⟨benzene ring⟩—CH_2CHNH_2COOH

Tyrosine

White[14] has studied the fluorescence of tyrosine and has found that it is maximal between pH 4 and 9 and falls off sharply at extreme pH values. In various derivatives of tyrosine, including esters, peptides, and tyrosine-o-phosphate, the effect of pH is altered so that fluorescence extends into the alkaline region. This is consistent with the increase in pK values of the esters and peptides and the inability of the phenolic group to dissociate in the O-phosphate linkage. Some sort of interaction of the side chain with the phenolic group is indicated by these studies since dissociation of the side chain has a marked effect on the quantum yield of fluorescence.

The native fluorescence of tyrosine is not sufficiently sensitive or spe-

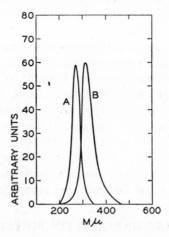

FIG. 2. Excitation (curve A) and fluorescence (curve B) spectra of tyrosine obtained with a commercial fluorescence spectrometer (uncorrected) (after Duggan and Udenfriend[15]).

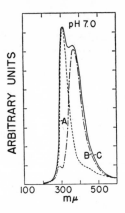

Fig. 3. Resolution of tyrosine and tryptophan fluorescence with a commercial fluorescence spectrometer (after Duggan and Udenfriend[15]). Curve A, 1.0 μg. of tyrosine; curve B, 0.1 μg. of tryptophan; curve C, curve A plus curve B.

cific to utilize it for tyrosine assay in crude tissue extracts. However, it has been found practical for measurement of tyrosine in hydrolyzates of purified proteins. The fluorescence characteristics of tyrosine are shown in Fig. 2. Duggan and Udenfriend[15] demonstrated that with commercially available fluorescence spectrometers the fluorescence of tyrosine could be resolved from that of tryptophan (Fig. 3); furthermore, in an acid hydrolysate, the tryptophan is largely destroyed. All this made it possible to devise the following simple and specific procedure for measuring the tyrosine content of proteins:

Hydrolysis of the protein sample (20–25 mg.) is carried out in 7 N H_2SO_4, in an open tube, in an autoclave, for 20 hours at 2 atmospheres. The hydrolysate is then neutralized by the dropwise addition of ammonium hydroxide, and diluted to 25 ml. with water. An aliquot of the neutralized hydrolysate is diluted tenfold with phosphate buffer, pH 8.0, and its fluorescence is measured in a fluorescence spectrometer. Known amounts of tyrosine, when added to mixtures of amino acids made up to simulate a protein hydrolysate, were quantitatively recovered. The excitation and fluorescent spectra of acid hydrolysates of proteins prepared in this manner were identical with those of authentic tyrosine. The tyrosine contents of several purified proteins, as determined by the direct fluorometric method, are shown in Table III. Velick[18] has used this procedure to determine the tyrosine content of triose-phosphate dehydrogenase. The value obtained was essentially the same as the previously reported[19] value of 4.57%.

Although acid hydrolysates of purified proteins lend themselves to the

TABLE III

TYROSINE CONTENTS OF PROTEINS AS DETERMINED BY SPECTROFLUOROMETRIC ASSAY[a]

| | Tyrosine | |
Protein	Found[b]	Given in literature[b]
Insulin_____	11.6	12.25 (8)[c]
Bovine serum albumin_____	6.0	5.5 (16)
β-Lactoglobulin_____	3.75	3.69 (17)

[a] Data from Duggan and Udenfriend.[15]
[b] The values are given in grams per 100 g. of protein.
[c] Bibliographic references in parentheses.

direct fluorometric assay of tyrosine, most biological materials contain many interfering phenols and substances which absorb in the region of tyrosine excitation and fluorescence. Several years ago a colorimetric method was developed for tyrosine based on its reaction with 1-nitroso-2-naphthol in the presence of nitric acid and traces of nitrite.[20] Upon heating there is first formed a red product which is gradually converted to a stable yellow product. The excess reagent can then be removed by extraction with an organic solvent. The nitrosonaphthol products of tyrosine and tyramine have been studied but their compositions are not known. As a colorimetric method this proved fairly specific and reasonably sensitive. More recently, however, Waalkes and Udenfriend[21] found that the nitroso-naphthol product of tyrosine is strongly fluorescent (excitation, 460 mμ; fluorescence, 570 mμ; uncorrected). Measurement of the fluorescence of this derivative endows the original procedure of Udenfriend and Cooper[20] with much greater sensitivity and specificity than can be obtained with the original colorimetric method. The procedure for fluorometric measurement of tyrosine in plasma and tissue homogenates is described:

Reagents. 1-Nitroso-2-naphthol: 0.1% 1-nitroso-2-naphthol (practical grade) in 95% alcohol. Nitric acid reagent: 24.5 ml. of 1:5 nitric acid is mixed with 0.5 ml. of 2.5% $NaNO_2$.

One milliliter of plasma is diluted to 4 ml. with water, and 1 ml. of 30% trichloroacetic acid is added. After 10 minutes the mixture is centrifuged. Homogenates are adjusted to 6% trichloroacetic acid and centrifuged. To 2 ml. of deproteinized plasma or homogenate, in a glass-stoppered centrifuge tube, are added 1 ml. of the nitrosonaphthol reagent and 1 ml. of the nitric acid reagent. The tube is stoppered, shaken, and placed in a water bath at 55° for 30 minutes. After cooling, 10 ml. of ethylene dichloride are added, and the tube is shaken to extract the unchanged nitrosonaphthol reagent. After the tube is centrifuged, the supernatant aqueous layer is transferred to a cuvette and its fluorescence is measured. The fluorescence of the tyrosine derivative, resulting from its excitation at 460 mμ, is measured at 570 mμ.

Duplicate tissue samples and a tyrosine recovery sample are run through the entire procedure with tyrosine standards and a reagent blank. Recovery of tyrosine from plasma and tissues, by this procedure, is essentially quantitative. The sensitivity of the nitrosonaphthol method for tyrosine, as described here, is greatly increased in comparison to its colorimetric assay. For example, to read the samples in the plasma filtrates (10–15 μg./ml.) the spectrophotofluorometer was employed at a sensitivity setting which is actually a fraction of what could be utilized. It has been found that 1 μg. or less of tyrosine can be measured by the procedure outlined in this method. In comparison, for a sample containing 10–15 μg. of tyrosine the optical density obtained in the Beckman spectrophotometer is only about 0.050, near the limits of its useful range. Moreover, the greater specificity of the fluorometric method is demonstrated by the similarity of the excitation and fluorescence spectra of the nitrosonaphthol derivatives obtained with plasma filtrates and with authentic tyrosine (Fig. 4). In contrast, the absorption spectrum of the nitrosonaphthol chromophore from plasma differs appreciably from that of the standard

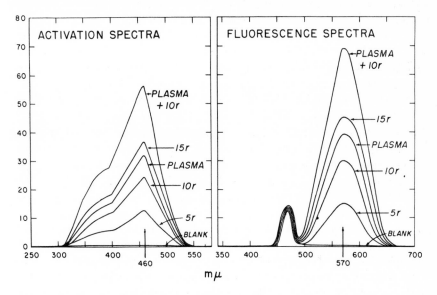

Fig. 4. Excitation and fluorescence spectra for the nitrosonaphthol derivative of tyrosine from standard solutions, from plasma, and from plasma plus 10 μg. of added tyrosine (after Waalkes and Udenfriend[21]).

TABLE IV

NONPROTEIN TYROSINE CONTENT OF SOME TISSUES DETERMINED BY THE
NITROSONAPHTHOL FLUOROMETRIC PROCEDURE[a]

Tissue	Tyrosine concentration μg./g. or μg./ml.
Plasma—human	10–13
Plasma—rat[b]	8–9
Brain—rat	14–15
Muscle—rat	16–18
Liver—rat	23–27
Heart-rat	16

[a] All tissues were obtained in the fasting condition.
[b] Sprague-Dawley male rats were used.

tyrosine derivative.[20] The colorimetric method, as applied to plasma, is not entirely specific although it has been shown to be more specific than other chemical methods for tyrosine assay.

The fluorometric method does not distinguish between tyramine and tyrosine; nor does the colorimetric assay. However, the amount of tyramine normally present in tissues is relatively small and none is found in plasma. In recent studies in this laboratory it has been shown that other metabolites of tyrosine, p-hydroxyphenylacetic acid, p-hydroxyphenylpyruvic acid, and homogentesic acid do not interfere with the nitrosonaphthol fluorescent analysis because the condensation products, being acidic, are extracted from the acidified solution into ethylene dichloride along with the excess nitrosonaphthol.

The fluorescent modification of the nitrosonaphthol procedure is applicable to many types of biologic materials. It has been used to determine the tyrosine levels of a variety of tissues; some are shown in Table IV.

B. Tyramine

Tyramine is the decarboxylated derivative of tyrosine and possesses the same fluorescence characteristics. However, being a basic compound it can be separated by solvent extraction or by chromatographic procedures and then determined separately. Tyramine also yields a fluorescent derivative with 1-nitroso-2-naphthol. This reaction coupled with an extraction into ethyl acetate is the basis for a fluorometric procedure for measuring the amine in urine.[22]

Ten milliliters of urine are adjusted to pH 10 using a pH meter or narrow-range pH paper. This sample is transferred to a glass-stoppered bottle containing 4 g. of NaCl plus 5 ml. of 0.5 M sodium borate buffer, pH 10, saturated with NaCl. Thirty milliliters of ethyl acetate are added and extraction is carried out by shaking for 15 minutes. The bottle is then centrifuged and 25 ml. of the ethyl acetate layer is transferred to a similar bottle containing 10 ml. of salt-saturated 0.5 M sodium borate buffer, pH 10. After shaking and centrifuging, 20 ml. of the ethyl acetate layer are removed to a 35-ml. glass-stoppered tapered centrifuge tube containing 3 ml. of 0.2 N HCl. The tyramine is extracted into the acid aqueous phase by shaking and the tube is then centrifuged. The ethyl acetate phase is removed by aspiration and discarded. A 2-ml. aliquot of the acid phase is transferred to a 15-ml. glass-stoppered centrifuge tube to which 1 ml. of nitrosonaphthol reagent and 1 ml. of nitric acid reagent are added.* After mixing, the tube is heated in a water bath at 55° for 30 minutes. Following this, the excess nitrosonaphthol is removed by extraction with 5 ml. of ethylene dichloride. After centrifugation this should leave an almost colorless aqueous phase from which approximately 2 ml. is removed for fluorometric assay (excitation, 465 mμ; fluorescence 565, mμ; uncorrected).

This assay measures only unconjugated tyramine in urine. Following hydrolysis in 1 N HCl for 1 hour at 100° the values may rise two- to three-fold. The assay does not measure any of the tyrosine in urine, as this is removed in the initial extraction and wash. p-Hydroxyphenylethanolamine and its N-methyl derivative, which are present in urine,[23, 24] have only one-tenth the fluorescence of an equivalent amount of tyramine. The small quantities present in unhydrolyzed urine do not contribute appreciably to the measured tyramine. Following hydrolysis, however, the hydroxyphenylethanolamines in the urine do reach significant proportions. Recovery of tyramine added to normal urine is 95–100%. The range of urinary tyramine concentrations which may be accurately measured is from 30 to about 2000 μg./liter.

Tyramine is present in some tissues and in urine. Normal human urine usually contains about 200 to 300 μg./24 hours. Administration of drugs which inhibit the enzyme, monoamine oxidase, markedly increase tyramine excretion, and measurement of the amine in urine has been used to follow the efficiency of drug treatment in patients.[25] A drop in tyramine excretion is observed when inhibitors of the enzyme aromatic L-amino acid decarboxylase are administered[26] and this assay can also be used to follow the effectiveness of treatment with drugs such as α-methyl dopa.

---

* The reagents are prepared as described in the procedure for tyrosine assay.

C. Phenolic Acids

p -Hydroxyphenylpyruvic acid

Homogentisic acid

Quantitatively, most of the tyrosine in animals is metabolized via these two phenolic acids. Although fluorometric procedures for their assay have not yet been reported, they should be feasible since both compounds are fluorescent (excitation, 290 mμ; fluorescence, 340–345 mμ; uncorrected),[27] and they can be separated from the parent amino acid by simple solvent extraction procedures. Although urine contains too many phenolic acids to permit such assay, it should be possible to utilize the native fluorescence of these compounds in studies with the partially purified enzymes along the homogentisic acid pathway of tyrosine metabolism.

D. 3,4-Dihydroxyphenylalanine

3,4-Dihydroxyphenylalanine
(dopa)

Further oxidation of the benzene ring of tyrosine yields the amino acid, 3,4-dihydroxyphenylalanine (dopa), which is the parent substance of melanin, on the one hand, and of the physiologically important epinephrines on the other. The latter are present and act in such small amounts that only fluorometric and biological procedures can be used for their assay. Dopa is itself fluorescent (excitation, 285 mμ; fluorescence, 325 mμ; uncorrected).[27] However, this fluorescence is characteristic of all catechols and phenols and cannot be used for quantitative assay in tissue extracts. Dopa does, however, yield a highly fluorescent derivative with the reagents used for assay of the epinephrines by the trihydroxyindole procedure (see below). The product has its excitation maximum at 365 mμ and its fluorescence maximum at 495 mμ.[28] Dopa can be separated from the catecholamines by passage over a Dowex 50 (Na⁺) column. The amines are adsorbed and the amino acid passes through into the effluent.[28] Dopa in the eluate can then be adsorbed onto alumina and assayed in the same manner as norepinephrine or epinephrine (see below). Ethylene diamine can also be used to detect the amino acid on paper chromatograms fluorometrically.[29]

E. 3,4-Dihydroxyphenylethylamine

HO⟨⟩CH₂CH₂NH₂
HO

3,4-Dihydroxyphenylethylamine
(dopamine)

Decarboxylation of dopa to dopamine is a key step in the formation of the epinephrines. A procedure for following the enzymatic reaction fluorometrically is described in the section on enzymes (Chapter 9). Dopamine possesses the characteristic fluorescence of a catechol and can condense with ethylenediamine to yield a highly fluorescent product. The latter reaction has been used for dopamine assay in urine by differential procedures.[30]

Dopamine reacts slowly with most agents which are used for oxidation of the epinephrines and so yields little fluorescence under conditions devised for assaying the latter compounds. However, dopamine can be converted to a highly fluorescing dihydroxyindole by oxidation with iodine and subsequent rearrangement, in the manner shown in Fig. 5.[30, 31] This reaction has been used as the basis for the measurement of dopamine in tissues. The procedure described below is a modification[32] of the one originally described by Carlsson and Waldeck.[30]

Adjust the pH of the sample to about 6.5 preferably with the aid of a pH meter. To a test tube add 1–3 ml. of sample (0.2–2 μg. dopamine), 0.5 ml. of 0.1 M phosphate buffer, pH 6.5, water to give a total volume of 3.8 ml., and 0.05 ml. of iodine solution.* After 5 minutes add 0.5 ml. of alkaline sulfite solution.† After another 5 minutes add 0.6 ml. of 5 N acetic acid (the pH drops to about 5.3) and heat the sample at 45° under standard laboratory lighting conditions for 30 minutes to produce the fluorophore.‡ The fluorescence is stable for 24 hours and is measured in a fluorescence spectrometer. Excitation and fluorescence peaks are at 345 and 410 mμ, respectively (uncorrected instrumental values) (Fig. 6).

A standard and a reagent blank are run together with the sample. When tissue extracts are analyzed, a "tissue blank" and an "internal standard"

---

* Dissolve 0.254 g. of iodine and 5 g. of potassium iodide in 5 ml. of water and dilute to 100 ml.

† The alkaline sulfite solution is prepared by dissolving 5.04 g. of the $Na_2SO_3 \cdot 7H_2O$ in 10 ml. of water and diluting with 5 N sodium hydroxide to 100 ml.

‡ In the original procedure of Carlsson and Waldeck,[30] photochemical means were used to bring about the rearrangement to the fluorophore. The samples, in silica test tubes, were irradiated with a mercury lamp (peak emission at 254 mμ) for 10 minutes. The heating procedure in ordinary light is simpler and more reproducible; nor does it require special silica test tubes.

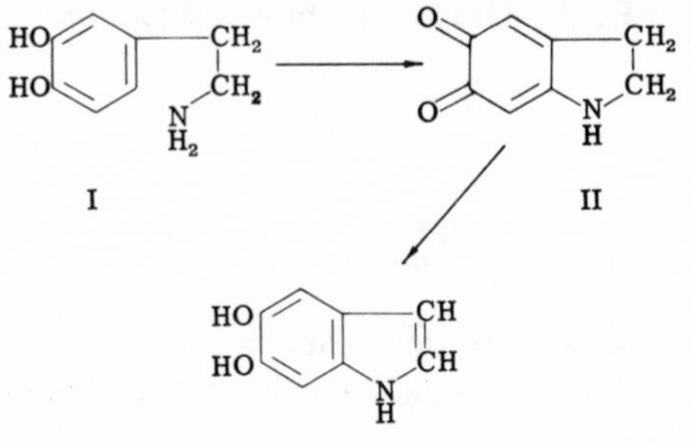

HO—CH₂ ... O—CH₂

[figure: chemical structures]

I   II

III

FIG. 5. Oxidation and rearrangement of dopamine to yield 5,6-dihydroxyindole (III).

are also run together with the sample. The "internal standard" is a sample treated as above, except that a known amount of dopamine has been added. Before plasma or tissue extracts are assayed, they must, of course, be purified in order to remove interfering substances. A Dowex 50 column is used for this purpose as described by Bertler et al.[28]

There is a linear relation between the concentration of dopamine and the fluorescence intensity up to about 0.4 μg./ml. Epinephrine, norepinephrine, and epinine (N-methyl dopamine) added in amounts equal to that of dopamine give readings which hardly differ from that of the reagent blank (Table V). This is due partly to the differences in fluorescence characteristics between dopamine and the epinephrines and partly to low yields of adrenolutine and noradrenolutine which are obtained under the experimental conditions employed in the dopamine assay. Dopa, when treated as described above, yields a compound with fluorescence charac-

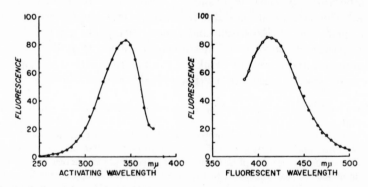

FIG. 6. Excitation and fluorescence spectra of the dopamine fluorophor according to Carlsson[31]).

TABLE V

FLUORESCENCE INTENSITIES OF SOME CATECHOLS IN THE DOPAMINE ASSAY[a]

Compound	Fluorescence intensity arbitrary units	Excitation peak mμ	Fluorescence peak mμ
Dopamine	50[b]	345	410
Epinephrine	4	—	—
Norepinephrine	4	—	—
Epinene	2	—	—
Dopa	56	345	410
Reagent blank	3	—	—

[a] In each case 0.2 μg./ml. was used; excitation was at 345 mμ, fluorescence at 410 mμ.
[b] Set to 50 (after Carlsson and Waldeck[30]).

teristics indistinguishable from those of the fluorophore of dopamine. However, when the purification procedure described by Bertler et al.[28] is used, dopa will not interfere with the estimation of dopamine, since it is not taken up by the sodium form of Dowex 50. Tyramine and other substituted phenylalkylamines give no fluorescence in this assay.

On using the foregoing procedure, Carlsson et al.[33] found dopamine to be present in brain in as high an amount as norepinephrine. In the area of the caudate nucleus it is apparently the only catecholamine present.[34] Holtz[35] has reported that, in the lung and intestinal tract of the sheep and ox, dopamine is also the only catecholamine present. Holzer and Hornykiewicz[36] related drug effects to dopamine levels in brain. It may be that instead of being only an intermediate in epinephrine formation, dopamine possesses physiologic functions of its own. This procedure for its assay should prove useful in evaluating what these functions may be.

F. Epinephrine and Norepinephrine

HO—⟨ ⟩—CHOHCH$_2$NH$_2$
HO

HO—⟨ ⟩—CHOHCH$_2$NHCH$_3$
HO

Norepinephrine   Epinephrine

The sympathetic nervous system synthesizes, stores, and secretes hormonal agents which are involved in its physiologic function. Epinephrine and norepinephrine are secreted by the medulla of the adrenal gland, whereas other sympathetically innervated tissues contain mainly the latter compound. These catecholamines are among the most potent agents known and are present in most tissues in submicrogram amounts. They were first detected by their pharmacological actions which led to biological methods of quantitative assay. Much of our present knowledge in this area was attained by such bioassay procedures. Bioassay of catecholamines can be remarkably sensitive and does not require elaborate apparatus. However, it is not suited to the modern biochemical laboratory. Furthermore, more and more physiologically active agents are being discovered which can interfere with these simple biological tests. The pathways of catecholamine biosynthesis and metabolism are now fairly well established. It is apparent that many of the intermediate compounds are structurally related and so require specific procedures for their assay. The methods for dopa and dopamine (above) were devised to detect them in the presence of the epinephrines. Procedures for other metabolites will be discussed later in this section.

Catecholamines, in their native form, do fluoresce. Epinephrine, norepinephrine, dopa, and dopamine are all maximally excited at 285 mμ and fluoresce at 325 mμ (uncorrected), the fluorescence being maximal at pH 1.[27] The native fluorescence of the catecholamines is due to their phenolic structure and is therefore not specific. The fluorescence is also the same for all the catechol derivatives. Because of this lack of specificity and requirement for the quartz ultraviolet, the native fluorescence of the epinephrines has not been applied to their measurement in blood or urine. However, it has been applied to measure total catechols in the adrenal gland.[28]

The application of chemical procedures to change the fluorescence characteristics of the catecholamines was originally necessary because there were no instruments available to detect fluorescence in the ultraviolet region. There are presently available two types of chemical procedures for measuring epinephrine and norepinephrine in body fluids. One involves oxidation and cyclization leading to fluorescent hydroxyindoles, the other involves oxidation and condensation with ethylenediamine to yield fluorescent quinoxaline derivatives.

1. TRIHYDROXYINDOLE REACTION

The fluorescence of oxidation products of epinephrine in alkaline solutions was first observed by Loew in 1918.[37] The phenomenon was further studied by Paget in 1930,[38] and Gaddum and Schild[39] described a fluorometric method for epinephrine in 1934. In 1940 Hueber[40] reported making use of the evanescent yellow-green fluorescence of the oxidation products, which appeared when epinephrine solutions were made alkaline, for assay in blood. Heller et al.[41] attempted to use similar procedures but concluded that they were not sufficiently sensitive or reproducible. As used by Heller

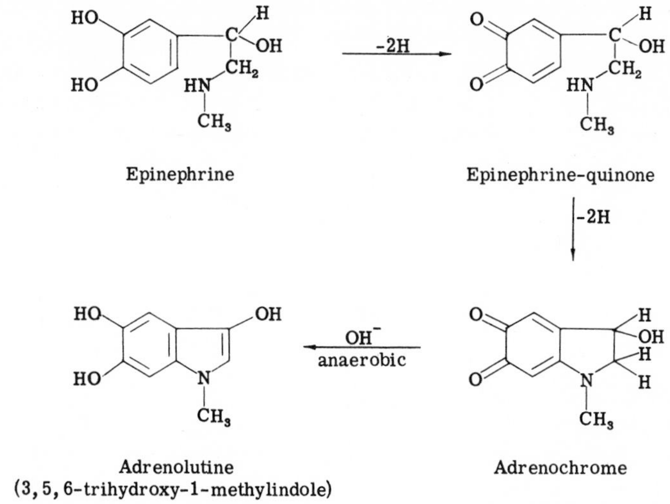

Epinephrine   Epinephrine-quinone

Adrenolutine   Adrenochrome
(3, 5, 6-trihydroxy-1-methylindole)

FIG. 7. Conversion of epinephrine to the highly fluorescent trihydroxyindole, adrenolutine.

and his predecessors, the method required the measurement of a fluorescence which persisted for seconds and the duration of which was sensitive to pH, dissolved oxygen, and the presence of reducing agents. It was not until the mechanism of the oxidation and rearrangement of the epinephrines to trihydroxyindoles was elucidated[42-45] that satisfactory applications of this procedure to tissue assay became possible. A summary of the over-all reaction for epinephrine is shown in Fig. 7. The product with norepinephrine is the same except that it lacks the methyl group on the indole nitrogen.

The trihydroxyindole method has been applied in a variety of modifications, all of which involve quite similar manipulations; isolation of the epinephrines with an adsorbent, elution, treatment with an oxidant to form the aminochrome, addition of alkaline ascorbic acid to yield the trihdroxyindole, and measurement of fluorescence. The large number of modifications which have been published attest to the fact that the assay, at its limits of sensitivity, is subject to many variables and interferences. Since the lowest levels are found in normal plasma, assay of epinephrines in this tissue is most exacting. Measurement in sympathetically innervated tissues and urine are simpler. Some of the modifications which have been more widely used are presented in Table VI. Crout[51] utilized the contributions of Price and Price,[46] of Cohen and Goldenberg,[47] and of many others to devise a procedure which has been used successfully for the simultaneous estimation of epinephrine and norepinephrine in urine and tissues. The procedure

Investigators and reference No.	Protein precipitant	Adsorbent	Eluting fluid
Price and Price[46]	None	Alumina column, pH 8; pressure.	0.3 N acetic acid
Cohen and Goldenberg[47]	None	Alumina batch process	0.2 N acetic acid
Bertler et al.[28] and Vendsalu[48]	Perchloric acid	Dowex 50 × 4, H$^+$ form; column	1 N HCl; pressure
Shore and Olin[49]	Extraction from 0.01 N HCl into n-butanol; noradenraline extracted back into 0.01 N HCl after addition of 2 volumes of heptane to the n-butanol extract		
DuToit[50]	None	Amberlite XE-64, sodium form; column	0.5 N HCl
Crout[51]	None for urine and plasma; trichloroacetic acid for tissues	Alumina batch adsorption and column elution	0.2 N acetic acid
Euler and Lishajko[52]	Acidified urine heated and filtered	Alumina column, from pH 8.2 to 8.3	0.25 N acetic acid

[a] Values on the left are filters or wavelength settings for epinephrine; those on the

makes use of the observation that both amines are oxidized to the aminochrome stage at pH 6.5 whereas only epinephrine undergoes significant oxidation at pH 3.5. The fluorometric assay of the catecholamines is of such interest that Crout's procedure[51] will be described in some detail. For those with more than a casual interest in this field the original papers should be consulted.

VI

Trihydroxyindole Procedure

Oxidant and pH	Measurement of fluorescence[a]		Applications
	Farrand fluorometer		
Ferricyanide	Excit. 436 mμ	400 mμ	Human and dog
pH 6.0	Wratten 35	Wratten 35	plasma
	Fluor. 500 mμ and Wratten 57		
	Farrand fluorometer		
MnO₂	Excit. 436 mμ	405 mμ	Human plasma and
pH 6.5	Fluor. Corning 3486	Ilford-Bight 623	urine
	(540 mμ)	(490 mμ)	
	Aminco-Bowman spectrophotofluorometer		
Ferricyanide	Excit. 455 mμ	410 mμ	Heart, brain; hu-
ph 6.0–6.5 with zinc	Fluor. 540 mμ	540 mμ	man plasma
	Aminco-Bowman and Farrand spectrofluorometers		
Iodine	Excit.	400 mμ	Norepinephrine
pH 5.0	Fluor.	520 mμ	only, in brain and heart
	Farrand fluorometer		
Ferricyanide	Excit. 436 mμ	Corning 4308 and 9863: Wratten 36	Human urine
pH 6.1–6.2 with zinc	Fluor. Corning 3486 (540 mμ)	Kodak 75	
	Aminco-Bowman spectrophotofluorometer		
Iodine			Human urine and
pH 6.5, both amines;	Excit. 410 mμ	395 mμ	pheochromocy-
pH 3.5, epinephrine only	Fluor. 520 mμ	505 mμ	toma plasmas; brain and heart
	Coleman Model 12C photofluorometer		
Ferricyanide	Excit. 436 mμ	395 mμ	Human urine
pH 6.2–6.3 (zinc has no effect)	Fluor. Corning 3486 (540 mμ)	Ilford-Bight (490 mμ)	

right are for norepinephrine. Excit. = for excitation; Fluor. = for fluorescence.

Twenty-four-hour urine specimens are collected in glass bottles containing 15 ml. of 6 N HCl; the final pH should be 3 or less. At this pH the catecholamines are stable in urine for at least several days at room temperature and for years at 2°. At neutral or alkaline pH values, however, their deterioration may be rapid. Volumes greater than 2500 ml./day are difficult to handle technically, and it may be necessary to impose some water restriction on the patients. Norepinephrine and epinephrine are present in urine as the free amines and as acid-labile conjugates. The method as presented is for the determination of free urinary catecholamines. Total (free plus conjugated) catecholamines may be determined by hydrolyzing the urine sample prior to assay (100° for 20 minutes at pH 0.5–1).

Reagents

Application of the trihydroxyindole method at its full sensitivity requires great care in the preparation of reagents. All solutions of mineral acids, bases, and buffers are prepared from reagent grade chemicals with water that has been distilled and passed through a commercial deionizer. Water purified in this manner is used throughout the procedure. Acid-washed aluminum oxide (for chromatographic use) is prepared in the following manner.

Boil 200–300 g. of alumina in 1 liter of 2 N HCl for 30 minutes in a reflux apparatus. Pour off the cloudy supernatant, stir the alumina briefly in 1 liter of distilled water, allow to settle for 5 minutes, and again decant the supernatant. Repeat this washing and decanting process with distilled water 12–15 times until the wash water clears after 5 minutes of settling and is at pH 4–5. Collect the alumina in a large suction funnel, allow to dry overnight in an open pan at room temperature, and then heat in an oven at 200° for 2 hours.

Alkaline ascorbic acid solution. A 1% ascorbic acid solution is prepared and 3 volumes are immediately mixed with 7 volumes of 5 N NaOH. This solution must be used immediately.

Epinephrine and norepinephrine working standards, 1 μg./ml. in 0.01 N HCl, are prepared by suitable dilutions of more concentrated standards. When stored at 2° they are stable indefinitely.

Procedure

An aliquot of urine equivalent to 10% of the 24-hour volume (50–250 ml.) is filtered through two sheets of Whatman No. 1 filter paper on a suction funnel to remove any sediment. The filtrate is transferred quantitatively to a 400-ml. beaker (smaller aliquots are diluted to approximately 200 ml. with water) and 5 ml. of 0.2 M EDTA* and 3 g. of alumina are

---

* EDTA (ethylenediamine tetraacetate, disodium salt) is added to the urine and the eluate to prevent the formation of a gelatinous calcium-magnesium-phosphate precipitate in the final reaction mixture which is read in the fluorometer. The presence of such a precipitate will cause light scattering and absorption and will markedly diminish the fluorescence.

added. With constant stirring on a magnetic stirrer and with constant monitoring of pH, bring the mixture to pH 7–7.5 by the dropwise addition of 5 N NaOH. Adjust to pH 8.4 by the further dropwise addition of 0.5 N NaOH. Stir for 7 minutes, adding 1–2 drops of 0.5 N NaOH, as necessary, to keep the solution at pH 8.4. The stirrer should be run fast enough to keep the alumina suspended and to minimize the grinding of alumina by the stirring bar. After 7 minutes of stirring, rinse the electrodes and stirring bar with a few ml. of water. Allow the alumina to settle for 3 to 5 minutes, and carefully decant the clear supernatant. Transfer the alumina† quantitatively, with water, to a glass column approximately 1.2 cm. in diameter. Allow the water to drain through and wash with two 10-ml. portions of water. Flow will stop when the water reaches the top of the alumina column. The water should flow by gravity at a rate of 1 to 2 ml./minute. If slower than this, it indicates the presence of fine granules of alumina in the column, the result of inadequate preparation of the alumina or too vigorous grinding during stirring. After washing with water the alumina should be white and the effluent about pH 6. Add exactly 3 ml. of 0.2 N acetic acid to the column and discard the effluent; this volume is sufficient to displace water from the alumina. Following this, add 9.4 ml. of 0.2 N acetic acid (in two 4.7-ml. portions) and collect the eluate in a calibrated centrifuge tube; it should be water clear and colorless. Dilute the eluate to 9.5 ml., add 0.5 ml. of 0.2 M EDTA, centrifuge to remove occasional granules of alumina which may pass through, and transfer to a test tube. The catecholamines are stable in this eluate for at least several weeks when stored at 2°.

For measurement of epinephrine plus norepinephrine, the trihydroxy-indole reaction is carried out at pH 6.5. The additions made to a series of tubes are given in the tabulation. Following these additions add to each tube 1.0 ml. of 1 M acetate buffer, pH 6.5. Add 0.10 ml. of 0.1 N iodine

Tube	Additions
1	0.20 µg. norepinephrine (standard)
2	0.20 µg. epinephrine (standard)
3	0.20 ml. H$_2$O (reagent blank)
4	0.20 ml. eluate
5	0.20 ml. eluate (for sample blank)
6	0.10 ml. eluate plus 0.10 µg. norepinephrine (internal standard)

† Adsorption is performed using a batch technique rather than a column because this exposes the catecholamines to an alkaline pH for a much shorter period of time. For elution the column is preferable to a filter funnel as its shape permits more quantitative elution with a smaller volume of acid.

solution to each tube, wait exactly 4 minutes, then add 0.50 ml. 0.05 N sodium thiosulfate to each tube to destroy the excess iodine. The iodo-aminochromes of both amines have now been formed. Prepare the 5 N NaOH-1% ascorbic acid solution and add 1.0 ml. to each of the tubes, except the sample blank (tube 5). To the latter tube add 0.70 ml. of 5 N NaOH, wait 15 minutes for any fluorescent indoles formed to deteriorate, then add 0.30 ml. of 1% ascorbic acid to complete the blank. Dilute all the tubes to 5.0 ml. with deionized water and allow them to stand for an additional 30–45 minutes before reading. When ferricyanide is used as the oxidizing agent, the base catalyzed isomerization of the aminochrome to the trihydroxyindoles occurs rapidly (1–2 minutes). With iodine, however, the process requires 45–60 minutes and is light-sensitive. Shielding the tubes from room light or reading prior to 30 minutes will produce erratic results. Fluorescence is stable for at least another hour.

For measurement of epinephrine the trihydroxyindole reaction is carried out at pH 3.5. Prepare a second series of tubes exactly as described above, except that 0.10 μg. of epinephrine is used in place of the norepinephrine in the internal standard (tube 6). Add 1.0 ml. of 1 M acetate buffer, pH 3.5, to each of the tubes and treat them exactly the same as for the pH 6.5 series. At pH 3.5 epinephrine is oxidized quantitatively but norepinephrine remains almost unchanged.

Fluorometric Assay

With a fluorescence spectrometer the instrument is set at the peak wavelengths for norepinephrine for the pH 6.5 series (excitation peak at 395 mμ; fluorescence peak at 505 mμ) and at the peak wavelength for epinephrine for the pH 3.5 series (excitation peak at 410 mμ; fluorescence peak at 520 mμ). With filter instruments use the 405-mμ mercury line for excitation and a complementary filter which transmits in the yellow-green region. A satisfactory primary filter combination consists of Corning glass filters Nos. 4308, 3060, and 5970; this passes nearly monochromatic light at 405 mμ. The secondary filter system may be composed of Corning glass filters Nos. 3385 and 4305; this passes light between 480 and 580 mμ, with peak transmittance at 505 mμ.

For both the pH 6.5 and pH 3.5 series set the meter with the corresponding standard and read the entire series of tubes. Samples which give readings which are off the scale (too concentrated) should be diluted appropriately with water and reassayed. Blank values are subtracted from the readings before calculation.

Calculations

For the determinations of free catecholamines in terms of norepinephrine (NE) equivalents, only the pH 6.5 figures are used. The catecholamine excretion is

$$\mu g./day = \frac{\text{reading of 0.1 ml. eluate at pH 6.5}}{\text{reading of 0.1 } \mu g. \text{ NE internal standard}}$$

$$\times 10 \times \frac{\text{urine volume}}{\text{volume of aliquot analyzed}}$$

Differential estimation of norepinephrine and epinephrine can be made from simultaneous equations based upon readings obtained at the two pH values. It is apparent that the greatest resolution is obtained with a procedure which distinguishes most between the two amines. Table VII shows the ratios of fluorescence intensity of epinephrine/norepinephrine obtained by this procedure and by other methods. Iodine oxidation at two different pH values resolves the two amines more completely (100:1) than does any

TABLE VII

COMPARISON OF METHODS FOR THE DIFFERENTIAL ESTIMATION OF EPINEPHRINE (E)
AND NOREPINEPHRINE (N)

Principle of method	Oxidizing agent	Ratios of fluorescence Intensities—E/N		Reference
		pH 6.5	pH 3.5	
Oxidation of eluates at	Iodine	0.5/1	50/1	(51)
two different pH values;	Ferricyanide	1.6/1	25/1	(53)
fluorescence measured	Manganese dioxide	1.4/1	20/1	(54)
at one set of wave-lengths.				
		Filter (wavelength) combination[a]		
		A	B	
Oxidation of eluates at	Ferricyanide	1/1	3/1	(52)
pH 6–6.5; fluorescence	Ferricyanide	1.2/1	5.7/1	(28)
measured with two	Manganese dioxide	1.1/1	5.8/1	(47)
different filter (wave-length) combinations.				

[a] Filter (or wavelength) combinations are described in detail in the individual references (after Crout[51]).

other fluorometric technique and is preferred for this reason. The resolution which is obtained by measurement at two sets of excitation and fluorescence wavelengths is relatively small (at best 5:1), so that the procedure is not precise with disproportionate amounts of the two amines. Of course, the latter procedure is simpler and requires less sample. It also gains in precision because both measurements are made on the same solution.

The recovery of added norepinephrine averages 91% (range 79–99%) and the recovery of added epinephrine 83% (range 68–103%). Duplicate samples analyzed on the same day will usually vary from their mean by no more than ±10%. If analyzed on separate days, the deviation may be as much as ±20%.

Dopa and dopamine are the only naturally occurring catechols which produce significant interference. The former is usually not present in detectable amounts but may be removed by isolating the epinephrines on a Dowex 50 (Na^+) column.[28] The normal urinary dopamine excretion of 100 to 200 μg./day may be estimated to produce a fluorescence equivalent to 3–6 μg. of norepinephrine per day. This would be significant only in patients with low norepinephrine excretion. The error due to dopamine is somewhat less with the ferricyanide oxidation technique and with bioassay procedures. The dopamine can, of course, be assayed independently (see above) and an exact correction applied. A frequent source of error in the trihydroxyindole assay for catecholamines is due to interference with fluorescence due to light scattering in the final sample. Most of the light scattering results from precipitates of calcium magnesium phosphate, in the absence of added ethylene diamine tetraacetate or when too little of the latter is added. Turbidity which is not apparent to the naked eye can produce appreciable light scattering and the scatter peaks should be checked (see Chapter 4) to make certain that scattering in the samples does not differ from that in the standards by more than ±50%. Even in the absence of turbidity, diminution of fluorescence may still occur owing to the presence in the eluates of contaminating compounds which absorb the excitation energy. The effects of such substances are largely diluted out by using small aliquots of eluate (0.1–0.2 ml.). The internal standard is run to correct for any remaining unpredictable effects of this type. With occasional urines the fluorescence of the internal standards is so markedly diminished as to make assay of those urines no more than semiquantitative.

Crout[51] reported that the excretion of free norepinephrine in hospitalized ambulatory patients was 30 ± 13 μg./day; epinephrine excretion was 5.6 ± 3.1 μg./day. The latter is barely significant. The total catecholamine excretion in hypertensive patients was 32 ± 18 μg./day, not significantly different from the nonhypertensives. In 90–95% of patients with pheochromocytoma (a catecholamine secreting tumor), the free catecholamine excretion was greater than 200 μg./day. Most patients with this disease can therefore be distinguished from other hypertensives. Crout and Sjoerdsma[55] consider daily urine excretion of 0 to 100 μg./day to be normal, 100–200 μg. to be suggestive of, and above 200 μg. to be diagnostic of pheochromocytoma. It should be pointed out that certain fluorescent drugs such as the tetracyclines or quinine and quinidine may interfere with catecholamine assay in urine. Metabolites of the latter two compounds have been reported[56] to yield abnormally high catecholamine values. Such interferences are, however, detectable by high sample blanks and by the characteristics of the excitation and fluorescent spectra.

As shown in Table VI, the procedure has also been widely applied to animal tissues and blood plasma. Bertler et al.[28] precipitate plasma proteins with perchloric acid and then absorb the epinephrines from the neutralized, desalted filtrate onto Dowex 50. The remainder of the procedure is outlined in Table VI. Shore and Olin[49] extract the norepinephrine from acidified brain and heart homogenates into butanol, wash the butanol free of interfering substances, then re-extract it into aqueous acid by adding heptane to the butanol. The norepinephrine (little or no epinephrine is present in these tissues) is then oxidized with iodine at pH 5. Excess iodine is destroyed with thiosulfate and the fluorescence is developed with alkaline ascorbic acid as described previously. Osinskaya[57] has used a procedure similar to Crout's[51] to measure epinephrine and norepinephrine in many organ tissues. Applications of the trihydroxyindole method to the assay of norepinephrine in tissues such as heart, brain, sympathetic nerves, etc., are satisfactory since these tissues contain relatively large amounts of the material. However, measurements in plasma present many problems. Norepinephrine content of normal plasma is less than 1 μg./liter and the normal epinephrine content is so low that it is difficult to be certain that any is present. In attempting to measure plasma catecholamines, one is actually using the fluorometric assay at the limits of its sensitivity. Under such conditions all the problems encountered in fluorometry become most prominent, i.e., reagent impurities, dirty glassware, light scattering, light absorption, etc. In Fig. 8 the spectral properties of the product obtained from plasma with this method are compared with a comparable assay for urine. It is apparent that whereas urine contains sufficient material so as to yield a proper fluorescence peak the material from plasma (at 10 times the instrumental sensitivity) is barely higher than the blank and cannot really be identified as norepinephrine fluorescence. The light scattering produced by the plasma sample is also disproportionately large. The mere increase of instrumental sensitivity by use of a more intense light source or a more sensitive photodetector cannot in-

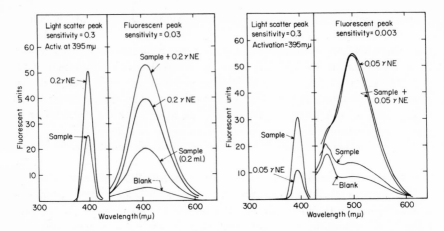

FIG. 8. Spectral characteristics of fluorophors obtained from urine, left (iodine oxidation) and from plasma, right (ferricyanide oxidation). Note that the light scatter is, in each case, so high that it must be measured at a lower instrumental sensitivity (after Crout[58]).

crease the real sensitivity of the method. The application of polarizing filters can eliminate much scattering including Raman scatter[58] (see Chapter 4). Perhaps they can effect a real increase in sensitivity. It is obvious from a survey of the literature that plasma norepinephrine levels, as measured by the trihydroxyindole method, differ appreciably from laboratory to laboratory. All that can be said is that the lowest values may be correct, assuming adequate controls have been run. In view of the extremely low instrumental readings (over the blank) obtained, the specificity of the assay must be seriously questioned. Small changes in apparent norepinephrine levels must certainly be interpreted with caution. It goes almost without saying that normal epinephrine levels, which are even lower, have little meaning although some individuals have developed instrumentation and techniques so as to yield, what appear to be, reasonable values (see Table VIII). So many divergent reports have appeared that in many instances the application of fluorescence assay to the determination of plasma catecholamines may be considered as the measurement of insignificant quantities with great precision. An excellent discussion and evaluation of the trihydroxyindole method as applied to plasma has been presented by Crout.[58] The criticisms do not apply to the markedly elevated levels such as are found after nicotine or related drugs, in some experimental conditions, or in patients with pheochromocytoma.

Adrenochrome, which is an intermediate in the conversion of epinephrine to the trihydroxyindole product, has been claimed to be a hallucino-

TABLE VIII

EPINEPHRINE AND NOREPINEPHRINE LEVELS REPORTED FOR NORMAL PLASMA IN MAN[a]

	Plasma levels (μg./liter) ,	
Investigators	Norepinephrine	Epinephrine
Price and Price[46]	0.34 ± 0.15	0.01 ± 0.07
Cohen and Goldenberg[47]	0.30 ± 0.07	0.06 ± 0.05
Vendsalu[48]	0.35 ± 0.10	0.07 ± 0.06
Holzbauer and Vogt[59]	<1.0	<0.06
Weil-Malherbe[b]	0.58 ± 0.11	0.22 ± 0.04

[a] The values are presented $\pm$ the standard error. References 46–48 represent trihydroxyindole assays. Reference 59 is a bioassay.

[b] These values were obtained with the ethylenediamine procedure (personal communication).

genic agent.[60] This has yet to be corroborated.[61] A procedure for the assay of adrenochrome in plasma was developed by the former group.[62] This depends on the same conversion to the trihydroxyindole product followed by fluorometric assay. Rather high levels of adrenochrome were reported in normal plasma and significant elevations in schizophrenia. However, the methodology has been criticized,[63] and Szara et al.[64] were unable to find any evidence of adrenochrome in the plasma of either normal or schizophrenic plasma using a comparable method.

2. CONDENSATION WITH ETHYLENEDIAMINE

The second reaction which has been widely used in the fluorometric assay of the catecholamines is the condensation with ethylenediamine. The reaction of diamines with oxidized catechols was first described by Wallerstein et al.[65] Natelson et al.[66] were the first to demonstrate the formation of fluorescent products from epinephrine and ethylenediamine and to recognize the importance of the reaction for analytical purposes. Shortly thereafter, Weil-Malherbe and Bone[67] published their well-known procedure for measuring norepinephrine and epinephrine, based on this reaction. Harley-Mason and Laird[68] later identified the fluorescent product formed with epinephrine. Their studies and those of Weil-Malherbe[69, 70] provide the basis for our current concept of the mechanism of the reaction with the catecholamines (Fig. 9). Epinephrine reacts with one molecule of ethylene diamine and maintains the N-methylethanolamine side chain. By contrast, norepinephrine reacts with two molecules of the reagent and loses the side chain during the reaction. Norepinephrine yields two major products which are chromatographically and fluorometrically identical

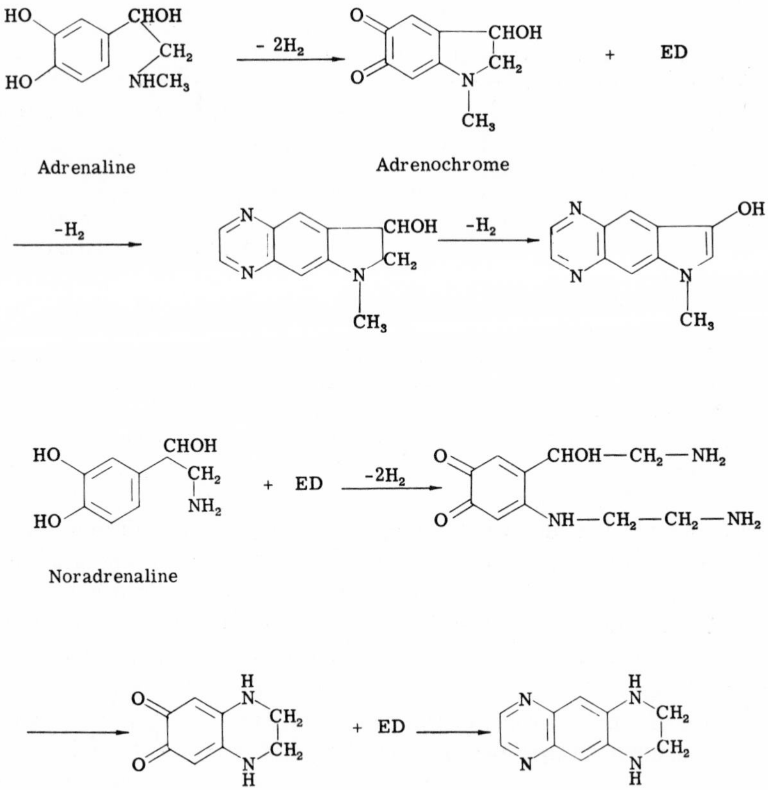

Adrenaline   Adrenochrome

Noradrenaline

FIG. 9. Reaction of ethylenediamine with epinephrine (adrenaline) and norepinephrine (noradrenaline) (according to Weil-Malherbe[70]).

with those formed from catechol and dihydroxymandelic acid. Only one product is shown in Fig. 9. The nature of the second product is not yet known. Although there have been reports that the ethylenediamine condensation leads to a large number of compounds,[71-74] Weil-Malherbe[70] has investigated this problem carefully and has shown that under the conditions of the assay, and with isolations carried out *in vacuo* and in the dark, epinephrine gives only one product whereas norepinephrine gives two major products. Nadeau and Joly[75] have proposed an entirely different mechanism from the one shown in Fig. 9. Nevertheless, it is fairly clear from the fluorescence characteristics (Table IX) that epinephrine and norepinephrine products are not the same.

As for its analytical applications, ethylenediamine, since it reacts with all catechols is inherently not as specific as the trihydroxyindole method.

TABLE IX

FLUORESCENCE CHARACTERISTICS OF EPINEPHRINE AND NOREPINEPHRINE PRODUCTS
WITH ETHYLENEDIAMINE

Investigator	Instrument	Fluorescence maximum (mμ)	
		Epinephrine	Nor-epinephrine
Weil-Malherbe[70]	Aminco-Bowman spectro-photofluorometer	525	485[a]
Mangan and Mason[76]	Farrand spectrofluorometer	525	495
Nadeau and Joly[73]	Modified "Model B" Beck-man spectrophotometer	530	490
Crout[58]	Aminco-Bowman spectro-photofluorometer	—	480[a]

[a] The reported excitation maximum is at 420 mμ. Similar excitation was apparently used for epinephrine fluorescence although this was not stated to be its maximum.
The excitation wavelength used in references 73 and 76 was 436 mμ, but this was not stated to be the maximum.
All the data were presumably uncorrected instrumental values.

On the other hand, following suitable separations it is possible to use the ethylenediamine reaction as a relatively specific and highly sensitive procedure. By introducing an anion exchange column to remove acidic and nonbasic catechols and with a fluorescence spectrometer for isolating the most suitable spectral regions for excitation and fluorescence detection, Weil-Malherbe* has been able to increase the specificity of the original procedure and to obtain plasma levels of norepinephrine and epinephrine consistent with the lowest values reported by the trihydroxyindole method (see Table VIII). The newly modified procedure for measurement of plasma catecholamines is presented in some detail since, as in the previous method, purification of reagents is very important in attaining the limits of sensitivity required to even detect the small amounts of catecholamines present in plasma.

Water, redistilled in an all-glass apparatus, is used throughout. Where the water supply is heavily chlorinated, a third distillation may be advisable. Anticoagulant solution is prepared by dissolving 1 g. of disodium ethylenediaminetetraacetate and 2 g. of sodium thiosulfate ($Na_2S_2O_3$ + $5H_2O$) in about 80 ml. of water. The mixture is titrated to pH 7.4 and the volume is made up to 100 ml. The 0.2 M sodium acetate solution is purified by passage through a column (25 cm. long, 2 cm. i.d.) of chelating

* H. Weil-Malherbe, personal communication.

resin (Dowex A-1, 50–100 mesh, Na$^+$ form; The Dow Chemical Company) adjusted to pH 8.4 with 0.5 N Na$_2$CO$_3$ and stored in a polyethylene bottle. Aluminum oxide, acid-washed, the product supplied by British Drug Houses, Ltd. "for chromatographic adsorption analysis," is recommended as it gives consistent results and is practically free of fluorescent impurities. Before use, it is heated, in batches of 100 g., with 500 ml. of 2 N HCl for 20 minutes at about 80°. Owing to the tendency of the heavy powder to "bump," rapid and efficient stirring is essential. The powder is filtered on a sintered glass funnel and washed, first with 500 ml. of hot 2 N HCl, then with water. It is then freed from dust by stirring it up repeatedly in about 500 ml. of water and decanting the supernatant. Finally, it is dried at 300° for 3 hours. Amberlite CG-50, Type 2 (Rohm and Haas Company) is prepared in the following manner. The resin is first soaked overnight in 2 N HCl. Then, after washing by decantation, it is stirred for 30 minutes with about 10 volumes of 2 N NaOH and washed again. Alternating treatments with acid and alkali, each lasting about 30 minutes, are repeated three times. Finally, the resin, (Na$^+$ form) is stirred with 1 M sodium acetate buffer pH 6.0; the pH of the mixture is adjusted to 6.0 with dilute acetic acid. The resin is stored in 1 M acetate buffer pH 6.0. Before use, the required amount is washed with water and sucked free of adhering moisture. Commercial ethylenediamine is redistilled in batches of 200 ml. in an all-glass apparatus, initial and final portions of the distillate being rejected. The reagent is sufficiently stable for several weeks when stored in a dark glass-stoppered bottle in the refrigerator. Commercial isobutanol is refluxed over NaOH (10 g./liter) for 3 hours and distilled through a fractionating column in an all-glass apparatus. The solvent is stored in a dark glass-stoppered bottle in the refrigerator.

Preparation of plasma: Twenty milliliters of blood are run through plastic tubing from an antecubital vein into a glass-stoppered 25-ml. graduated cylinder containing 5 ml. of anticoagulant solution. The contents are inverted gently, the total volume is noted and a sample is removed for the hematocrit determination. The remainder is centrifuged for 20 minutes at 1000 r.p.m. (250 × g.). Plasma is separated and its volume measured. After adding an equal volume of sodium acetate solution, the pH is adjusted to 8.4 by the cautious addition of 0.5 N Na$_2$CO$_3$, using a pH meter and automatic stirrer.

After placing a small wad of glass wool over the constriction, 0.5 g. of alumina is introduced into the bulb of an adsorption tube. The stem is then filled with sodium acetate solution after closing the bottom end of the tube. The alumina in the bulb is covered with about 5 ml. of sodium acetate solution and swirled gently around to facilitate the escape of air bubbles. The adsorbent is finally allowed to sink to the bottom of the stem. When the meniscus of the sodium acetate solution is near the top of the alumina column, the plasma is poured in and allowed to flow through by gravity. The column is washed with 3 ml. of sodium acetate solution, followed by 3 ml. of water. Filtrates and washings are rejected and the receiver is replaced by a 10-ml. beaker. The column is then eluted with 3 ml. of 0.3 M acetic acid, followed by 3 ml. of water. The eluate (6 ml.) is mixed with 0.6 ml. of 1% EDTA and titrated to pH 6.0. It is then passed over a column of 0.3 g. Amberlite CG-50, Type 2 (weighed moist). The column is washed with 20 ml. of water and is eluted with 5 ml. of 1 M acetic acid, the eluate being collected in a 10-ml. tube of "low actinic" glass* with ground stopper. The rate of flow should not exceed 0.5 ml./ minute. The eluate is transferred to a stoppered tube* containing 0.8 ml. of ethylenediamine and the tube is shaken and placed in a water bath at 55° for 25 minutes. After cooling, about 1.5 g. of NaCl is added through a funnel, followed by 3 ml. of isobutanol. The mixture is shaken mechanically for 4 minutes and the organic layer is transferred with a "low-actinic" pipette to a "low-actinic" test tube which is then covered with tin foil.

Reagent blanks are carried through the entire procedure ("column blanks"). Standards containing 0.05 μg. of epinephrine and norepinephrine, respectively, and a second reagent blank are prepared in 5 ml. of 1.0 M acetic acid and carried through the condensation procedure along with the plasma eluates.

Fluorescence is measured with a fluorescence spectrometer (using the widest slits). The wavelength of excitation is set at 420 mμ and fluorescence is measured at two wavelengths, 510 and 580 mμ. The readings of the plasma eluates are corrected for the "column blank," those of the standards for the appropriate reagent blank. The latter is usually slightly less fluorescent than the "column blank" but the difference should not be large.

Calculations: The fluorescence of the norepinephrine derivatives at 510 mμ is approximately twice that of the epinephrine derivative. At 580 mμ the epinephrine derivative is more fluorescent, also by a factor greater than 2 (Fig. 10). Simultaneous equations set up from the two readings and known concentrations of the two standards and of the readings of an unknown permit calculation of the proportions of each of the two amines present. Recoveries vary from 70 to 100%. It should be pointed out that the differences in fluorescence intensity between the two amines, at the two wavelengths settings, are not so great as to permit precise determina-

---

* The "low-actinic" tubes with ground-glass stoppers may be obtained from Corning Glass Works.

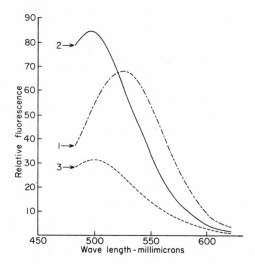

FIG. 10. Fluorescence spectra of ethylenediamine condensation products of epinephrine (1), norepinephrine (2), and the reagent blank (3). In each instance 0.2 μg./ml. of the base was used and measurements were made in isobutanol. Excitation was at 436 mμ (after Mangan and Mason[76]).

tion of the composition of a mixture when one or the other amine predominates. Furthermore, 580 mμ is far removed from the fluorescence peak where the fluorescence intensities of both compounds are about 10% of what they are at their maxima. This limits the sensitivity of the differential assay.

There have been many applications of the ethylenediamine method to the assay of the epinephrines in tissues and with each improvement in the method the measured values for human plasma have been found lower and lower. Weil-Malherbe's first values[67] for norepinephrine were about 15 to 16 μg./liter.* Subsequent values by Gray and Young[77] were 3.4–4.9 μg./liter. In a more recent report by Weil-Malherbe[69] the values for plasma norepinephrine were 1.5 μg./liter. The values obtained with the method presented above are 0.58 μg./liter (Table VIII). Obviously, the specificity of the plasma assay is far from established. As with the trihydroxyindole method, the plasma increment in fluorescence intensity over reagent blanks is not very large. A spectrophotometric study of a typical plasma assay with the ethylenediamine method is shown in Fig. 11. It is the opinion of this author that in none of the methods currently available has

---

* Total epinephrines were measured and the values were recalculated in terms of norepinephrine.

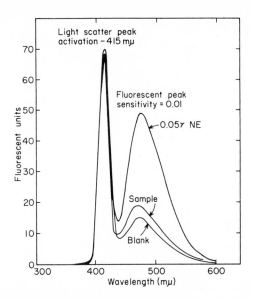

FIG. 11. Fluorescence spectra and light scatter of 0.05 μg. of authentic norepinephrine, a sample of plasma "norepinephrine," and a reagent blank assayed by the ethylenediamine method (after Crout[58]).

it been possible to establish the identity of the apparent norepinephrine in normal plasma. Reported changes in the circulating norepinephrine due to diseases, drugs, or experimental conditions must also be viewed with skepticism unless these are rather large (several times the controls) and, when increased, can be shown to yield characteristic excitation and fluorescence maxima. The ethylenediamine method, like the trihydroxyindole method, can be used diagnostically in pheochromocytoma and in situations where the plasma norepinephrine levels are markedly elevated. One important limitation of the ethylenediamine method is that it yields fluorescence of comparable intensity with dopamine. Tissues which contain appreciable amounts of this relatively inactive precursor of norepinephrine cannot be assayed for norepinephrine by this method. A summary of some catecholamine levels in various animal tissues is shown in Table X.

G. Catechol Acids

One of the important routes of catecholamine metabolism involves oxidative deamination catalyzed by the enzyme, monoamine oxidase. The intermediate aldehydes are further oxidized to the dihydroxy acids.

TABLE X

CATECHOLAMINE LEVELS IN VARIOUS TISSUES[a,b]

Tissue	Dopamine μg./g.	Norepinephrine μg./g.	Epinephrine μg./g.	Reference
Rat				
Brain	1.0	0.44	0	[c]
Heart	0	1.1	0	[c]
Spleen	—	0.4	—	(79)
Adrenal gland	—	130	1100	(79)
Guinea Pig				
Brain	0.55	0.28	0	[c]
Heart	0.02	2.0	0	[c]
Spleen	—	1.5	—	(78)
Adrenal gland	—	68	560	(78)
Rabbit				
Brain	0.2	0.5	0	(49)
Heart	0	2.0	0	(49)
Adrenal gland	—	10	470	(78)
Cattle				
Lung	1.1	0.02	0	(35)
Splenic nerve	4.5	4.5	0	(35)
Stellate ganglion	1.5	1.5	0	(35)
Small intestine	17	0.4	0	(35)
Adrenal medulla	80	1200	3000	(78)
Sheep				
Brain	0.1	0.1	—	(35)
Lung	25.0	0	0	(35)
Small intestine	3.5	0	—	(35)
Adrenal medulla	28	540	1100	(78)
Banana				
Peel	700	122	0	(79)
Pulp	8	2	0	(79)
Human				
Plasma	0	0.0003[d]	0.00001[d]	[e]

[a] Wet weight.

[b] The values in the table are to be taken only as indicative of the range of values to be expected in a given organ. Factors such as diet, sex, age, strain, etc., can influence the catecholamine levels significantly.

[c] These values represent the mean of observations in the author's laboratory.

[d] μg./ml.

[e] Values taken from Table VIII.

Both dihydroxyphenylacetic acid and dihydroxy mandelic acid are found in urine[80] and have also been demonstrated in plasma. Both acids possess native fluorescence characteristic of phenols.[27] A more characteristic fluorescence is that formed by condensation with ethylenediamine. It is possible to determine dihydroxy acids by extraction from acid solution into ether, re-extraction into pH 7.0 buffer and treatment with ethylenediamine as described for catecholamines. Such a procedure would, of course, determine all catechol acids. More specific procedures for the catechol acids have been developed by Rosengren.[81]

H. O-Methyl Catechol Metabolites

The recent finding that a major portion of the circulating catecholamines are metabolized by methylation on the meta hydroxy group has made it important to develop assays for normetanephrine, metanephrine,

$$CH_3O-C_6H_3(OH)-CHOHCH_2NHCH_3$$

Metanephrine

vanillyl mandelic acid, and related compounds. The following method of assay, based on the native fluorescence of the O-methyl catecholamines has been developed by Axelrod and Tomchick[82] and applied to enzyme studies.

Metanephrine, normetanephrine, and 3-methoxytyramine are extracted from borate buffer (pH 10) into 20 volumes of ethylene dichloride containing 2% isoamyl alcohol. A 15-ml. aliquot of the solvent is shaken with 1.5 ml. of 0.1 N HCl and the acid extracts are subjected to fluorometric assay, excitation being at 285 mμ and fluorescence at 335 mμ (uncorrected). Known amounts of the synthetic compounds are carried through the entire procedure to correct for the partition of the methylated metabolites in the two-phase system. While this procedure is adequate for studies with purified enzymes or for identification of material obtained by macroisolation procedures,[83] it is not sufficiently specific for measurements in tissue or urine extracts.

A fairly sensitive and specific spectrofluorometric method has been developed for the assay of metanephrines by Bertler et al.[84] They have reported that the metanephrines undergo oxidation and rearrangement to the same trihydroxyindole products as the parent amines. Oxidation of the metanephrines requires a more alkaline pH and more drastic conditions and the yield of fluorophor is somewhat less. Smith and Weil-Malherbe[85] have developed a fluorometric method for the differential estimation of metanephrine and normetanephrine in urine, based on their oxidation to fluorescent indoles at two different pH values. 3-Methoxytyramine gives no fluorescence under similar conditions. Urine is hydrolyzed in acid, passed over alumina to remove catechols, and desalted. The metanephrines are adsorbed on Amberlite CG-50 resin at pH 6.0 and are eluted with 1 N formic acid. As little as 0.25 to 0.5 μg. of each base may be measured with recoveries of 80 to 90%. Preliminary results of studies on normal humans indicate hourly excretion rates of 1 to 4 μg. for metanephrine and 3–12 μg. for normetanephrine. Brunjes et al.[86] have utilized a very similar procedure and reported uncorrected spectral data; for metanephrine 390 mμ excitation and 480 mμ fluorescence and for normetanephrine 430 mμ excitation and 520 mμ fluorescence.

IV. TRYPTOPHAN AND ITS METABOLITES

Indoles are among the most highly fluorescent of the biologically occurring compounds. The intense fluorescence makes these products readily detectable in extracts and on chromatograms and as a result they have received a great deal of attention. There have been countless indole derivatives reported in nature. Some of the more important metabolic products will be discussed.

In animals and plants, tryptophan is converted to tryptamine and indole acetic acid. It is also hydroxylated to 5-hydroxytryptophan which is decarboxylated to 5-hydroxytryptamine (serotonin). The latter compound is an important physiologic agent. It, in turn, is oxidatively deaminated to 5-hydroxyindoleacetic acid. A most important pathway of tryptophan metabolism in animals involves rupture of the ring by tryptophan pyrrolase. This reaction yields a number of fluorescent aromatic amines and quinoline derivatives. Fluorometric assay procedures have been devised for many of these metabolites.

A. Indoles, General

Indoles which are not substituted on the ring system fluoresce at 348 mμ when excited at 287 mμ. These are corrected values reported by Teale and Weber[10b] for tryptophan in neutral aqueous solution. Under such conditions the amino acid fluoresces with an efficiency of about 20%. Congeners of tryptophan with different substituents on the 3 position of the indole ring fluoresce at the same wavelengths, but with different efficiencies. For all these compounds the fluorescence yield varies with pH and is lowest in

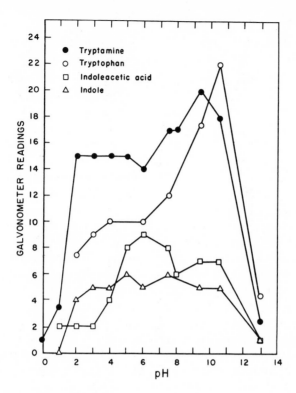

FIG. 12. Effect of pH on the fluorescence of tryptophan and related compounds.

strong acid and base.[14] The data presented in Fig. 12 indicate that dissociation of the side chain has an appreciable effect on the fluorescence yield. This is not only of theoretical interest but of practical value as well. Thus, enzymatic interconversions, as from tryptamine to indoleacetic acid, should be detectable by a large change in fluorescence intensity at pH 3. Binding or interaction of such a molecule in a manner which influences the charge of the side chain will also influence fluorescence yield. Many of the fluorescent studies on proteins depend on the fluorescence of tryptophan and effects on the fluorescence yield of tryptophan produced by molecular interactions within the protein molecule. These are discussed in the section on proteins (Chapter 6).

The native fluorescence of indoles is sufficiently intense and specific to permit its use for quantitative assay purposes. Procedures for many of these metabolites have been developed based on their native fluorescence.

B. Tryptophan

Protein hydrolysates[15]: Of all the naturally occurring amino acids in proteins, only tryptophan and tyrosine exhibit appreciable fluorescence in

$$\text{CH}_2\text{CHNH}_2\text{COOH}$$

Tryptophan

aqueous media, with presently available instrumentation. The fluorescence intensity of tryptophan is greatest at about pH 11 (Fig. 12) and at this pH is approximately 100 times greater than that of a comparable concentration of tyrosine. Furthermore, the fluorescence maximum of tryptophan (350 mμ) occurs about 40 mμ from that of tyrosine (310 mμ). With a fluorescence spectrometer one can also resolve the tryptophan and tyrosine fluorescent bands (Fig. 3) so that it is possible to measure tryptophan in the presence of a large excess of tyrosine.

Protein samples as small as 100 μg. may be used. These are subjected to hydrolysis in test tubes of Corning brand No. 7280, alkali-resistant glass to reduce light scattering due to colloidal silicate. The samples, in 1 ml. of 5.0 N NaOH, are heated in unstoppered tubes in the autoclave for 20 hours at 2 atmospheres. The cooled hydrolysates are acidified with 1.5 ml. of 5 M H$_2$SO$_4$ and are made up to 25 ml. with water. It may be necessary to clarify this acidified solution by centrifugation. One milliliter of the clear solution is diluted with 5 ml. of 1 M sodium carbonate and the fluorescence is assayed using the excitation and fluorescence maxima of tryptophan (see above). Tryptophan standards are carried through the entire procedure, including hydrolysis, to correct for losses during alkaline hydrolysis (5–10%). A reagent blank is also carried through the over-all procedure.

Recoveries of tryptophan from known amino acid mixtures are quantitative and the values for the tryptophan contents of insulin, bovine serum albumin, and β-lactoglobulin have been found to be in excellent agreement with values obtained by other procedures (Table XI). It should be pointed out that in insulin the presence of 11.6% tyrosine did not interfere with the tryptophan assay. The fluorometric assay of tryptophan is now the simplest and most sensitive of all available procedures. Because of its simplicity it is also precise. It should therefore be an excellent tool for studies on protein structure. Velick[18] has used this procedure for the determination of the tryptophan content of glyceraldehyde-3-phosphate dehydrogenase.

TABLE XI

TRYPTOPHAN CONTENT OF PROTEINS

Protein	Tryptophan	
	Found[a]	Given in literature[a]
Insulin	0	0 (87)[b]
Bovine serum albumin	0.71	0.6 (16)
β-Lactoglobulin	1.75	1.92 (17)

[a] The values are given in grams per 100 g. of protein.
[b] Bibliographic reference in parentheses.

Plasma: Although biological materials, such as plasma, contain many substances which can fluoresce, it was shown by Duggan and Udenfriend[15] that tryptophan is the only component normally present in tungstic acid filtrates of plasma to yield measurable fluorescence under the conditions of tryptophan assay. As shown by countercurrent distribution, fluorescence pH curves, and studies with various quenching agents, the fluorophor in such filtrates is identical with tryptophan.

To 1.0 ml. of plasma are added 4 ml. of water and 0.5 ml. of 0.6 N H_2SO_4. Addition of 0.5 ml. of 10% sodium tungstate, with constant stirring, precipitates the proteins which are removed by centrifugation. Tungstate ion, which has a slight quenching effect on tryptophan fluorescence, is precipitated by the addition of 1 ml. of 0.25 M barium chloride to 3.0 ml. of the clear tungstic acid filtrate. After centrifugation, to remove the resulting precipitates of barium sulfate and barium tungstate, a 2.0-ml. aliquot of the solution is transferred to another centrifuge tube containing 0.5 ml. of 2 M sodium carbonate. The barium carbonate precipitate is removed by centrifugation for 10 minutes at 3000 r.p.m. A portion of the supernatant fluid, which should be clear and colorless, is transferred to a cuvette and fluorescence is assayed using the appropriate wavelength settings for excitation and fluorescence (see above). Standards for comparison may be prepared by appropriate dilutions of tryptophan in 0.3 M sodium carbonate. As an additional check, internal standards (tryptophan added to plasma) should be used. A reagent blank, substituting water for plasma is carried through the entire procedure. The fluorescence spectra of plasma, standards, and blank, in a typical determination are shown in Fig. 13.

The procedure has been applied to plasma of various species of laboratory animals and man. Fasting levels in man are about 10 μg./ml. Tissues

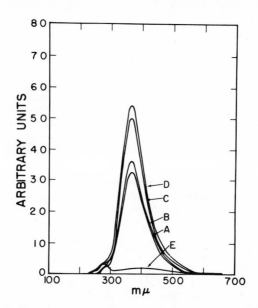

FIG. 13. Fluorescence spectra of tryptophan standards (curves A and C), plasma tryptophan (curve B), plasma plus added tryptophan, and reagent blank (Duggan and Udenfriend[15]).

other than plasma contain absorbing substances which make it impossible to utilize the procedure for tryptophan exactly as presented above. It may be that an intermediary isolation of tryptophan, perhaps by chromatographic means, would permit direct fluorometric assay of the tryptophan.

Another procedure for fluorometric assay of tryptophan involves condensation with formaldehyde followed by oxidation to yield norharman.[88] The details of this reaction will be presented in the following section on tryptamine. The reaction has been applied to trichloracetic acid filtrates of plasma and tissues. Values obtained on normal human plasma were found to be comparable to those obtained by the previously described procedure.[15] The method is highly specific and its sensitivity is so great that it can be applied to the assay of tryptophan in fingertip blood.

C. Tryptamine and Indoleacetic Acid

Tryptamine is derived from tryptophan by decarboxylation. The amine is widely distributed in the plant kingdom including ordinary fruits and vegetables.[81] It is present in animal tissues, including brain, and in urine where its daily excretion serves as an index of amine formation[26] and metabolism.[89]

The following simple method for measurement of tryptamine in human urine[89] involves isolation of the amine from alkalinized urine by extraction into benzene, return to an acid aqueous phase, and fluorometric assay. Fifteen milliliters of urine in a 50-ml. glass-stoppered centrifuge tube are adjusted to pH 11–14 with 10 N NaOH. Thirty milliliters of benzene are added and the tube is shaken for 20 minutes and then centrifuged. The benzene is transferred to a similar tube and washed once by shaking with 5 ml. of 0.1 N NaOH. After centrifugation, a 25-ml. aliquot of the benzene layer is transferred to a 35-ml. glass-stoppered tapered centrifuge tube containing 0.7 ml. of 0.1 N HCl. The tube is shaken for 15 minutes, centrifuged, and the supernatant benzene layer is removed by aspiration and discarded. A 0.5-ml. aliquot of the acid layer is transferred to a test tube containing 1 ml. of 0.13 M borate buffer, pH 10, and the tryptamine in this buffered solution is assayed in a fluorescence spectrometer; 285-mμ excitation and 360-mμ fluorescence (uncorrected).

A tryptamine standard (about 2 μg.) and a reagent blank are carried through the entire extraction procedure. As little as 0.1 μg. in 15 ml. of urine can be measured by this assay; fluorescence is proportional to trypt-

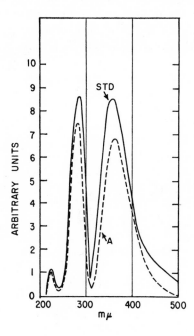

Fig. 14. Excitation and fluorescence spectra of authentic tryptamine (curve STD) and apparent tryptamine isolated from urine (curve A) by the procedure of Sjoerdsma et al.[89]

amine concentration from 0.1 to 8 μg. The average tryptamine excretion in man is about 75 μg./day. The identity of the extracted urinary tryptamine was established by comparison of its spectral characteristics (Fig. 14) and countercurrent distribution pattern with those of the authentic compound.

Some drugs produce extractable fluorescent material which interferes with this procedure. However, the presence of such foreign fluorophors is readily detected by their excitation and fluorescence spectra which differ from those of tryptamine. Examination of the excitation and fluorescence spectra of the urine extracts should be included in the assay procedure when drugs are included in a patient's regimen or when a new patient's urine is assayed for the first time.

The method described above is excellent for human urine, but is not applicable to the assay of tryptamine in the tissues or urine of some laboratory animals. Another fluorometric procedure for measuring tryptamine, developed by Hess and Udenfriend,[88] is applicable to urines and tissues of laboratory animals and humans. The procedure involves extraction of the tryptamine from tissue with an organic solvent, condensation with formaldehyde, and oxidation of the tetrahydronorharman to the highly fluorescent norharman. The chemical steps in this conversion are shown in Fig.

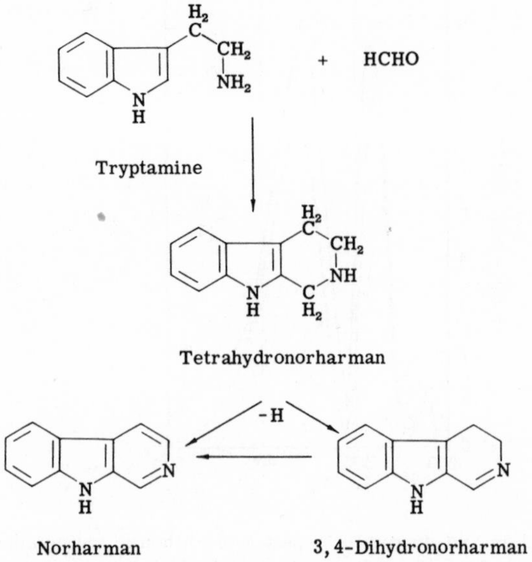

Tryptamine

Tetrahydronorharman

Norharman   3,4-Dihydronorharman

FIG. 15. Conversion of tryptamine to norharman.

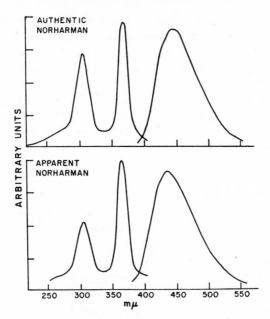

FIG. 16. Excitation and fluorescence spectra of authentic norharman and of apparent norharman prepared from tryptamine extracted from guinea pig liver following the administration of tryptophan and a monoamine oxidase inhibitor (after Hess and Udenfriend[88]).

15. That the final product is norharman is shown by the characteristic excitation and fluorescent spectra (Fig. 16).

The procedure for tissues is as follows: To 3 to 5 ml. of tissue homogenate or urine, containing 0.1–5 μg. of tryptamine, in a glass-stoppered centrifuge tube, are added 0.3 ml. of 10 N NaOH to bring the pH to about 11. Twenty milliliters of benzene are added and the tube is shaken for 15 minutes on an automatic shaker. The tube is centrifuged and as much of the benzene as possible is transferred to another tube containing 5 ml. of 0.1 N NaOH. The tube is shaken for 5 minutes, centrifuged, and 15 ml. of the benzene is transferred to another tube containing 4 ml. of 0.1 N H$_2$SO$_4$. This tube is also shaken for 5 minutes, centrifuged, and 3 ml. of the acid layer are transferred to a test tube containing 0.1 ml. of 18% formaldehyde. The tube is covered with a glass marble and placed in a boiling water bath for 20 minutes following which 0.1 ml. of 5% aqueous H$_2$O$_2$ is added and heating is continued for another 20 minutes. After cooling to room temperature, fluorescence is measured in a fluorescence spectrometer: 365-mμ excitation, 440-mμ fluorescence (uncorrected). Known amounts of tryptamine added to tissue yield recoveries which are essentially quantitative, 90–100% based on extracted standards. When the reagents themselves are carried through the entire extraction procedure and compared to 0.1 N H_2SO_4, the resulting fluorescence is equivalent to about 0.2 to 0.3 μg. of tryptamine. This blank fluorescence limits the sensitivity of the method. The amounts of tryptamine normally present in the tissues of several laboratory animals were found to be too small, 0.05–0.10 μg./g., to give significant readings above the reagent blank range. The normal values are presumably less than 0.08 μg./g. However, following the administration of L-tryptophan or of monoamine oxidase inhibitors, tissue levels of tryptamine rise to highly significant values (1–2 μg./g.).[90] When applied to human urine this procedure yielded the same values as were obtained by the method of Sjoerdsma et al.[89] Since the condensation procedure occurs only with indolethylamine derivatives, indoleacetic acid and 5-hydroxyindoleacetic acid do not interfere. Tryptophan and 5-hydroxytryptamine in concentrations over 500 times greater than tryptamine do not interfere in the over-all procedure, the former being removed by the solvent extraction and washing procedure.

Jepson and Stevens[91] have reported that under appropriate conditions tryptamines on paper chromatograms react with ninhydrin to yield highly fluorescent β-carbolines. Although the reaction takes place only on paper, the product can be eluted and its fluorescence assayed in solution.[92] The latter group has combined chromatographic separation with the ninhydrin fluorescence reaction into a procedure for the measurement of tryptamine in animal urine.

Like tryptophan and tryptamine, indoleacetic acid is highly fluorescent. It is present in animal urine as an end product of tryptamine metabolism and is widely distributed in the plant kingdom where it functions as a regulatory substance. Because it can be extracted from acidified solutions into nonpolar solvents, such as benzene, then returned to aqueous alkali, it is possible to separate this acid from related indoles and other interfering substances and assay it fluorometrically. Excitation and fluorescence maxima are the same as for tryptophan. Weissbach et al.[93] have used fluorometric assay of indoleacetic acid to follow the oxidative deamination of tryptamine, and N-methyl tryptamine. Details of this will be presented in the section on assay of enzymes (Chapter 9).

Many other indole derivatives are known in nature. Some will be discussed in the section relating to plants (Chapter 11) others in the section on drugs (Chapter 13). Procedures for detecting a number of indoles on paper chromatograms by the characteristic fluorescence produced by different spray reagents have been reported by Smith.[94]

D. *N*-Acyl Derivatives of Tryptophan

Uphaus *et al.*[95] noticed that proteins when dissolved in trifluoroacetic acid developed a green fluorescence of intensity roughly proportional to their tryptophan content. Upon further study they found that the reaction is characteristic of *N*-acyl derivatives of tryptophan; *N*-acetyl tryptophan, glycyl tryptophan, and tryptophan containing proteins. The reaction in oxygen free solutions of trifluoroacetic acids (and in other perfluoroacids) yields a product with an excitation maximum at 420 mμ and a fluorescence maximum at 530 mμ (uncorrected). The fluorophor from acetyl-DL, tryptophan, in a nitrogen atmosphere, was isolated and identified as 1-methyl-3-carboxyl-3,4-dihydro-β-carboline. In the presence of oxygen and light, the green fluorescent dihydro-β-carboline is oxidized to the β-carboline which absorbs at 280 mμ and fluoresces at 470 mμ.

1-Methyl-3-carboxyl-β-carboline

Tryptophan does not yield comparable fluorescence under these conditions. It should be pointed out that tryptophan on condensation with formaldehyde, as described by Hess and Udenfriend[88] (see above), yields a comparable dihydro-β-carboline, norharman. However, the fluorescence characteristics of this derivative in acid solution are 315- and 365-mμ excitation and 440-mμ fluorescence. Perhaps the trifluoroacetic interacts with the compound in some manner to alter its fluorescence. Although no analytical applications were suggested by the authors, the reaction in trifluoroacetic acid should be applicable as a specific method for assay of conjugated forms of tryptophan, including peptides, in the presence of free tryptophan and other indoles.

Freed and Salmre[96] have reported that the phosphorescence of indoles may be more specific than their fluorescence.

E. 5-Hydroxyindoles

An important route of tryptophan metabolism involves hydroxylation to 5-hydroxytryptophan followed by decarboxylation to 5-hydroxytryptamine. Although 5-hydroxytryptamine (serotonin) was isolated and identified only a little more than 10 years ago, it is now recognized as one of the most potent physiologically active amines. It is widely distributed in nature, both in plants and animals, and is currently receiving the attention of many physiologists, pharmacologists, and biochemists. Fortunately, serotonin and other 5-hydroxyindoles are highly fluorescent in aqueous

$$HO\diagup\!\!\diagdown\diagdown\diagdown CH_2CH_2NH_2$$

5-Hydroxytryptamine
(serotonin)

solution so that they can be assayed by this sensitive procedure. In neutral or slightly acid solutions, 5-hydroxyindoles fluoresce at 330 mμ when excited at 295 mμ[97] (uncorrected). When the acidity is increased, by addition of HCl, the 330-mμ fluorescence decreases and a new fluorescence peak appears at 550 mμ[98] (uncorrected). In 3 N HCl the 550-mμ fluorescence is maximal and the 330-mμ fluorescence is extremely low. This shift of fluorescence from the ultraviolet to the visible, with increasing acidity, is completely reversible and is not accompanied by a noticeable change in the absorption spectrum. Unsubstituted indoles do not exhibit such a shift, indicating that the phenolic group is involved in some manner. The energy difference between the exciting and emitted light (about 47 kcal./mole) is unusually high for fluorescence, and would suggest, perhaps, other species being involved in an energy transfer mechanism. However, no other absorption band is apparent. Further studies on the fluorescence of 5-hydroxyindoles in strong acid are warranted.

1. 5-HYDROXYTRYPTAMINE

There have now been published a number of methods for serotonin assay based on the above fluorometric properties. Each of these procedures was devised for a specific purpose. The solvent extraction method of Udenfriend et al.[97] was designed to measure serotonin in relatively pure extracts; for enzyme studies and platelets. For these purposes the 330-mμ fluorescence in dilute acid is adequate. Bogdanski et al.[99] utilized a modification of this procedure, along with 550-mμ fluorescence in 3 M HCl, to assay serotonin in brain and other tissues. The fluorometric measurement of serotonin in whole blood requires precipitation of proteins with zinc sulfate and NaOH.[100] Where the blood levels are sufficiently high (above 1 μg./ml.) as in rabbits, chickens, and patients with malignant carcinoid,* fluorometric assay is carried out directly on the acidified protein free filtrate. Where blood levels are below 1 μg./ml., protein precipitation may be followed by solvent extraction as in the procedure of Udenfriend et al.[97] A combination of protein precipitation, solvent extraction, and differential

---

* A serotonin secreting tumor of enterochromaffin origin.

absorption has been devised to assay serotonin and histamine in the same sample of blood.[101] The procedure devised by Shore *et al.*[49] for norepinephrine assay permits the simultaneous assay of serotonin as well. Oates[102] has devised a chromatographic procedure for the fluorometric assay of serotonin in urine of patients with malignant carcinoid. Details of some of these modifications may be found in *Methods of Biochemical Analysis.*[100]

The most widely used procedure for measurement of serotonin in plant and animal tissues involves extraction with *n*-butanol from a salt-saturated alkaline homogenate and the return of the serotonin to an aqueous phase by the addition of heptane to the *n*-butanol and shaking the mixed solvents with dilute acid. The acid solutions are then assayed fluorometrically and compared with appropriate standards and blanks.

The reagents for the assay are relatively simple but great care must be taken to prepare some of them.

Borate buffer: To 94.2 g. of boric acid dissolved in 3 liters of water are added 165 ml. of 10 *N* NaOH. The buffer solution is then saturated with purified *n*-butanol and NaCl by adding these substances in excess and shaking. Excess *n*-butanol is removed by aspiration and excess salt is allowed to settle. The final pH should be about 10.0. Reagent grade *n*-butanol is purified by first shaking with an equal volume of 0.1 *N* NaOH, then with an equal volume of 0.1 *N* HCl, and finally twice with distilled water. Practical grade of heptane is treated in the same manner as the *n*-butanol.

One part of tissue is homogenized in 2 parts of 0.1 *N* HCl and an aliquot, containing 0.2–5.0 μg. of serotonin, is transferred to a 60 ml. glass-stoppered bottle containing 3 ml. of borate buffer pH 10.0. If necessary, the pH is adjusted to between 9 and 10 by the addition of anhydrous sodium carbonate, but the pH should not exceed 10.5. The solution is diluted to an appropriate volume (usually less than 10 ml.) and 3–4 g. of NaCl are added to saturate the aqueous phase with salt. Fifteen milliliters of butanol are added and the bottle is shaken for 10 minutes. After centrifugation the fluid is decanted from the solid material into another bottle and the aqueous phase is removed by aspiration. The butanol layer is then washed with an equal volume of borate buffer pH 10 to remove any neutral or acidic 5-hydroxyindoles, such as 5-hydroxytryptophan and 5-hydroxyindole-acetic acid. Three washings may be necessary when large amounts of 5-hydroxytryptophan are present. Ten milliliters of the butanol phase are then transferred to a 40-ml. glass-stoppered shaking tube containing 15 ml. of heptane and 2.5 ml. of 0.1 *N* HCl. The tube is shaken for 5 minutes, centrifuged, and 1 ml. of the dilute acid is transferred to a test tube containing 0.3 ml. of 12 *N* HCl. The solution is assayed fluorometrically in a fluorescence spectrometer; excitation, 295 mμ; fluorescence, 550 mμ (uncorrected). Standards, recoveries, and a reagent blank are carried through the entire procedure in the same way. Serotonin added to tissues is recovered to the extent of 92 to 106%.

Instrumentation: Since excitation of 5-hydroxyindoles is below 300 mμ, it is necessary to use quartz for the cuvettes and optics. Fluorescence spectrometers have been mainly used for this assay. These are to be preferred because they permit identification as well as assay. However, some of the commercially available fluorophotometers are so designed that it may be possible to detect and assay serotonin with relatively little modification.

Distribution of serotonin between the salt-saturated buffer, pH 10, and butanol is such that, with the volumes used in the procedure, only 90–95% is extracted into the *n*-butanol. Since as many as three equilibrations between solvent and buffer may be employed as much as 15 to 20% of the serotonin may be lost. When heptane is added to the water-saturated butanol, water separates out from the solvent, increasing the volume of the acid aqueous phase. It is for these reasons that standards are prepared by carrying known amounts of serotonin through the entire extraction procedure. Details of the extraction procedure can be varied when necessary provided such standards are carried through the procedure. Reagents contribute a small "blank" which may be equivalent to as much as 0.1 μg. of serotonin. Although other fluorescing material is frequently present in tissue extracts, there is seldom interference at the excitation and fluorescent wavelengths characteristic of serotonin. As little as 0.1 μg. of serotonin can be detected by this procedure, but at least 0.3 μg. are required for accurate assay.

The amounts of serotinin present in some plant and animal tissues are shown in Table XII. Although serotonin is sufficiently elevated in the blood of patients with malignant carcinoid as to be diagnostic of the disorder, the excretion of 5-hydroxyindoleacetic acid in the urine is so high that a simple colorimetric test for the urinary acid metabolite is preferable to the fluorometric assay of blood serotonin.

Other 5-hydroxyindoles, such as 5-hydroxytryptophan and 5-hydroxyindole-acetic acid, have fluorescence characteristics similar to those of serotonin and may be assayed fluorometrically following suitable isolation procedures. Details of such methods for animal tissues have been presented by Udenfriend et al.[100] 5-Hydroxytryptophan is found only in urine of certain patients with malignant carcinoid. A fluorometric assay for the amino acid in such urines has been reported.[103] It involves removal of the serotonin by a cation exchange resin, conversion of the 5-hydroxytrypto-

TABLE XII

5-HYDROXYTRYPTAMINE IN TISSUES

Tissue	5-Hydroxytryptamine content[a] μg./g.
Mouse	
Brain	0.8
Lung	2.0
Blood	0.2
Liver	0.2
Rat	
Brain	0.4
Lung	2.1
Stomach	3–5
Liver	0.2
Blood	0.2
Guinea Pig	
Brain	0.6
Lung	0.2
Blood	0.2
Rabbit	
Brain	0.5
Lung	2.1
Intestine	7–14
Stomach	25
Blood	4–6
Dog	
Hypothalamus	1.5
Blood	0.4
Cat hypothalamus	2.0
Human	
Hypothalamus	0.5
Carcinoid tumor	300–1500
Blood (normal)	0.1–0.3
Blood (carcinoid)	0.5–3.0
Bufo Marinus, paratoid gland	20–40[b]
Venus mercenaria	40[c]
Hydra oligactis	1.5[c]
Banana	
Peel	50–150
Pulp	20–30
Tomato	12
Pineapple	25

[a] Based on wet weight of tissue.

[b] The oxidized, N-methyl analog, dehydrobufotenine, is present in many times this amount.

[c] Welsh, J. H., and Moorehead, M., J. Neurochem. 6, 146 (1960).

TABLE XIII

Fluorescence Characteristics of Some Tryptophan Metabolites[a]

Compound	Structure	Excitation maximum mμ	Fluorescence maximum mμ	Medium[b]	Reference
Kynurenine		370	490	pH 11	(27)
3-Hydroxy-kynurenine		365	460	pH 11	(27)
5-Hydroxy-kynurenine		375	460	pH 11	(27)
Kynurenic acid		325 325 340	405 440 435	pH 7 pH 11 14 N H$_2$SO$_4$	(27) (27) (104)

TABLE XIII (*continued*)

Compound	Structure	Excitation maximum mμ	Fluorescence maximum mμ	Medium	Reference
Xanthurenic acid		350 370	460 530	pH 11 5 N NaOH	(27) (104)
Anthranilic acid		300	405	pH 7	(27)
3-Hydroxy-anthranilic acid		320	415	pH 7	(27)
5-Hydroxy-anthranilic acid		340	430	pH 7	(27)

[a] The measurements were made with an Aminco–Bowman spectrophotofluorometer calibrated against a quinine standard.

[b] In most instances this represents the medium in which fluorescence intensity is highest.

phan to serotonin by partially purified guinea pig kidney decarboxylase, and assay of the enzymatically formed serotonin as described in the preceding section. The method is applicable to urines containing 0.5 mg. of the amino acid per day, or more. It should also be applicable to studies on 5-hydroxytryptophan formation and metabolism.

On acidified paper chromatograms, serotonin and other 5-hydroxyindoles yield a pink fluorescence when irradiated with a mercury lamp (365 mμ). Although this is not too sensitive (5 μg./spot), it is fairly specific and is especially useful when chromatography or electrophoresis on paper are used as isolation procedures. A much more sensitive (0.01–0.1 μg./spot) but less specific procedure for the fluorometric detection of serotonin and other primary indoles alkylamines on paper chromatograms involves reaction with ninhydrin in acetic acid to yield 3,4-dihydro-β-carbolines.[91] These derivatives give an intense greenish-blue fluorescence under ultraviolet light (365 mμ) when adsorbed on paper.

F. Kynurenine Pathway of Metabolism

The major route of tryptophan catabolism proceeds via ring cleavage catalyzed by the enzyme tryptophan pyrrolase. The immediate product is N-formyl kynurenine which is split to kynurenine. From here there are produced a large number of substituted aminophenols and quinolines. Most of these compounds are highly fluorescent on paper chromatograms when exposed to an appropriate ultraviolet lamp. Kynurenine, 3-hydroxykynurenine, anthranilic acid, 3-hydroxyanthranilic acid, kynurenic acid, and xanthurenic acid also fluoresce in aqueous solution. Excitation and fluorescence spectra of these and related compounds are presented in Table XIII.

Both kynurenic acid and xanthurenic acid occur in normal human urine. However, the amounts of each compound which are excreted vary with dietary intake of tryptophan, vitamin deficiency, and disease. The following fluorometric assay for these two quinoline derivatives in urine has permitted Satoh and Price[104] to evaluate the various factors influencing this route of tryptophan metabolism in man as well as in experimental animals:

For isolation ion exchange columns consisting of a glass tube (1.2-cm. o.d. and 15 cm. long) attached to the bottom of a 250-ml. Erlenmeyer flask are used. A constriction near the bottom of the tube holds a plug of glass wool and a column of Dowex 50 (H$^+$) 3 cm. long. Before use, the column is washed with 50 ml. of 5 N HCl and 100 ml. of water. Duplicate samples consisting of from 2 to 5% of a 24-hour urine are added to each of two graduated cylinders and diluted to 120 ml. with water. To one of the samples, 1 ml. of each standard solution is added; 30 ml. of 1 N HCl are added to both cylinders. The mixed solutions are added to the Dowex 50 (H+) columns and the columns are washed with 50 ml. of 0.2 N HCl, and 20 ml. of water. The kynurenic and xanthurenic acids are then eluted with 396 ml. of water. The entire procedure is carried out under gravity flow rates. Four milliliters of 0.5 M phosphate buffer (pH 7.4) are added to each sample to make a total volume of 400 ml. (pH 7.4 ± 0.1).

Kynurenic acid: One to 6 ml. of the effluent from the Dowex 50 column are transferred to a test tube and diluted to 6 ml. with 0.005 M phosphate buffer (pH 7.4). Then 4.0 ml. of concentrated H_2SO_4 are slowly added to each tube, mixed, and cooled in a water bath. The resulting fluorescence remains stable for 2 hours, and the samples are read in 1 hour. The excitation maximum is at 340 mμ and the fluorescence maximum at 435 mμ (see Table XII). Standards and blanks for comparison are prepared by comparable dilutions of reagents with and without added kynurenic acid.

Xanthurenic acid: Aliquots of the column effluent (0.1–5 ml.) are diluted to 5 ml. with 0.005 M phosphate buffer in small centrifuge tubes. Five milliliters of saturated NaOH are added to each tube and after mixing the tubes are centrifuged for 5 minutes in a clinical centrifuge. Fluorescence is measured (370-mμ excitation, 530-mμ fluorescence).

According to Satoh and Price,[104] the fluorometric procedure is simpler than other available procedures and, in the case of xanthurenic acid, is much more specific and precise than previously used colorimetric methods.

Roy *et al.*[105] have identified 8-hydroxyquinaldic acid in the urine of

8-Hydroxyquinaldic acid

pyridoxine deficient animals on a high tryptophan intake. On paper chromatograms this compound emits a red fluorescence when excited with the 254-mμ mercury line. It should also fluoresce in solution, either alone or in the form of a metal complex.

V. HISTIDINE AND HISTAMINE

A procedure which was developed for the fluorometric assay of histamine,[106] and which will be described in detail in the following section, has been applied to the assay of histidine in extracts and hydrolysates of peptides.[107]

Histidine is isolated from mixtures on a Dowex 50 (H⁺) column. After washing the column with water, the amino acid is eluted with 4 N ammonia. The eluate is then taken to dryness to remove all traces of ammonia and the residue is taken up in water. An aliquot, containing 0.01–0.06 μmole of histidine, is condensed with o-phthalaldehyde in dilute NaOH solution as described for histamine (below). The resulting fluorescence is at 480 mμ when excitation is at 340 mμ (uncorrected).

The mechanism of the condensation of o-phthaldehyde with histidine is probably similar to that with histamine. However, the fluorophor produced with histidine is much less fluorescent than the corresponding histamine product. Another limitation of this procedure as applied to histidine is that unlike histamine, the product cannot be extracted into organic solvents. This makes it difficult to separate interfering substances such as ammonia and other water soluble compounds. However, the o-phthalaldehyde procedure for histidine is more specific and sensitive than existing methods. It should be possible to develop simple isolation procedures which will also make it satisfactory in other respects.

Decarboxylation of histidine gives rise to histamine, another of the potent vasoactive amines which has been widely studied by physiologists and pharmacologists. Many bioassay and colorimetric procedures have been used to measure histamine in tissues. However, the fluorometric procedure recently developed by Shore et al.[106] combines the sensitivity of bioassay with the convenience of a chemical procedure and is, in addition, highly specific.

The method involves condensation of histamine with o-phthalaldehyde to yield a highly fluorescent product. Although the product itself has not been identified, corresponding types of condensation between histamine and pyridoxal have been reported. These led Shore et al.[106] to suggest the reaction shown in Fig. 17. They suggest, however, that subsequent rear-

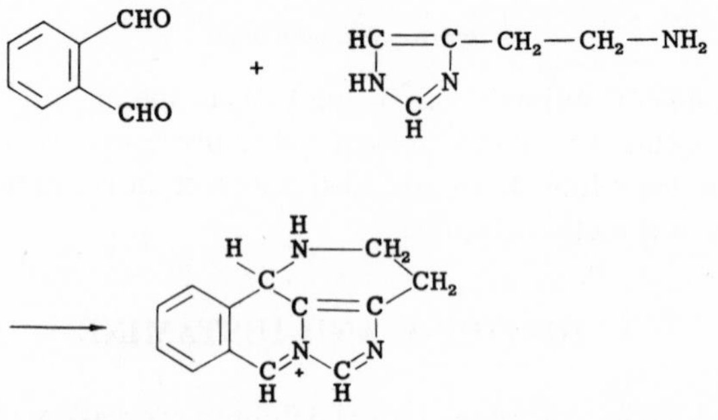

Fig. 17. Reaction between o-phthalaldehyde and histamine as suggested by Shore et al.[106]

rangement and oxidation may yield a fluorophor of somewhat different structure than the initial condensation product.

The over-all procedure for assay of histamine in tissues, as reported by Shore et al.[106] involves extraction of histamine into n-butanol from alkalanized perchloric acid tissue extracts and return of the histamine to an aqueous solution before condensation with o-phthalaldehyde. Weissbach[101] et al. have used chromatographic procedures to isolate histamine and perhaps this may also be used prior to the condensation.

The purest solvents are used and are further purified if necessary. Commercially obtained o-phthalaldehyde, if not pure, may be recrystallized from ligroin. It is prepared as a 1% solution in methanol which is stable for at least 2 weeks when kept in a dark bottle under refrigeration. To prepare salt-saturated 0.1 N NaOH, sufficient solid NaCl is added to 0.1 N NaOH so that an excess of solid NaCl remains.

Tissues are homogenized in 9 volumes of 0.4 N perchloric acid using a motor-driven glass homogenizer. The homogenate is allowed to stand for 10 minutes and is then centrifuged. A 4-ml. aliquot of the supernatant fluid is transferred to a 25-ml. glass-stoppered shaking tube containing 0.5 ml. of 5 N NaOH, 1.5 g. of solid NaCl, and 10 ml. of n-butanol. The tube is shaken for 5 minutes to extract the histamine into the butanol and, after centrifugation, the aqueous phase is removed by aspiration. The organic phase is then shaken for about 1 minute with 5 ml. of salt-saturated 0.1 N NaOH. This wash removes any residual amounts of histidine which may be present. The tube is then centrifuged and an 8-ml. aliquot of the butanol is transferred to a 40-ml. glass-stoppered shaking tube containing 2.5–4.5 ml. of 0.1 N HCl and 15 ml. of n-heptane. After shaking for about 1 minute, the tube is centrifuged and the histamine in the aqueous phase is assayed fluorometrically as described below.

For the analysis of blood in species, such as the dog, which contain only small amounts of circulating histamine, 5 ml. of oxalated blood are hemolyzed by the addition of 4.5 ml. of water and 0.5 ml. of concentrated (10–12 N) perchloric acid is added. The tube is then shaken for 10 minutes. The subsequent steps are similar to those described above.

The solubility characteristics of histamine are such that in the initial extraction about 92% of the histamine is extracted into the butanol. After the additional wash of the butanol with salt-saturated 0.1 N NaOH, a total of about 85% of the histamine remains in the butanol phase. There are also small volume changes in the various phases owing to the relative solubility of water in butanol. To correct for these factors, standards are prepared by carrying known amounts of histamine, added to 0.4 N perchloric acid, through the procedure. Recoveries of histamine added to tissue homogenates range from 90 to 100%.

After extraction from tissues the histamine is condensed with *o*-phthalaldehyde in strongly alkaline solution. The resulting, rather labile fluorescent product, is stabilized upon acidification. To estimate histamine in the acid extracts, a 2-ml. aliquot containing 0.005–0.5 μg. histamine/ml. is transferred to a test tube and 0.4 ml. of 1 *N* NaOH is added followed by 0.1 ml. of *o*-phthalaldehyde reagent. After 4 minutes, 0.2 ml. of 3 *N* HCl is added. The contents of the tubes are thoroughly mixed after each addition. The solution is then transferred to a cuvette and the fluorescence at 450 mμ resulting from excitation at 360 mμ (uncorrected) is measured in a fluorescence spectrometer. The fluorescence of the acidified solution is stable for at least ½ hour and the intensity is proportional to histamine concentration over the range 0.005–0.5 μg./ml.

Corrections may be made for the small fluorescence blanks in tissues or reagents by omission of the condensation step. This is accomplished by adding all the reagents to a separate aliquot of tissue extract, but reversing the order of addition of *o*-phthalaldehyde and 3 *N* HCl. The resulting solution contains tissue and reagent blanks, but not the fluorophor resulting from the condensation of the reagent with histamine, since the condensation reaction does not take place in acid.

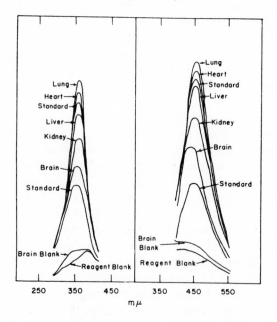

FIG. 18. Excitation (left) and fluorescence (right) spectra of the fluorophors produced by treatment of authentic histamine and apparent histamine from various rabbit tissues with *o*-phthalaldehyde (after Shore et al.[106]).

The following compounds do not produce significant fluorescence at the spectral peaks (excitation, 360 mμ; fluorescence, 450μ) of the histamine fluorophor: acetylhistamine, 1-methyl-4-(β-aminoethyl) imidazole, (1,4-methylhistamine), N-alkyl histamines (side chain), n-propyl, methylhistidine, 1,4-methylimidazolacetic acid, urocanic acid, carnosine, anserine, serotonin, 3,4-dihydroxyphenylethylamine, norepinephrine, spermine, spermidine, and, in concentrations lower than 4 μg./ml., ammonia. Histidine and histidylhistidine produce fluorophors spectrally similar to that of histamine, but they do not interfere with the histamine assay since they are not extracted by butanol from an alkaline solution. Ammonia, at concentrations greater than 4 μg./ml. produces some fluorescence, but it is poorly extracted into butanol and consequently does not interfere in the

TABLE XIV

Tissue Levels of Histamine as Determined by the o-Phthalaldehyde-Fluorometric Procedure[a]

Tissue	Histamine level (μg./g.)
Rat	
Brain	0.2–0.4
Kidney	0.5–1.5
Liver	1.1–2.0
Heart	3.8–5.2
Lung	1.7–6.9
Guinea pig	
Brain	0.2–0.3
Kidney	2.4–4.5
Liver	1.2–2.2
Heart	5.8–6.4
Lung	19–35
Rabbit	
Brain	0.2–0.4
Kidney	0.5–0.9
Heart	0.5–1.2
Lung	1.9–4.1
Blood	1.8–4.9
Dog	
Brain	0.2–0.4
Blood	0.05

[a] After Shore et al.[106]

analysis of tissue or blood. Fluorescent readings no greater than reagent blanks are obtained when as much as 150 µg. of ammonia are run through the entire procedure. The reaction thus appears to be remarkably specific for N-unsubstituted imidazolethylamines in most mammalian tissues.

The material extracted from tissues and condensed with o-phthalaldehyde to form a fluorophor was identified as histamine by (1) the characteristic excitation and fluorescent spectra (Fig. 18), (2) comparison with bioassay procedures, (3) comparison of solubility characteristics with those of authentic histamine, and (4) demonstration of its metabolism by diamine oxidase. The amounts of histamine found in various animal procedures by this procedure are shown in Table XIV.

In human urine there are present sufficient substances which interfere with the method as presented for tissues. However, Oates et al.[108] have introduced a chromatographic isolation step in the procedure which removes interfering substances. About 15 to 25 ml. of urine (representing a definite fraction of the daily excretion) is adjusted to pH 6.5, diluted to 30 ml. with 0.25 M phosphate buffer, and passed over an IRC-50 Type I column Na$^+$ buffered to pH 6.5; 10 × 55 mm. The column is then washed with two portions of 0.5 M acetate buffer, pH 6.5 (10 and 5 ml.). The histamine is eluted with 5 ml. of 1 N HCl and the eluate is assayed as described previously. Values obtained for adult human urine with this method are approximately 15–65 µg./day.*

VI. ARGININE AND RELATED GUANIDINES

Conn and Davis[109] have reported fluorescence methods for arginine, creatine, and related guanidinium compounds which are more specific and sensitive than existing methods. They found that ninhydrin, in strongly alkaline media, forms highly fluorescent products with guanidine and N-substituted guanidines. To produce the fluorophor, 2 ml. of 0.5% ninhydrin in water are added to 2 ml. of the solution of guanidinium compound. After mixing, 2 ml. of 1.0 N KOH are added and the solution is again shaken and stored at room temperature, shielded from direct light, for 9 to 14 minutes. With creatine, concentrations as low as 0.07 µg. can be determined and fluorescence is proportional to concentration from 0.5 to 75 µmole/liter. The sensitivity for arginine is about one-third that of creatine.

As shown in Fig. 19, the condensation products of guanidinium compounds with ninhydrin exhibit two excitation peaks, at 305 and 390 mµ.†

---

* J. A. Oates and A. Sjoerdsma, personal communication.

† The instrument was calibrated against a quinine standard.

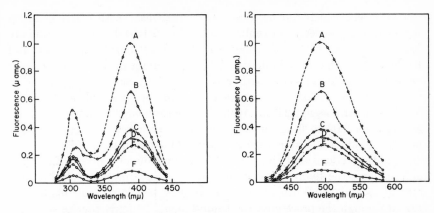

FIG. 19. Excitation (left) and fluorescence (right) spectra of the condensation products of various guanidinium compounds with ninhydrin. In each instance there were 50 μmoles of guanidinium compound per liter of 1.0 N KOH containing 0.5% ninhydrin. Curve A, creatine; curve B, dimethylguanidine; curve C, guanidoacetic acid; curve D, methylguanidine; curve E, arginine; curve F, guanidine (after Conn and Davis[109]).

With the Farrand spectrofluorometer, fluorescence excited at 390 mμ was twice as intense as that at 305 mμ. Fluorescence is at 495 mμ. The spectra of all compounds tested were the same. However, the fluorescence intensity varied, N,N-disubstituted compounds being most highly fluorescent and unsubstituted guanidine being least fluorescent. Creatine phsophate, creatinine, guanine, urea, and ammonia exhibited negligible fluorescence under the conditions of the analytical procedure. As to the mechanism, the authors suggest that in alkaline solution the ninhydrin ring is opened to form o-carboxyphenylglyoxal which combines with the guanidine group to give the fluorescent product. Since o-carboxyphenylglyoxal, which is present in excess, strongly absorbs light in the region of the required excitation, it interferes with fluorescence measurement. Fortunately, upon standing in alkali the phenylglyoxal is converted to the corresponding mandelic acid which no longer absorbs at the wavelength of excitation. The 9–14 minute wait following addition of the reagents is not only to permit condensation with guanidines to go to completion, but to ensure complete destruction of the interfering o-carboxyphenylglyoxal.

Conn[110] has made use of the ninhydrin reaction as the basis for a sensitive method for the estimation of creatine in serum, blood, and urine. With serum and blood a barium hydroxide-zinc sulfate filtrate is prepared using 1.0 ml. of the tissue, 15.0 ml. of water, 2.0 ml. of 0.3 N barium hydroxide, and 2.0 ml. of 5% zinc sulfate. The solution is carefully mixed and filtered through fine filter paper. Urine contains interfering fluorescent materials which can be removed by treatment with a strong anion exchange resin. Fifteen to twenty milliliters of a 1:50 dilution of urine are placed in a 50-ml. flask with about 3 g. of IRA-401 resin (Na$^+$ form), and the flask is agitated for 30 minutes. The supernatant urine dilution is used for assay. Blanks and standards are prepared by substituting water or standard solutions for the equivalent amount of either blood or urine. For fluorometric assay, 2.0 ml. of filtrate or resin supernatant fluid (containing from 0.03 to 6 μg. of creatine) are transferred to a cuvette containing 1.0 ml. of 1% ninhydrin in alcohol. One milliliter of 10% alcoholic KOH is then added and the contents are immediately mixed. Exactly 5 minutes later the fluorescence is measured, using a quinine solution as a reference standard. When a series of samples, standards, and blanks are carried through the procedure, the timing should be fairly precise since the fluorescence takes 3 or 4 minutes to develop and then fades gradually. Fluorescence is measured using 390 mμ for excitation and 495 mμ for fluorescence. In this region of the spectrum, filters are practical and these are discussed in the paper by Conn.[110] Recoveries are essentially quantitative.

The compounds which can interfere with the reaction in vertebrate tissues are arginine, guanidoacetic acid, guanidobutyric acid,[111] methylguanidine, and guanidine. The latter two compounds, if present at all, occur only in trace amounts. The bulk of the arginine in blood and serum is lost through the precipitation with barium hydroxide and zinc sulfate. Arginine in the urine is very low except in patients with certain aminoacidurias. If, for any reason, excessive amounts of arginine should be

TABLE XV

CREATINE IN HUMAN SERUM AND URINE

Tissue	Value	Method	Reference
Serum	0.19–0.79^a	Fluorescence—ninhydrin	(110)
	0.19–0.48	Picric acid (by difference)	(113)
	0.28–0.62	Bacterial enzyme+picric acid	(114)
	0.14–0.44	Ion exchange; diacetyl	(112a)
Urine	18.6–58.5^b	Fluorescence—ninhydrin	(110)
	4 –40^b	Bacterial enzyme; diacetyl	(115)
	22.2–74.6^c	Ion exchange; diacetyl	(112a)
	14.6–34.4^c	DPN–linked enzyme assay	(116)

^{a} mg. % (all serum values).
^{b} mg./liter.
^{c} mg./day.

present, the amino acid may be removed by treatment with an ion exchange resin as described by Anderson *et al.*[112a] Guanidoacetic and perhaps guanidobutyric acids are interferences which cannot be separated. Fortunately, the fluorescence formed with creatine is several times as intense as that formed with either arginine or the guanidino acids; creatine is also present in much larger quantities. A comparison of creatine levels in serum and urine with those obtained by other methods is shown in Table XV.

Another sensitive fluorometric procedure for arginine involves condensation with *o*-phthalaldehyde* under the same conditions as were described for histamine[106] (see above). Unlike the histamine fluorophore the arginine product is stable in the alkaline solution in which condensation takes place. Agmatine, the decarboxylation product of arginine (1-amino-4-guanidobutane), can be readily separated from interfering substances (including arginine) and a fluorometric assay has been developed for it.[112b] Maximal excitation is at 355 mμ and fluorescence at 470 mμ (uncorrected). Although the compound has not been found in mammalian tissues, it is a metabolic product in bacteria, plants, and invertebrates.

VII. CYSTEINE AND GLUTATHIONE

Sulfhydryl compounds are, themselves, not fluorescent. However, they do react with a variety of reagents which should permit the application of fluorometry to this class of compounds. Reaction with aldehydes has long been known and used and Cohn and Shore* have shown that *o*-phthalaldehyde does yield fluorescent derivatives with sulfhydryl compounds in general. Quite unexpectedly, however, they found that the product with glutathione has many times the fluorescence intensity of cysteine or other simple sulfhydryl compounds (nanogram quantities can be detected). Since glutathione is also present in higher concentrations in tissues, it should be possible to utilize this reaction for a fairly sensitive and specific procedure for the assay of this important peptide. The major excitation peak of the glutathione product is at 490 mμ, and the fluorescence peak is at 545 mμ (uncorrected).

VIII. γ-AMINOBUTYRIC ACID

γ-Aminobutyric acid is widely distributed in nature. It is found in plants, invertebrates, and mammals. In the latter it is limited almost entirely to the central nervous system. Since enzymes for its synthesis and metabolism

---

* V. H. Cohn, Jr., and P. A. Shore, personal communication.

are also uniquely found in the brain, γ-aminobutyric acid has become of great interest to the neurochemist. A sensitive and highly specific method for its assay was developed by Scott and Jakoby.[117] This procedure makes use of a specific enzyme for this amino acid and is coupled with the reduction of pyridine nucleotide. Although spectrophotometric assay has been utilized, it should be possible to apply fluorometric procedures to this coupled enzyme assay and increase its sensitivity manyfold (see Chapter 9).

Lowe et al.[12] have found that in aqueous solution at about pH 9.5 to 10 ninhydrin reacts with many amino acids to yield fluorescent condensation products.* Of the naturally occurring compounds γ-aminobutyric acid and phenylalanine (see above) yield the most intense fluorescence. They have utilized this reaction to develop a fluorometric assay for γ-aminobutyric acid which they have applied to the measurement of the enzyme, glutamic decarboxylase, in brain. The reaction of ninhydrin with γ-aminobutyric acid is rather complicated and the fluorescent product has not been identified. To obtain maximal amounts of fluorescence and proportionality between γ-aminobutyric acid concentration and fluorescence, it is necessary to have present in the reaction mixture rather large amounts of glutamic acid, which by itself, reacts only slightly with the reagent. To form the fluorophore in optimal yield, 0.1 ml. of solution, containing 3–100 μg. of γ-aminobutyric acid and 1.5 μmoles of glutamic acid, is transferred to a tube containing 0.2 ml. of ninhydrin solution (14 mM ninhydrin in 0.5 M carbonate buffer, pH 9.4). The tube is capped and heated at 60° for 30 minutes following which 5 ml. of copper tartrate reagent† are added. After standing at room temperature for 15 minutes, fluorescence is measured. Excitation is maximal at 375 mμ and fluorescence at 485 mμ (uncorrected).

IX. CHOLINE AND ACETYLCHOLINE

Although no fluorometric procedure for the assay of these important compounds has yet been applied to biological material, fluorometric tests for the pure compounds are available. Neu[118, 119] has reported a particularly sensitive method for the detection of choline, acetylcholine, and other onium compounds on paper chromatograms and in solution, as a spot test. For a spot test add several drops of a solution of tetraphenyl-

---

* The reaction of ninhydrin with argininine and creatine (see above) does not take place at this pH but requires strong alkali.

† Copper tartrate reagent: 1.6 g. Na_2CO_3, 329 mg. tartaric acid, and 300 mg. of $CuSO_4 \cdot 5H_2O$ are dissolved in 1 liter of water and stored at room temperature. Each should be dissolved separately then mixed and made up to volume.

diboroxide or 2-aminoethyl diphenylborate (50 mg. in 100 ml. of methanol) to a watch glass. Then add several drops of a quercitin solution (25 mg. of quercitin in 100 ml. of methanol) and mix. Upon the addition of microgram quantities of the amines a red-orange fluorescence appears. At higher concentrations the fluorescence appears red-brown; at still higher concentrations a precipitate appears. The same reagents can be used for spraying chromatograms. Although the mechanism of the reaction is not explained and not too many details of the reaction are presented, this sounds like a fairly sensitive reaction. Studies on its applications are certainly warranted.

REFERENCES

1. Coons, A. H., *Intern. Rev. Cytol.* **5**, 1 (1956).
2. Keston, A. S., Udenfriend, S., and Levy, M. *J. Am. Chem. Soc.* **72**, 748 (1950).
3. Patton, A. R., Foreman, E. M., and Wilson, P. C., *Science* **110**, 593 (1949).
4. Woiwood, A. J., *Nature* **166**, 272 (1950).
5. Shore, V. G., and Pardee, A. R., *Anal. Chem.* **28**, 1479 (1956).
6. Levy, A. L., and Chung, D., *Anal. Chem.* **25**, 396 (1953).
7. Wadman, W. H., Thomas, G. J., and Pardee, A. B., *Anal. Chem.* **26**, 1192 (1954).
8. Brand, E., *Ann. N. Y. Acad. Sci.* **47**, 187 (1946).
9. Stein, W. H., and Moore, S., *J. Biol. Chem.* **178**, 79 (1949).
10a. Von Horst, H., Tang, H., and Jurkovich, V., *Anal. Chem.* **31**, 135 (1959).
10b. Teale, F. W. J., and Weber, G., *Biochem. J.* **65**, 476 (1956).
11. Vladimirov, Yu. A., *Doklady Akad. Nauk S.S.S.R.* **116**, 780 (1957).
12. Lowe, I. P., Robins, E., and Eyerman, G. S., *J. Neurochem.* **3**, 8 (1958).
13. Chirigos, M. A., and Guroff, G., *Anal. Biochem.* **3**, 330 (1962).
14. White, A., *Biochem. J.* **71**, 217 (1959).
15. Duggan, D. E., and Udenfriend, S., *J. Biol. Chem.* **223**, 313 (1956).
16. Moore, S., and Stein, W. H., *J. Biol. Chem.* **178**, 53 (1949).
17. Holiday, E. R., *Biochem. J.* **30**, 1795 (1936).
18. Velick, S. F., *J. Biol. Chem.* **233**, 1455 (1958).
19. Velick, S. F., and Udenfriend, S., *J. Biol. Chem.* **203**, 575 (1953).
20. Udenfriend, S., and Cooper, J. R., *J. Biol. Chem.* **196**, 227 (1952).
21. Waalkes, T. P., and Udenfriend, S., *J. Lab. Clin. Med.* **50**, 733 (1957).
22. Oates, J. A., *in* "Methods of Medical Research" (J. H. Quastel, ed.), Vol. IX, p. 169, Year Book Med. Publ., Chicago, Illinois, 1961.
23. Kakimoto, Y., and Armstrong, M. D., *Federation Proc.* **19**, 295 (1960).
24. Pisano, J. J., Oates, J. A., Karmen, A., Sjoerdsma, A., and Udenfriend, S., *J. Biol. Chem.* **236**, 898 (1961).
25. Sjoerdsma, A., Lovenberg, W., Oates, J. A., Crout, J. R., and Udenfriend, S., *Science* **130**, 225 (1959).
26. Oates, J. A., Jr., Gillespie, L., Udenfriend, S., and Sjoerdsma, A., *Science* **131**, 1890 (1960).
27. Duggan, D. E., Bowman, R. L., Brodie, B. B., and Udenfriend, S., *Arch. Biochem. Biophys.* **68**, 1 (1957).
28. Bertler, Å., Carlsson, A., and Rosengren, E., *Acta. Physiol. Scand.* **44**, 273 (1958).
29. Weil-Malherbe, H., and Bone, A. D., *J. Clin. Pathol.* **10**, 138 (1957).
30. Carlsson, A., and Waldeck, B., *Acta. Physiol. Scand.* **44**, 293 (1958).

31. Carlsson, A., *Pharmacol. Revs.* **11**, 300 (1959).
32. Drujans, H., Sourkes, T. L., Layne, S., and Murphy, G. F., *Can. J. Biochem. Physiol.* **37**, 1154 (1959).
33. Carlsson, A., Lindqvist, M., Magnusson, T., and Waldeck, B., *Science* **127**, 471 (1958).
34. Bertler, Å., and Rosengren, E., *Experientia* **15**, 10 (1959).
35. Holtz, P., *Pharmacol. Revs.* **11**, 317 (1959).
36. Holzer, G., and Hornykiewicz, O., *Arch. exptl. Pathol. Pharmakol. Naunyn-Schmideberg's* **237**, 27 (1957).
37. Loew, O., *Biochem. Z.* **85**, 295 (1918).
38. Paget, M., *Bull. sci. pharmacol.* **37**, 537 (1930).
39. Gaddum, J. H., and Schild, H., *J. Physiol. (London)* **80**, 9P (1934).
40. Hueber, E. F., *Klin. Wochschr.* **19**, 664 (1940).
41. Heller, J. H., Setlow, R. B., and Mylon, E., *Am. J. Physiol.* **161**, 268 (1949).
42. Ehrlén, I., *Farm. Revy* **47**, 242 (1948).
43. Lund, A., *Acta. Pharmacol. Toxicol.* **5**, 75, 121 (1949).
44. Fischer, P., *Bull. soc. chim. Belges* **58**, 205 (1949).
45. Harley-Mason, J., *J. Chem. Soc.* p. 1276 (1950).
46. Price, H. L., and Price, M. L., *J. Lab. Clin. Med.* **50**, 769 (1957).
47. Cohen, G., and Goldenberg, M., *J. Neurochem.* **2**, 58 (1957).
48. Vendsalu, A., *Acta. Physiol. Scand.* **49**, *Suppl.* 173 (1960).
49. Shore, P. A., and Olin, J. S., *J. Pharmacol. Exptl. Therap.* **122**, 295 (1958).
50. DuToit, C. H., Wright Air Development Center Tech. Rept. 59–175 (1959).
51. Crout, R. J., *in* "Standard Methods of Clinical Chemistry" (D. Seligson, ed.), Vol 3, p. 62. Academic Press, New York.
52. Euler, U. S. v., and Lishajko, F., *Acta Physiol. Scand.* **45**, 122 (1959).
53. Euler, U. S. v., and Floding, I., *Scand. J. Clin. & Lab. Invest.* **8**, 288 (1956).
54. Lund, A., *Acta Pharmacol. Toxicol.* **6**, 137 (1950).
55. Crout, J. R., and Sjoerdsma, A., *Circulation* **22**, 516 (1960).
56. Sax, S., Waxman, H. E., Aarons, J. H., and Lynch, H. J., *Clin. Chem.* **6**, 168 (1960).
57. Osinskaya, V. O., *Biokhimiya* **22**, 537 (1957).
58. Crout, J. R., *Pharmacol. Revs.* **11**, 296 (1959).
59. Holzbauer, M., and Vogt, M., *Brit. J. Pharmacol.* **9**, 249 (1954).
60. Hoffer, A., Osmond, H., and Smythies, V., *J. Mental Sci.* **100**, 29 (1954).
61. Sobotka, H., Barsel, N., and Chanley, J. D., *Fortschr. Chem. org. Naturstoffe* **14**, 217 (1957).
62. Payza, A. N., and Mahon, M., *Anal. Chem.* **32**, 17 (1960).
63. Feldstein, A., *Science* **128**, 28 (1958).
64. Szara, S., Axelrod, J., and Perlin, S., *Am. J. Psychiat.* **115**, 162 (1958).
65. Wallerstein, J. S., Alba, R. T., and Hale, M. G., *Biochim. et Biophys. Acta* **1**, 175 (1947).
66. Natelson, S., Lugovoy, J. K., and Pincus, J. B., *Arch. Biochem.* **23**, 157 (1949).
67. Weil-Malherbe, H., and Bone, A. D., *Biochem. J.* **51**, 311 (1952).
68. Harley-Mason, J., and Laird, A. H., *Biochem. J.* **69**, 59P (1958).
69. Weil-Malherbe, H., *Pharmacol. Revs.* **11**, 278 (1959).
70. Weil-Malherbe, H., *Biochim. et Biophys. Acta* **40**, 351 (1960).
71. Burn, G. P., and Field, E. O., *Nature* **178**, 542 (1956).
72. Young, J. G., and Fischer, R. L., *Science* **127**, 1390 (1958).
73. Nadeau, G., and Joly, L. P., *Nature* **182**, 180 (1958).
74. Yagi, K., and Nagatsu, T., *Nature* **183**, 822 (1959).
75. Nadeau, G., and Joly, L. P., *Can. J. Biochem. Physiol.* **37**, 231 (1959).

76. Mangan, G. F., Jr., and Mason, J. W., *Science* **126**, 562 (1957).
77. Gray, I., and Young, J. G., *Clin. Chem.* **3**, 239 (1957).
78. Euler, U. S. v., "Noradrenaline," p. 60. C. C Thomas, Springfield, Illinois, 1956.
79. Udenfriend, S., Lovenberg, W., and Sjoerdsma, A., *Arch. Biochem. Biophys.* **85**, 487 (1959).
80. Euler, C. V., Euler, U. S. v., and Floding, I., *Acta Physiol. Scand.* **33**, 32 (1955).
81. Rosengren, E., *Acta Physiol. Scand.* **49**, 370 (1960).
82. Axelrod, J., and Tomchick, R., *J. Biol. Chem.* **233**, 702 (1958).
83. Axelrod, J., Senoh, S., and Witkop, B., *J. Biol. Chem.* **233**, 697 (1958).
84. Bertler, Å., Carlsson, A., and Rosengren, E., *Clin. Chim. Acta.* **4**, 456 (1959).
85. Smith, E. R. B., and Weil-Malherbe, H., *Federation Proc.* **20**, 182 (1961).
86. Brunjes, S., Blaylock, A., Wybenga, D., and Johns, V. J., Jr., *Clin. Research* **9**, 78 (1961).
87. Velick, S. F., and Ronzoni, E., *J. Biol. Chem.* **173**, 627 (1948).
88. Hess, S., and Udenfriend, S., *J. Pharmacol. Exptl. Therap.* **127**, 175 (1959).
89. Sjoerdsma, A., Oates, J. A., Zaltzman, P., and Udenfriend, S., *J. Pharmacol. Exptl. Therap.* **126**, 217 (1959).
90. Hess, S. M., Redfield, B. G., and Udenfriend, S., *J. Pharmacol. Exptl. Therap.* **127**, 178 (1959).
91. Jepson, J. B., and Stevens, B. J., *Nature* **172**, 772 (1953).
92. Davis, E. J., and de Ropp, R. S., *Biochem. Biophys. Research Communs.* **2**, 361 (1960).
93. Weissbach, H., Smith, T. E., and Udenfriend, S., *Biochemistry* **1**, 137 (1962).

94. Smith, I., "Chromatographic Techniques," William Heinemann Medical Books Ltd., London, 1958.
95. Uphaus, R. A., Grossweiner, L. I., and Katz, J. J., *Science* **129**, 641 (1959).
96. Freed, S., and Salmre, W., *Science* **128**, 1341 (1958).
97. Udenfriend, S., Weissbach, H., and Clark, C. T., *J. Biol. Chem.* **215**, 337 (1955).
98. Udenfriend, S., Bogdanski, D. F., and Weissbach, H., *Science* **122**, 972 (1955).
99. Bogdanski, D. F., Pletscher, A., Brodie, B. B., and Udenfriend, S., *J. Pharmacol. Exptl. Therap.* **117**, 82 (1956).
100. Udenfriend, S., Weissbach, H., and Brodie, B. B., *Methods of Biochem. Anal.* **6**, 95 (1958).
101. Weissbach, H., Waalkes, T. P., and Udenfriend, S., *J. Biol. Chem.* **230**, 865 (1958).
102. Sjoerdsma, A., Lovenberg, W., Oates, J. A., Crout, J. R., and Udenfriend, S., *Science* **130**, 225 (1959).
103. Sjoerdsma, A., Oates, J. A., Zaltzman, P., and Udenfriend, S., *New Engl. J. Med.* **263**, 585 (1960).
104. Satoh, K., and Price, J. M., *J. Biol. Chem.* **230**, 781 (1958).
105. Roy, J. K., Brown, R. R., and Price, J. M., *Nature* **184**, 1573 (1959).
106. Shore, P. A., Burkhalter, A., and Cohn, V. H. Jr., *J. Pharmacol. Expt. Therap.* **127**, 182 (1959).
107. Pisano, J. J., Wilscn, J. D., Cohen, L., Abraham, D., and Udenfriend, S., *J. Biol. Chem.* **236**, 499 (1961).
108. Oates, J. A., Marsh, E. B., and Sjoerdsma, A., *J. Clin.Chim. Acta* **7**, 488 (1962).
109. Conn, R. B., Jr., and Davis, R. B., *Nature* **183**, 1053 (1959).
110. Conn, R. B., Jr., *Clin. Chem.* **6**, 537 (1960).
111. Pisano, J. J., Mitoma, C., and Udenfriend, S., *Nature* **180**, 1125 (1957).
112a. Anderson, D. R., Williams, C. M., Krise, G. M., and Dowben, R. M., *Biochem. J.* **67**, 258 (1957).

112b. Cohn, V. H., Jr., and Shore, P. A., *Anal. Biochem.* **2,** 237 (1961).
113. Sandberg, A. A., Hecht, H. H., and Tyler, F. H., *Metabolism, Clin. and Exptl.* **2,** 22 (1953).
114. Allinson, M. J. C., *J. Biol. Chem.* **157,** 169 (1945).
115. Ennor, A. H., and Stocken, L. A., *Biochem. J.* **55,** 310 (1953).
116. Tanzer, M. L., and Gilvarg, C., *J. Biol. Chem.* **234,** 3201 (1959).
117. Scott, E. M., and Jakoby, W. B., *J. Biol. Chem.* **234,** 932 (1959).
118. Neu, R., *Z. anal. Chem.* **143,** 30 (1954).
119. Neu, R., *Naturwissenschaften* **44,** 181 (1957).

6

PROTEINS

I. GENERAL REMARKS

FLUORESCENCE assay has been applied in many ways to studies concerning (1) the chemical composition of proteins, (2) the physicochemical properties of proteins and other macromolecules, and (3) the interaction of proteins with one another and with smaller molecules. The aromatic amino acids tyrosine and tryptophan are fluorescent and methods for their fluorometric assay in protein hydrolysates have been reported[1] and were discussed in Chapter 5. A more specific method for tyrosine assay was also discussed[2] as were methods for histidine[3] and arginine.[4] It is, therefore, possible to utilize microfluorometric procedures to determine the amino acid composition of proteins. It may even be possible to devise less specific fluorometric reagents for assay of amino acids, which will be applicable to all the amino acids in a hydrolysate and thereby require much less sample than is presently needed for determination of the total amino acid composition of a protein.

Proteins possess native fluorescence. In most instances this is due to the presence of the aromatic amino acids, tyrosine and tryptophan. However, it may also result from other fluorescent structures bound to the protein other than in peptide linkage. Enzymes which firmly bind reduced pyridine nucleotides and flavines will be discussed later in this chapter. Proteins can also be combined chemically with fluorescent compounds to yield fluorescent protein derivatives. In many instances the properties of the "labeled" protein are sufficiently similar to the parent protein to permit its use in various studies concerning the free protein.

The aromatic amino acids, coenzymes, and fluorescent dyes combined with a protein will not necessarily possess the same fluorescence characteristics as they do in the free form. Thus, in the combined form the intensity of fluorescence may be higher or lower, the excitation and fluorescence spectra may shift, and the degree of fluorescence polarization will usually increase. When more than one fluorescing moiety is present in a macromolecule, it may also be possible to detect a transfer of excitation energy from one to the other. All these factors can be measured with a fair degree of precision and can yield information concerning the size and shape of macromolecules and their interaction with other molecules. Such studies have been applied to simple proteins, enzymes, toxins, and antigen-antibody complexes. In addition, the fluorescent labeling technique has made available immunological techniques of a high degree of sensitivity and specificity. The various uses of fluorescence as tools in studies with proteins will be discussed in this section.

Before discussing particular types of protein, we must discuss two phenomena which are particularly important when applying fluorescence assay to proteins. These are "energy transfer" and "fluorescence polarization." Although neither phenomenon is limited to proteins, it is in this field that they are most important in biology. For this reason they are discussed in this section.

II. ENERGY TRANSFER

In some of the previous sections we have occasionally referred to the term "energy transfer" in explaining certain fluorescent phenomena. We have used the term to signify mainly the transfer of radiation energy absorbed by one atom or molecule to another, the resulting fluorescence emission then being characteristic of the second species. This demonstration of energy transfer by fluorescence techniques is of importance in biology because it indicates that a system is capable of transferring electronic excitation energy no matter how it is induced. The proximity of light absorbing residues within a protein molecule makes energy transfer of such importance with regards to proteins that it must be discussed before turning to a general discussion concerning the fluorescence of intact proteins.

The transfer of electronic excitation energy from one molecule, or portion of a molecule, to another is now known to be such an efficient process that in many systems it may occur at every encounter. Furthermore, it is known to occur widely in nature even over distances as great as 50 Å or more. Largely through the efforts of Förster[5] it is now recognized that transfer can proceed by a radiationless mechanism whereby the molecules which are originally excited (primary oscillators) transmit the excitation energy to recipient molecules (oscillators) at a distance. This is somewhat like a vibrating tuning fork inducing vibration in a nearby fork having the same pitch. The interaction is due to a dipole-dipole coupling and not to the overlap of electron orbitals (see Chapter 2). The latter process can only act over very short distances. The two processes may also be compared in the same manner as wireless transmission (former) to telegraphy (latter).

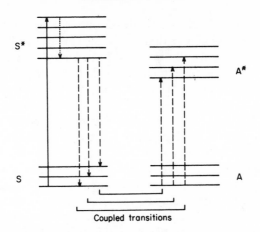

Coupled transitions

FIG. 1. A simplified energy level diagram showing the relationship between a sensitizer (S) and an acceptor (A). —, radiative transitions; ..., nonradiative transitions; —, transfer transitions (after Förster[6]).

The following explanation of electronic energy transfer between molecules has been presented by Förster.[6] Energy level diagrams of two molecules are shown in Fig. 1. The molecule S (sensitizer, donor) absorbs radiation and is excited to a higher vibrational level of its excited electronic state. It then returns by a nonradiative transfer to the lower vibrational level of the excited electronic state. In the absence of other molecular species, the electron would return to the ground state with the emission of fluorescence. However, when there is present another species, then the energy difference of the deactivating process (fluorescence or other) in the sensitizer molecule may correspond exactly to that of an absorption transition of a nearby molecule. When this happens and there is sufficient energy available, then both processes may occur simultaneously so that there results a transfer of the excitation from the sensitizer to the acceptor. These may be separate molecules or different parts of a single molecule. Coincidence between sensitizer and acceptor transitions occur when the absorption spectrum of the acceptor overlaps the fluorescence spectrum of the sensitizer. According to Förster[6] the distance, R_0, between sensitizer and acceptor at which an excited electron may transfer its energy or, with equal probability emit its own fluorescence, is a function of the lifetime of the excited state, the orientations of the two species and the degree of overlap of the absorption and fluorescence spectra. Conditions for the process of resonance energy transfer, frequently referred to as the Förster mechanism, are highly favorable in biological systems because of the presence of high local concentrations of absorbing matter. Transfer of radiation energy is, of course, of direct importance in photosynthesis. However, it is now recognized that resonance energy transfer (of chemical origin) may be more generally involved in biological systems. Such mechanisms may play a role in cytochrome oxidation, in pyridine nucleotide reduction, in the action of myosin and in many other biological systems. A summary of some applications of energy transfer in biology, based on a symposium on the subject held at the University of Nottingham in 1959, was published by Porter and Weber.[7] The entire symposium appeared in the *Discussions of the Faraday Society.*[8] In this area the suggestions of Szent György are that in living matter energy transfer is so great that it exhibits semiconductivity. A thorough discussion of the implications of energy transfer in biology is outside the scope of this book. A list of references on this subject is presented at the end of this chapter for those having more than a casual interest in this field. Some relatively simple examples of energy transfer in fluorescing systems are presented to help the reader understand the phenomenon and its applications.

First of all, energy transfer can occur between two different molecules in solution, provided the concentrations are sufficiently high. An example of this is shown in Fig. 2.[9] It can be seen that p-terphenyl emits appreciable

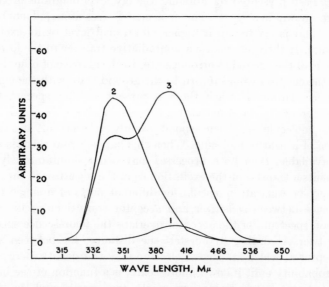

FIG. 2. Fluorescence of p-terphenyl, 1-naphthylphenyloxazole, and of the combined solutions (after Gemmil [9]).

Curve 1, 1-naphthylphenyloxazole, 0.000083%; curve 2, p-terphenyl, 0.27%; curve 3, 0.27% p-terphenyl and 0.000083% 1-naphthylphenyloxazole. The compounds were dissolved in toluene. The excitation monochromator was set at 270 mμ.

fluorescence at about 340 mμ when excited by light at 270 mμ. On the other hand, 1-naphthylenephenyloxazole shows no fluorescence at 340 mμ and emits only faintly at 400 mμ when excited by light at 270 mμ. However, when the two are mixed (curve 3) the fluorescence of the p-terphenyl decreases while that of the 1-naphthylenephenyloxazole increases. Obviously then, radiation absorbed by the p-terphenyl (donor or sensitizer) is converted to electronic energy and transferred to the 1-naphthylenephenyloxazole (acceptor) which is in turn excited and therefore emits its characteristic fluorescence. The concentrations of reactants in this instance are quite high (0.27% p-terpheryl), the amount of transfer diminishing greatly with dilution. According to our current concepts of this process, this must be so because with dilution the average intermolecular distances are increased beyond the range of dipole-dipole coupling. It should be pointed out that energy transfer does not always lead to an increase in fluorescence. The acceptor molecule in such a system may emit the transferred electronic energy as heat rather than as fluorescence. In such instances the transfer appears as a quenching of the fluorescence of the species absorbing the exciting radiation. It has been suggested that in many instances an unexpected quenching associated with increasing concentration of a solute may be associated with transfer of energy to a nonfluorescent dimer of the solute.[5, 10] One important aspect of energy transfer, whether it leads to fluorescence or quenching of fluorescence, is that the efficiency is greater the more closely the absorption band of the second component coincides with the fluorescence band of the first. It is also frequently observed that some molecules emit more intense fluorescence when they absorb energy by transfer from another absorbing species, either from another solute or from the solvent (usually solid) than they do when excited directly with light at their own absorption maxima. Such cases are referred to as "sensitized fluorescence." This phenomenon is discussed in detail by Pringsheim.[11a] In scintillation counting β or γ radiation excites the solvent. Electronic excitation energy is then transferred from the solvent to a solute, which either fluoresces or transfers its energy to a second solute which has more desirable emission characteristics with respect to the detecting system.[11b]

When two absorbing species are combined into a single molecule, the probability of energy transfer is increased enormously. When the combination is such that the two parts of the molecule are within the range of dipole-dipole coupling, then the phenomenon of energy transfer will always be observed even in dilute solution. A most appropriate example of intramolecular transfer in biological systems is that reported by Weber[12] between the purine and dihydro-pyridine portions of reduced pyridine nucleotides. Dihydro-diphosphopyridine (DPNH) in water, at pH 7–9,

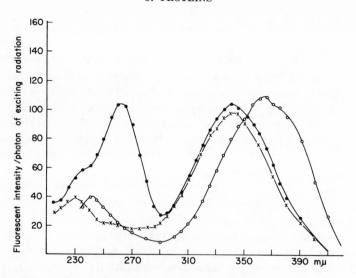

Fig. 3. Excitation spectra of dihydro-diphosphopyridine nucleotide and of its components (after Weber[12]). ○ — ○, 4-hydro-N-methyl nicotinamide in 0.1 M sodium carbonate; ● — ●, dihydro-diphosphopyridine nucleotide in 0.1 M tris buffer, pH 8.3; ×—×, dihydro-diphosphopyridine nucleotide in 0.1 M tris buffer, pH 8.3, after incubation for a half-hour at 38° with nucleotide pyrophosphatase.

emits fluorescence with a maximum at 468 mμ when excited at its absorption maximum 340 mμ. The quantum yield is about 2%. Similar fluorescence characteristics are shown by many reduced N-alkyl nicotinamides. However, whereas the excitation spectra of reduced N-alkyl nicotinamides are characterized by only one peak, coinciding exactly with their absorption spectra, the excitation spectrum of DPNH shows, in addition to the 340-mμ peak, a second peak of excitation at 260 mμ. This is a region where 80–90% of the absorption is due to the purine part of the molecule (Fig. 3). It was shown that 30% of the photons absorbed by the adenine appear as nicotinamide fluorescence. That the integrity of the DPNH is required for this energy transfer to occur was shown by recording excitation spectra before and after treatment with nucleotide pyrophosphatase. The enzyme splits the P—O—P linkage and liberates adenine nucleotide and nicotinamide nucleotide as independent molecules, which at the concentrations employed ($9 \times 10^{-6} M$) are too far apart (300 Å average distance) to exhibit energy transfer. As shown in Fig. 3, pyrophosphatase has no effect on the 340-mμ excitation but almost completely abolishes the 260-mμ excitation characteristic of purines. According to Weber[12] and Shore and Pardee,[13] purines are not fluorescent, so that the transfer from adenine must compete with a normally radiationless transition. The observed high probability of the transfer in the face of the radiationless transition indicates to Weber short-range interaction between the two parts of the molecule and suggests that DPNH exists in solution as an intramolecular complex. The presence of such a complex is also shown by the changes in absorption spectra before and after treatment with the nucleotide pyrophosphatase. Before treatment, E260/E340 is 2.85, after treatment it rises to 3.3. Weber[14] and Whitby[15] have suggested similar complexes to explain the absorption and fluorescent properties of flavin adenine dinucleotide.

Of course, unusual internal complexes must be assumed only if the adenine moiety is nonfluorescent. However, Duggan et al.[16] have reported that, in acid solution, adenine nucleotides do yield appreciable fluorescence (fluorescence, 390 mμ, uncorrected). It is conceivable, therefore, that in the dinucleotide excitation at neutral pH may lead to the same excited state as that of the ionized purine. This would have a high probability of energy transfer since there is then an appreciable lifetime of excitation and considerable overlap of the absorption band of the dihydro-pyridine with the observed fluorescence band of the adenine nucleotide.

The examples of energy transfer above were of relatively simple molecules as compared to proteins. It should be apparent that even in unconjugated proteins many of the absorbing residues must be sufficiently close to one another so as to result in considerable intramolecular energy transfer. Depending on the types of amino acids, their sequence and on their spatial configuration in the native protein, energy transfer may exert increases or decreases in quantum yield of fluorescence of a given fluorescent residue. The interaction is not limited to aromatic amino acids but can occur with other light absorbing groups, either as part of the peptide chain, such as cystine or peptide bonds, as prosthetic groups, heme or chlorophyll, as bound coenzymes, flavine nucleotides, and reduced pyridine nucleotides or as fluorescent dyes conjugated to proteins. As will be shown in subsequent sections, energy transfer is required to explain many of the observed interactions between proteins and other molecules.

III. FLUORESCENCE POLARIZATION

Before getting on with the discussion of protein fluorescence, it is also necessary to discuss fluorescence polarization, one of the more important tools which make fluorescence studies of proteins so meaningful. Detailed reviews of this subject have been presented by Weber[17] and Laurence.[18] As is well known, the size and shape of macromolecules, such as proteins, influence their Brownian motion, the latter involving both translational and rotational movements. In the case of large colloids it is possible to make direct visual observations of the Brownian motion. Since molecules of the size of the proteins cannot be observed, it is necessary to use indirect means for measurement. When a macromolecule is fluorescent, a study of the polarization of the emitted fluorescence is a measure of the Brownian motion which can, in turn, yield information concerning the size and shape of the molecule. The relationship between Brownian motion and fluorescence polarization has been most lucidly presented by Weber.[17]

The molecules of a homogenous protein have a definite size and shape which may be represented by a circle with a line through it to indicate a rigidly bound parameter or direction (Fig. 4). At any instant all possible directions are represented equally so that the molecules are randomly oriented, as in Fig. 4a. If, at a given time, we restrict our attention to those molecules oriented in the same direction (Fig. 4b) and observe only these we find that in a short time they too have become randomly disposed. However, if the time interval is sufficiently short randomization may be only partial (Fig. 4c). State c differs from state b in the degree of disorientation which has been introduced. The angle defined by the original direction in b and the new direction in c may be called θ and the mean value of its cosine, cos θ, can be used as a measure of the degree of orientation. Initially, when there is perfect orientation (Fig. 4b) cos θ is unity; it falls to a value of zero when orientation is completely random. The time necessary to pass from the state of complete orientation to a given state of disorientation is termed the *relaxation time*. An intermediate state of disorientation is selected since it is difficult to measure the time required for complete disorientation. The relaxation time of a protein varies with temperature and viscosity.

Perrin[19] devised a simple and elegant procedure for determining the disorientation caused by the Brownian motion in the case of fluorescent

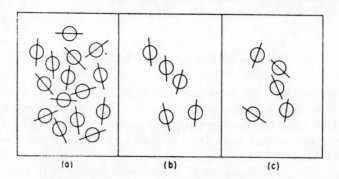

(a)   (b)   (c)

FIG. 4. Randomization brought about by Brownian motion (after Weber[17]).

molecules based on excitation with polarized light and measurement of the polarization of the emitted fluorescence. When polarized light is used for excitation, only those molecules will absorb appreciable energy that have the direction of their absorption oscillator parallel to the plane of vibration of the exciting wave. Thus, excitation with polarized light selects those molecules oriented around a particular direction as in state b, Fig. 4. Since each molecule emits a plane-polarized wave parallel to its emission oscillator, the emitted fluorescence should also be polarized. However, there is another factor which determines the observed polarization, this is the interval between excitation and emission, τ. For many molecules in aqueous solution, τ is about 10^{-8} second. For small molecules the relaxation time is much smaller than τ, so that by the time emission takes place the excited molecules are already randomly oriented and the resulting fluorescence is unpolarized. However, with large molecules, such as proteins, relaxation times are quite long compared to τ; this means that Brownian motion will have produced little disorientation during the interval between excitation and emission and the fluorescence will be partially polarized. In media of high viscosity and at low temperatures Brownian motion is lowest and fluorescence polarization is highest. Measurement of fluorescence polarization as a function of temperature and viscosity permits the estimation of the intensity of rotary Brownian movement and therefore of the relaxation time of a molecule. These terms can be utilized to yield information concerning molecular size and shape. The equation of Perrin shows the relationships between fluorescence polarization and many of these factors:

$$\frac{1}{p} = \frac{1}{p_0} + \left(\frac{1}{p_0} - \frac{1}{3}\right)\left(\frac{RT}{V}\right)\frac{\tau}{\eta}$$

where p = observed polarization.

p_0 = a constant (maximal value of p obtained in a rigid medium).

R = gas constant (8.314×10^7 ergs/degree centigrade/mole).

T = absolute temperature (°K).

η = viscosity (poise).

τ = interval between excitation and emission (a characteristic of the molecular species).

V = molecular volume of the fluorescent rotational unit (cubic centimeters).

Polarization, p, is defined by the following equation:

$$p = \frac{F_{||} - F_{\perp}}{F_{||} + F_{\perp}}$$

where $F_{||}$ is the fluorescence intensity measured with polarizing and analyzing prisms oriented with their electric vectors parallel and $F_\perp$ is the fluorescence intensity measured with the electric vectors of the two prisms crossed (See Chapter 2).

Most proteins possess some degree of fluorescence by virtue of their aromatic amino acid residues. Those which do not may be conjugated to fluorescent dyes and thereby made fluorescent. Furthermore, the value τ can be determined by independent means. Fluorescence polarization can therefore be used to study relatively small and symmetric proteins over wide ranges of pH and buffer concentrations. Steiner and McAlister[20] have determined mean relaxation times of many globular proteins with fluorescence polarization and the former[21] has utilized the technique to investigate the combination of trypsin with soybean inhibitor.

There is one additional way in which fluorescence polarization can provide useful information. As was pointed out earlier, the fluorescence emitted by small molecules in aqueous solution is itself randomly oriented. However, low molecular weight substrates, coenzymes, prosthetic groups, or haptenes, when they interact with their specific proteins, may alter the fluorescence polarization of the protein as well as its excitation and fluorescence spectra and quantum yield of fluorescence. Each of these changes provides additional information concerning the combination of the smaller group with the protein. Furthermore, when the smaller group is itself fluorescent, upon combination with the protein in a rigid manner, it can assume the long relaxation time of the macromolecule and thereby exhibit a high degree of fluorescence polarization. Less rigid bonding will yield smaller degrees of polarization. It should also be pointed out that even small molecules may exhibit appreciable fluorescence polarization. Many do so in viscous media at low temperatures. However, if the average lifetime of the excited state is sufficiently short, then a relatively small molecule can exhibit fluorescence polarization even in aqueous solution at room temperature. This is the case with DPNH. Weber[12] has estimated that $\tau = 5 \times 10^{-10}$ second for the reduced coenzyme in distilled water. Velick[22] estimates it to be even smaller. These low values are consistent with the observed low quantum yield, about 2%. Because of this low value for τ, DPNH solutions in water yield relatively high fluorescence polarization.

Fluorescence polarization studies of proteins give reliable estimates of molecular size when account is taken of molecular symmetry. The advantages it offers over other methods are:

(1) High sensitivity; microgram quantities of protein may be used.
(2) It is independent of protein and salt concentration.
(3) Applicability over a wide temperature range.
(4) Applicability over a wide range in hydrogen ion concentration.

(5) Conjugated proteins may be studied in the presence of other non-fluorescent proteins.

(6) Measurements are simple and the equipment is generally available and relatively inexpensive.

IV. FLUORESCENCE OF SIMPLE PROTEINS

That proteins absorb in the region 270–300 mμ has long been known. In fact, this is made use of by the enzymologist during purification of enzymes. The absorption in the far ultraviolet is due mainly to tyrosine and tryptophan, and it is therefore not surprising that proteins should fluoresce. However, even as recent as 1953 so little had been done in this area that Weber[17] could only predict that "A study of the ultraviolet fluorescence of aromatic amino acids and proteins would no doubt provide interesting details as regards the interaction of the aromatic residues in the protein molecules . . . ." Previous reports of visible fluorescence in simple proteins[23] were undoubtedly artifacts most likely due to impurities.

Shore and Pardee[13] in 1956, were about the first to report detailed studies on the native fluorescence of unconjugated proteins. These authors utilized a Beckman spectrophotometer with an RCA 1P28 photomultiplier attachment and measured fluorescence in the same direction as the transmitted excited light. They did this by first measuring absorption and then utilizing a filter to remove most of the exciting light. Corrections for photomultiplier response were also made. With this procedure they first measured the fluorescence efficiencies of a number of dyes of known efficiency. However, as was pointed out earlier (Chapter 5), the fluorescence efficiencies of the aromatic amino acids reported by Shore and Pardee[13] were later shown to be in error by Teale and Weber.[24] Nevertheless, Shore and Pardee[13] were the first to report the actual fluorescence of purified proteins and to demonstrate that this fluorescence is due almost entirely to the aromatic amino acids. They found the following proteins to be fluorescent when excited at 280 mμ; bovine plasma albumin, ribonuclease, lysozyme, tobacco mosaic virus protein, gramicidin, chymotrypsinogen, and glyceraldehydephosphate dehydrogenase (rabbit muscle). They were not able to detect appreciable fluorescence in purified nucleic acids, nucleotides, and nucleosides. Following the studies of Shore and Pardee on proteins, Teale and Weber[24] reported a most careful and elegant study of the fluorescence properties of the free aromatic amino acids. A summary of some of their findings was presented in Chapter 5, Table II. Although tryptophan and tyrosine have similar fluorescence efficiencies in water, the former contributes more fluorescence because of its higher extinction coefficient. The fluorescence contribution of phenylalanine is extremely low. They reported that the fluorescence spectra of the three aromatic amino acids in neutral water solutions consist of single bands in the ultraviolet with maxima of 282 mμ for phenylalanine, 303 mμ for tyrosine, and 348 mμ for tryptophan. The excitation spectra were also found to correspond to the known absorption spectra of the amino acids and in each case there was shown to be a constant quantum yield of fluorescence over the entire spectrum; 4% for phenylalanine, 21% for tyrosine, and 20% for tryptophan. Teale and Weber[24] also pointed out that the excitation spectrum, fluorescence spectrum, and quantum yield can be used to characterize a molecular species in solution.

Shortly thereafter Konev[25a] reported studies pertaining to the "centers responsible for fluorescence in proteins." The proteins investigated were serum albumin, egg albumin, gourd γ-globulin, serum γ-globulin, arachin, zein, gliadin, casein, sturnine, clupein, pepsin, and bromelin. For all of these proteins he observed qualitatively identical fluorescence spectra with two maxima at 313 and 350 mμ. The first maximum coincided with tyrosine, the second with tryptophan.* A mixture of aromatic amino acids in the proportions in which they were known to be present in a given protein were shown to yield a fluorescence spectrum qualitatively the same as that of the protein. A similar fluorescence spectrum was given by a hydrolyzate of the protein. Proteins without tryptophan (histones) and peptides such as glycyltyrosine gave fluorescence spectra with a single maximum at 313 mμ. The two protamines, clupein and sturnine, which do not contain aromatic amino acids showed no fluorescence in the region 250 to 400 mμ. Konev[25a] also compared the intensities of fluorescence of intact proteins with their hydrolyzates and with simulated amino acid mixtures. The data, reported as the ratio of intensity at 350 mμ (tryptophan) over intensity at 313 mμ (tyrosine) (I 350/I 313), indicated that the relative fluorescence of tryptophan falls with liberation of the free amino acids during hydrolysis. The higher fluorescence intensity of tryptophan in the intact protein indicated to Konev[25a] transfer of energy between tyrosine and tryptophan within the protein molecule.

In this same paper Konev[25a] also reported data on the excitation spectra of several proteins and again compared them with comparable amino acid mixtures and with the denatured protein.† He was able to show that in a

---

* It should be pointed out that Teale and Weber[24] reported values of 303 mμ for tyrosine and 348 mμ for tryptophan. Whereas the two values for tryptophan are within experimental error, those for tyrosine differ by 10 mμ. Values for tyrosine excitation closer to those of Konev have been obtained by others using commercial fluorescence spectrometers although these may require additional correction.

† To measure excitation spectra, a spectrophotometer was used as the excitation monochromator and a filter to cut off the exciting radiation. The observed excitation spectra are therefore for all fluorescence above about 300 mμ. Excitation energy was calibrated with a 0.02% solution of fluorescein in ammonia.

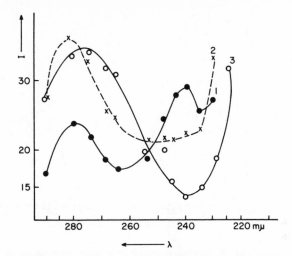

Fig. 5. Excitation spectra: (1) native γ-globulin, rabbit blood serum; (2) γ-globulin denatured at 60°; (3) γ-globulin hydrolyzate. All solutions adjusted to pH 7.5 (after Konev[25a]).

neutral medium (pH 6.7–8.2), protein solutions of comparable absorbance, at the wavelength of excitation (280 mμ), yielded spectral curves of an identical character. These had a clearly defined maximum at 280 mμ, a minimum at 255 mμ, and a second maximum at 240 to 245 mμ (Fig. 5). Following heat denaturation or hydrolysis, the 240- to 245-mμ excitation maximum disappeared, although there was little or no change in the absorption spectrum. Konev suggested that the 240- to 245-mμ excitation in the native protein is due to absorption of energy by a component of the protein, other than the aromatic amino acids, and that transmission of this energy to the aromatic amino acids occurred. He further suggested that the 240- to 245-mμ absorbing component responsible for the energy transfer is the peptide bond, possibly in its enol form. It is of interest that in their earlier report Shore and Pardee[13] also observed excitation of proteins below 250 mμ and stated that it may be indicative of energy transfer. They, too, suggested the peptide bond as being one primary absorbing species at 250 mμ but also suggested cystine as another absorbing species which could contribute to low wavelength excitation.

Vladimirov[25b] has observed a transfer of excitation energy from phenylalanine to tryptophan in mixed crystals and suggests that this may occur in proteins. He has also reported that excitation spectra of intact yeast cells exhibit properties characteristic of protein except for a minimum at 260 mμ in the region of nucleic acid absorption. This would indicate that energy absorbed by nucleic acids is not transmitted to proteins.

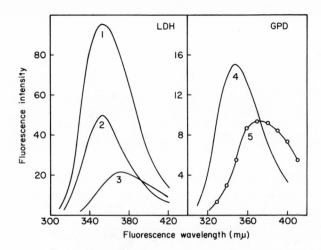

Fluorescence wavelength (mμ)

FIG. 6. Fluorescences of enzyme proteins compared with those of their alkaline hydrolysates.

Curves 1 and 2 are for intact lactic dehydrogenase in 0.1 M tris acetate buffer, pH 7.1, and in 1.0 N sodium carbonate, respectively. Curve 3 is the complete alkaline hydrolysate of the same concentration of lactic dehydrogenase in 1.0 N sodium carbonate. Curve 4 is the fluorescence of glyceraldehyde phosphate dehydrogenase in 1.0 N sodium carbonate. Curve 5 is the alkaline hydrolysate at the same concentration and in the same solvent. Excitation was at 290 mμ in all cases (after Velick[22]).

In a more recent study, Velick[22] reported data on the fluorescence of two purified enzymes, lactic dehydrogenase (beef heart; LDH) and glyceraldehyde-3-phosphate dehydrogenase (rabbit skeletal muscle; GPD) and compared this data with alkaline hydrolysates of the proteins and with simulated amino acid mixtures. These results are summarized in Fig. 6. It can be seen that both intact proteins yielded only one peak of fluorescence emission with a maximum lower than that of free tryptophan by about 10 mμ. These findings are in contrast to those of Konev[25a] who reported two distinct peaks for protein fluorescence, one at 313 mμ and one at 350 mμ. Conceivably, the latter investigator used an instrument with higher resolution. However, Teale and Weber[26] have also reported a failure to detect tyrosine fluorescence in tryptophan containing proteins. Nevertheless, both Velick[22] and Konev[25a] agree that there is considerable enhancement of the 350-mμ fluorescence in the intact proteins as compared to the component amino acids. Furthermore, the 350-mμ fluorescence of LDH was found to be 5 times as intense as that of GPD per unit weight of protein, whereas the tryptophan content was only twice as high. Clearly, these data, once more, indicate interaction of the aromatic amino acid residues within the protein molecule.

Teale and Weber[26] have presented an excellent discussion of the problem of native protein fluorescence. They point out that in the protein molecule the environment of the amino acid residues is modified in the following ways:

(a) the surrounding medium is less polarizable than water,

(b) rotational freedom is restricted,

(c) adjacent groups of the polypeptide framework may take part in specific quenching processes, and

(d) the average distance from one aromatic amino acid residue to another is about 10 Å; efficient transfer of energy has been shown to take place over such distances.

After examining a large number of proteins, including many that were globular, they divided them into two classes.[27] Class A proteins were those containing phenylalanine and tyrosine but no tryptophan. In these proteins phenylalanine fluorescence was absent, the fluorescence spectrum being characteristic of tyrosine, with a maximum emission at 303 mμ. However, the quantum yield of fluorescence in class A proteins, such as insulin, ribonuclease, and zein, was found to be only one-fifth to one-seventh of that of free tyrosine. Two trypsin inhibitors, which contained tyrosine as the sole aromatic amino acid were actually nonfluorescent. Class B proteins were those containing tryptophan as well as tyrosine and phenylalanine. In all these proteins only one fluorescence peak was observed, that due to the tryptophan. These findings are consistent with those of Velick[22] and again contradict those of Konev[25a] who reported peaks for both tyrosine and tryptophan in class B type proteins. The quantum yield of fluorescence of class B proteins varied widely from 0.04 to 0.3, compared to 0.2 for free tryptophan. The two enzymes studied by Velick[22] appeared to have quantum yields even higher than those studied by Teale and Weber.[26] The latter investigators also reported that the effects of various solvents on class B proteins differed from protein to protein. Thus, in viscous media (glycols) albumins showed no change in quantum yield whereas large increases were shown by trypsin and chymotrypsin. In a denaturing medium such as 8 M urea, the quantum yields of trypsin and chymotrypsin were increased, whereas values for the albumins were decreased. Class B proteins also exhibited a shift in maximum fluorescence emission; 320–350 mμ in the proteins as compared to 348 mμ for free tryptophan. There was also a shift in absorption maximum to 295–310 mμ as compared to 275–280 mμ for free tryptophan and tyrosine. The diminished[25a] or absent[22] tyrosine fluorescence in class B proteins was considered to be due to transfer of energy from the tyrosine to the tryptophan residues. However, Teale[28] showed that when these proteins were irradiated with a narrow band of energy specific for the excitation of protein bound tryptophan (295–310 mμ), the quantum yield actually increased appreciably due to the loss of quanta absorbed by tyrosine at the shorter wavelengths. Thus, it appears that tyrosine fluorescence energy in class B proteins is not transferred to tryptophan but is quenched by processes arising from secondary bonding of the phenolic group. Just what role the tryptophan plays in the quenching of tyrosine fluorescence was not at all made clear in these studies. Furthermore, Weber[29] had shown that in mixtures of phenols and indoles, the phenol was able to transfer excitation energy to the indole over a distance of 18.5 Å. The recent studies by Weber[27, 29] and by Teale[28] are the most comprehensive relating to the fluorescence of unsubstituted proteins. Although more definitive interpretation of their data is still required, it is apparent that quantitative methods of fluorescence are most useful in detecting the interaction of amino acid residues within a protein molecule and to demonstrate structural differences between molecules which, by amino acid assay alone, are not too different.

Boroff and Fitzgerald[30] and Boroff[31] have recently applied fluorometry to studies on the purified toxin of *Clostridium botulinum*. The toxin, which is an unconjugated protein, demonstrated one fairly intense fluorescence band with an excitation maximum at 287 mμ and a fluorescence maximum at 350 mμ (uncorrected). These data are again characteristic of tryptophan. The authors have utilized this fluorescence in experiments which purported to show that tryptophan is near or at the toxic center of the protein molecule. They did this by comparing toxicity and fluorescence following various chemical and physical inactivation procedures. However, another group[32a] reinvestigated the fluorescence of *botulinum* toxin and were unable to demonstrate a direct relationship between protein fluorescence and toxicity.

Fluorometric protein assay: The characteristic fluorescence of proteins is sufficiently sensitive and specific as to make it practical for use in quantitative assay. Neutral solutions of many tissue extracts contain few other materials which emit in the region 340–350 mμ when excited at 280 mμ. Under such conditions, nucleic acids, purines, and pyrimidines show no fluorescence, and it should therefore be possible to assay proteins during enzyme purification procedures without the need for nucleic acid correction as in the widely used spectrophotometric method. Of course, it is necessary to make comparisons with appropriate protein standards and to make allowance for the light scattering brought about by partially purified biological extracts. Konev and Kozunin[32b] have used the native fluorescence of protein to develop a rapid and precise method for measuring the protein content of milk. The milk is diluted about 1 to 20 and its native fluorescence is measured in a specially devised photofluorometer (Chapter 3). Arrangements are provided to determine and correct for the light scattering of the solutions. By using this procedure Konev and Kozunin[32b] were able to determine protein in milk with an accuracy comparable to that obtained with the Kjeldahl procedure and with a speed far greater than is available with any current method. They propose this procedure for general use in the dairy industry and also discuss its more general applications in biology.

V. FLUORESCENCE OF PROTEINS WITH PROSTHETIC GROUPS

A large number of proteins exist in nature in combination with smaller molecules which may be bound to the protein in some form of loose or tight linkage. The nonpeptide molecules or "prosthetic groups" may be, among other things, metals, porphyrins, carbohydrates, or any one of the numerous coenzymes. If the prosthetic group is itself capable of absorbing radiant energy, it can influence the fluorescence of the protein either by affecting the fluorescence of the aromatic amino acids in the peptide chain or by endowing the protein with its own fluorescence.

It is known that light absorbed by the aromatic amino acid residues in carbon monoxide compounds of heme proteins is active in promoting photochemical changes. For this to occur, it is necessary that there be a transfer of electronic energy from the oscillators originally excited (aromatic amino acids) to the heme residues. Weber and Teale[33] investigated this problem by measuring the quantum yields of fluorescence and lifetimes of the excited state of a number of intact heme proteins and then comparing these with the corresponding values observed when the heme was split off.* On utilizing the ultraviolet fluorescence due to the aromatic amino acids and the visible fluorescence obtained by conjugating the heme proteins with 1-dimethylamino naphthalene-5-sulfonyl chloride (see below), they were able to show a marked fluorescence quenching by transfer of the excitation energy to the heme. A summary of their findings is shown in Table I. It is evident from these data that with the exception of horse-radish peroxidase the heme containing proteins are essentially devoid of fluorescence. By contrast, the heme-free globins were all found to have fluorescence quantum yields comparable to that of free tryptophan (about 0.2). They calculated that the transfer of the excitation energy to

---

* The protein-heme linkage was broken either by the method of D. L. Drabkin, [*J. Biol. Chem.* **158**, 72 (1945)] or that of F. J. W. Teale (unpublished). The latter involves extraction of the heme by methyl-ethyl ketone from the acidified protein solution.

TABLE I

RATIO OF FLUORESCENCE EFFICIENCY OBSERVED AFTER REMOVAL OF HEME TO THAT
MEASURED IN THE INTACT PROTEIN $(F_0/F)^a$

	F_0/F	
Protein	Native heme protein	Dye conjugated protein
Myoglobin Fe^{2+}	—	18–32
Myoglobin Fe^{3+}	100	9–15
Horse-radish peroxidase Fe^{2+}	—	6–11
Horse-radish peroxidase Fe^{3+}	7	3.5–7
Cytochrome c Fe^{3+}	—	>100
Hemoglobin Fe^{2+}	—	54–60
Hemoglobin Fe^{3+}	>100	38–40
Catalase Fe^{3+}	>100	5–7

[a] After Weber and Teale.[33]

heme has a probability at least 100 times greater than fluorescence and
even 20 times greater than that of other competing radiationless processes.
Therefore, the quantum yield of the indirect chemical effects (photodecomposition) resulting from the transfer of radiation absorbed by the
aromatic amino acids to the heme should be as high as that produced by
excitation of the heme at its wavelength of maximum absorption. The data
were also used to estimate which energy transfers occurred between the
absorbing centers (tryptophan or conjugated dye) and the heme residues
within the heme-protein molecules. It should be pointed out that Shore
and Pardee[13] in examining a large number of proteins had found that
carbonmonoxymyoglobin was least fluorescent. There are other prosthetic
groups which absorb light but are not fluorescent. It will be of interest to
see how they interact with the fluorescence of the protein. However,
studies of this kind require highly purified proteins in order to be meaningful.

It has been possible to obtain many pyridine nucleotide dehydrogenases
in pure form and fluorometry has become one of the more widely used
tools with which to study enzyme-coenzyme-substrate interaction. Theorell
and Bonnichsen[34] showed that DPNH forms a complex with alcohol dehydrogenase in which the absorption maximum of the reduced coenzyme is
shifted from 340 mμ (free) to 325 mμ (bound). At a later date Boyer and
Theorell[35] showed that in the enzyme-coenzyme complexes there is also
an intensification of fluorescence and a shift in the fluorescence spectrum
(462 mμ, free; 440 mμ, bound) (Fig. 7). It is now recognized that binding

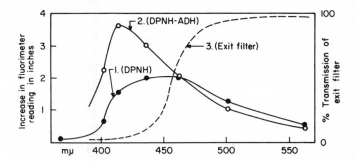

Fig. 7. Fluorescence spectra of free DPNH (1) and of DPNH combined with alcohol dehydrogenase, corrected for the fluorescence of the enzyme (2). This difference is not apparent in filter instruments which use exit filters of the type shown in curve 3 (after Boyer and Theorell[35]).

of reduced pyridine nucleotide coenzyme to a dehydrogenase generally leads to an alteration in the fluorescence intensity and a shift in excitation and fluorescence maxima towards the ultraviolet. Fisher and McGregor[36] have reported that they can obtain similar increases in fluorescence intensity and alteration of the spectra of DPNH following its adsorption onto nonspecific surfaces such as zinc and manganese phosphate precipitates and ion exchange resins. They point out, therefore, that the fluorescent changes can only be taken as an index of the nonspecific binding of the reduced coenzyme onto the protein enzyme. However, although the mere adsorption of the coenzyme onto the protein is responsible for fluorescence changes, more detailed studies have indicated that fluorometric studies can contribute substantially to our understanding of the specific interaction of pyridine nucleotides with individual dehydrogenases. Table II indicates the number of fluorescent studies recently undertaken in this area. A comprehensive review on coenzyme binding, stressing these and other fluorometric studies has appeared recently.[37] A most thorough report of the application of fluorescence to' enzyme-coenzyme complexes was presented by Velick.[22] Although portions of this paper have been discussed in previous sections, it will be reviewed here in more detail since it presents most clearly the theoretical and practical aspects of such studies.

Velick[22] compared a number of fluorescent properties of two different enzymes, lactic dehydrogenase (LDH) and triosephosphate dehydrogenase (TPD) and found the following:

(1) Excitation and fluorescence maxima of DPNH in the LDH complex are shifted towards the ultraviolet. The spectra of the GPD-DPNH complex coincide with those of free DPNH.

TABLE II

FLUOROMETRIC STUDIES ON DEHYDROGENASE-REDUCED PYRIDINE
NUCLEOTIDE COMPLEXES

| | | Fluorescence | | |
Enzyme	Coenzyme	Intensity	Direction of spectral shift 200–800 mμ	Reference
Alcohol dehydrogenase (liver)	DPNH	Increased	←	(35)
Alcohol dehydrogenase (yeast)	DPNH	Increased	←	(39)
Lactic dehydrogenase (heart)	DPNH	Increased	←	(22)
Lactic dehydrogenase (liver,	DPNH	Increased	←	(40)
Lactic dehydrogenase (heart and muscle)	DPNH	Increased	←	(38)
Lactic dehydrogenase (heart)	DPNH	Increased	←	(38)
Lactic dehydrogenase (heart)	APDH[a]	Increased	←	(38)
Triosephosphate dehydrogenase	DPNH	Decreased	←	(22)
Glutamic dehydrogenase	DPNH and TPNH	Increased	←	(41)
Malic dehydrogenase	DPNH	Increased	←	(42, 43)
Isocitric dehydrogenase	TPNH	—	←	(44)
Uridine diphosphogalactose-4-epimerase	DPNH	—	←	(45)

[a] APDH = acetylpyridine analog of DPNH.

(2) Whereas the fluorescence intensity of the LDH-DPNH complex is
higher than that of free DPNH, that of the TPD-DPNH complex is
lower. The latter represents the only instance of its kind (see Table II).

(3) Both proteins, in the free form, show appreciable fluorescence due
to the tryptophan residues. The combination with DPNH depresses the
native fluorescence of the proteins (Fig. 8). The diminution in tryptophan
fluorescence, when excited at 280 mμ, is balanced by an appearance of
DPNH fluorescence indicating transfer of excitation energy from the
tryptophan residues to the reduced pyridine nucleotide.

(4) The fluorescence of the free proteins is higher in intensity than that
of a corresponding hydrolyzate and the fluorescence maxima of the aromatic amino acids in the intact proteins is displaced towards the ultraviolet.

(5) Changes in fluorescence intensity and polarization are so high that
titrations between enzyme and coenzyme can be carried out at the microgram level to yield data concerning the number of coenzyme binding sites

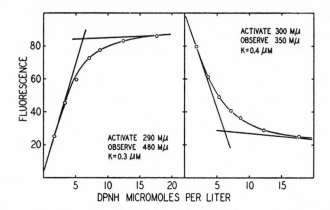

FIG. 8. Titration of lactic dehydrogenase with DPNH for stoichiometry and dissociation constants. The curve on the left is nucleotide fluorescence excited by energy transfer from the protein. The curve on the right represents quenching of protein fluorescence by DPNH. Combining weight of 38,700 and 36,300 g. of protein per mole of DPNH were calculated by the two methods (after Velick[22]).

and their dissociation constants. Even DPN, which is not itself fluorescent, can be titrated fluorometrically since it quenches the native protein fluorescence (as does heme) when combined in the complex (Fig. 8). Competition between oxidized and reduced coenzyme for the binding sites was also demonstrated. A summary of the data obtained by titration studies is shown in Table III.

(6) Demonstration of release of bound DPNH from the enzymes on titration with p-chloromercuribenzoate with no stoichiometry between the

TABLE III

COENZYME BINDING STUDIES WITH LACTIC DEHYDROGENASE AND
TRIOSEPHOSPHATE DEHYDROGENASE

Finding	Lactic dehydrogenase	Triosephosphate dehydrogenase
Minimum combining weight_____	37,500	45,000
Number of binding sites_____	4	3
KDPN_____	About $10^{-4}\ M^a$	$0.06 \times 10^{-6}\ M$
KDPNH_____	$0.35 \times 10^{-6}\ M$	$0.24 \times 10^{-6}\ M$

[a] The disparity between KDPN and KDPNH made this measurement susceptible to appreciable error due to trace contamination of DPN with DPNH (after Velick[22]).

thiol reagent and the reduced pyridine nucleotide, indicating that the inhibitor acts by a mechanism other than direct displacement.

(7) Demonstration of a competition between the two proteins for DPNH.

(8) Demonstration that phosphate increases the fluorescence intensity of the TPD-DPNH complex and that other anions have little effect. Since the phosphate effect is most pronounced at about pH 6.7 and almost disappears at pH 8.5, it would suggest that the phosphate effect involves histidine residues which may participate in the coenzyme binding. By contrast phosphate and other ions lower the fluorescence efficiency of LDH.

(9) The change in fluorescence polarization of DPNH on binding with the two enzymes is quite different (Fig. 9), LDH produces a much higher degree of polarization than TPD even though the molecular weights and axial ratios of the two proteins are similar. Velick suggests that these differences indicate different mechanisms of interaction of the two proteins with the reduced coenzyme. Furthermore in the LDH-DPNH complex there was no detectable transfer of purine exciting energy (260 mμ) to the pyridine as occurs in the free coenzyme,[12] indicating that the conformation of DPNH in the protein bound form is such that the purine and pyridine groups are further removed from one another than in the free coenzyme. Spatial models of the enzyme coenzyme complexes were constructed based on these data.

Many other interesting studies of enzyme-pyridine nucleotide interaction have been carried out (Table II). Thus, Shifrin et al.[38] showed that analogs of reduced pyridine nucleotides underwent similar fluorescent changes on binding with enzyme. In fact, the reduced acetylpyridine analog showed almost a tenfold increase in fluorescence on binding with

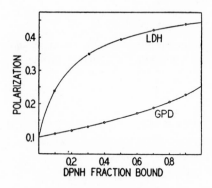

FIG. 9. Fluorescence polarization of DPNH as a function of the mole fraction of the coenzyme that is bound to lactic dehydrogenase and glyceraldehyde phosphate dehydrogenase (after Velick[22]).

either alcohol dehydrogenase or lactic dehydrogenase. Winer and Schwert[46] used fluorometric methods to demonstrate ternary complexes of lactic dehydrogenase, DPNH, and certain carboxylic acids such as oxamate and oxalate. Binary and ternary complexes with malic dehydrogenase, coenzyme, and L-malic acid were also demonstrated by Theorell and Langan.[43] Maxwell *et al.*[45] isolated uridine diphosphogalactose-4-epimerase and found it to contain, in its purest form, fluorescent material characteristic of reduced pyridine nucleotide. Further studies[47] have shown that the pyridine nucleotide is firmly bound in the epimerase molecule.

Although the flavin nucleotide coenzymes are also fluorescent there has not been as much use of fluorescence in studying their interaction with proteins. One reason is that, in general, the flavin fluorescence disappears when bound to proteins.[48] Notatin[49] and xanthine oxidase[50] are devoid of any flavin fluorescence. On the other hand, the fluorescence of diaphorase corresponds in intensity and emission spectrum to that of the free prosthetic group, flavin adenine dinucleotide.[48] Weber[17] has shown that solutions of diaphorase show strongly polarized fluorescence and that the fluorescence of the free prosthetic group is unpolarized. More recently, Walaas and Walaas[51] showed that purified D-amino acid oxidase is also fluorescent, indicating that the isoalloxazine is not bound to the protein in the same manner as in other nonfluorescent flavoproteins.

Hagins and Jennings[52] have used fluorometric methods to investigate the possibility of a highly efficient energy transfer between rhodopsin molecules (protein + vitamin A aldehyde). They found that vitamin A solutions show a high degree of fluorescence polarization and a high fluorescence efficiency even in the most concentrated solutions. These findings are not consistent with appreciable energy transfer from one molecule to another. They also failed to detect evidence of energy transfer in direct studies on the microfluorescence of rhodopsin in rods.

Other coenzymes are fluorescent and should lead to enzyme-coenzyme complexes amenable to fluorometric study. This is true of pyridoxal phosphate and pyridoxamine phosphate, yet there have been no studies with decarboxylases and transaminases comparable to those with the dehydrogenases. Many other prosthetic groups while not fluorescent absorb light. In such cases it should be possible to study their interaction with the enzyme by the effect on the fluorescence intensity of the aromatic amino acid residues in the protein. Even nonfluorescent, nonabsorbing prosthetic groups may influence the native fluorescence of an enzyme by increasing or decreasing energy transfer between residues. In plants, chlorophyll-protein complexes and other photosynthesizing protein pigments are fluorescent. These will be discussed in a later section (Chapter 11). As more proteins are purified and studied there will undoubtedly be found more normally occurring fluorescent prosthetic groups. As an example, purified bovine elastin when hydrolyzed by elastase yields peptides combined with a yellow fluorescent compound.[53]

VI. ANTIBODY-HAPTEN INTERACTION

When antibody formation is induced with certain conjugated antigens, the resulting antibodies can combine specifically, not only with the intact antigen conjugate but with the relatively simple conjugating molecule, or hapten. With antigens substituted on the ϵ-amino groups of the lysine side-chain with 2,4-dinitrophenyl groups, antibodies are produced which react with ϵ-N-dinitrophenyl lysine. In a comparable manner, antigens conjugated with azobenzene-p-arsonate yield antibodies which combine with p-phenylarsonate azotyrosine. Velick et al.[54] have shown that when either of these haptens combines with purified, specific antibody the tryptophan fluorescence of the protein is quenched. The absorption bands of the haptens are in the region 300–400 mμ which overlap the 350-mμ fluorescence band of tryptophan (Fig. 10). It is for this reason that tryptophan transfers its energy so efficiently to the hapten. The haptens themselves are not fluorescent and release the transferred energy in a form other than radiation. In this respect the haptens resemble the heme proteins. Diminution of fluorescence quantum yield was shown to be specific

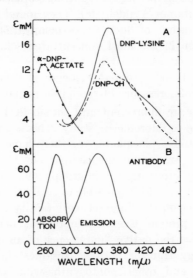

FIG. 10. A, absorption spectra of the 2,4-dinitrophenyl haptens; B, absorption and fluorescence emission bands of the antibody protein (after Velick et al.[54]).

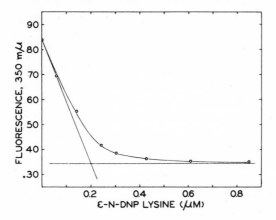

FIG. 11. The titration of 20 μg. of antibody with N-dinitrophenyl-lysine at 26°, pH 7.4 in 1.0 ml. of 0.01 M potassium phosphate buffer, 0.15 M sodium chloride. The intersection of the linear extrapolations is the antibody concentration in microequivalents per liter (after Velick et al.[54]).

in the case of ϵ-N-dinitrophenyl lysine and to occur to the extent of 70%. It is, therefore, a sensitive indicator of complex formation. Hapten-antibody titrations were carried out (Fig. 11) to obtain the stoichiometry of the reaction, and it was even possible to measure a dissociation constant of 2.4×10^{-9} for the ϵ-N-dinitrophenyl lysine haptene with its specific antibody. Few other methods could be used for such a small measurement. The study by Velick et al. was an extensive one, additional studies being carried out with antibody fragments. The data made it possible to locate the tryptophan residues with respect to the hapten and to show that each hapten must capture the excitation energy of eight tryptophan residues.

This approach to the study of antibody-hapten interaction and to antibody-antigen interaction provides the immunochemist with a most powerful tool which is remarkably sensitive and specific and can be made to yield precise data. There can be no doubt that the fluorescence quenching technique of studying antibody interaction will be used widely in the future.

VII. CHEMICALLY PREPARED FLUORESCENT PROTEIN CONJUGATES

The combination of proteins with colored dyes has long been used in immunology. In order to increase the sensitivity of the dye techniques,

Creech and Jones[55] introduced the use of fluorescent dyes. In these early studies they treated proteins with 1:2 benzanthryl isocyanate to form the corresponding carbamido conjugates. Since then, the use of fluorescent conjugates has become an important and widely used technique in protein studies. Although protein-dye conjugates are used in widely separated fields of study, immunology and protein structure by depolarization methods being the two most important, similar problems arise in all such studies. These include the choice of fluorescent reagents for labeling, the preparation and purification of these reagents, methods for treatment of the proteins with the reagent, purification of the protein dye conjugates, and fluorometric assay of the conjugates during a particular study. In addition to these problems, application of fluorescence microscopy in immunology requires that the fluorescent antibody have fluorescence which differs from that of the background tissues. For protein studies there is also the requirement of choosing a dye with an average lifetime of fluorescence, τ, sufficiently long to produce appreciable polarization with the size molecule under study.

A. Application to Studies on Protein Structure

The use of dye conjugates was utilized for such studies before instruments were available for detecting the native ultraviolet fluorescence of proteins. Since very few proteins fluoresced in the visible region, conjugation with a fluorescent dye was the only way in which fluorescence polarization studies could be applied to proteins in a general manner. Following formation of a protein-dye conjugate, fluorescence polarization can be measured as described earlier in this section. Substitution of this and other parameters in the Perrin equation permits calculation of molecular volumes. Further calculations can then be made of molecular weight. If molecular weight is known from other measurement, fluorescence polarization can provide data concerning molecular asymmetry. Determinations of intramolecular energy transfer can also be made. The various ways in which fluorescence polarization of dye protein conjugates can be utilized in studies of protein structure have been well reviewed by Weber[17] and Laurence.[18] Many individuals have applied this technique and a large number of dyes have been utilized. Investigations in this area by Weber,[17] Steiner,[20] Chadwick,[56] and their colleagues are among the more noteworthy. Shore and Pardee[57] utilized the method to study energy transfer in proteins and nucleic acids as have Weber and Teale.[33]

There are some problems concerning fluorescence studies on proteins which are unique to dye conjugates. The choice of an appropriate dye is one. In general, the dye should be one with a high fluorescence quantum yield so that only a small number of residues need be inserted into the protein to endow it with sufficient fluorescence intensity to permit measurements in dilute solution. Too many groups can alter the properties of the protein. The excitation and fluorescent spectra are not too important since generally the studies are carried out with pure materials in the absence of interfering fluorescence. The dye should be one that can be converted readily to a derivative such as a sulfonyl chloride, isocyanate, or isothiocyanate, in order to react with proteins under relatively mild conditions so as to avoid denaturation. The properties of the dye and its reagent form should be such that they can be separated from the conjugated protein. For energy transfer studies it is desirable that the dye (in conjugated form) have an absorption peak overlapping that of the protein fluorescence emission, 304 mμ for class A proteins (tyrosine only), 325–350 mμ for class B proteins (tryptophan containing). The dye should also have minimal absorption at the protein excitation maxima (270–280 mμ). For polarization studies it is also necessary that the average lifetime of the fluorescent state of the protein conjugate be compatible with the size of the protein under investigation. Thus, a dye which emits its fluorescence over a long interval of time will yield little fluorescence polarization when combined with a small protein or peptide which has a much shorter relaxation time. Conversely, a dye which emits fluorescence (as a conjugate) over a very short interval of time, when combined to proteins above a certain size, will remain almost completely polarized during the entire time of emission, independent of the size of the protein.

The dye reagent which has been used most widely in protein structural studies is 1-dimethylaminonaphthalene-5-sulfonyl chloride (Fig. 13). The corresponding sulfonic acid can be converted to the sulfonyl chloride by treatment with phosphorous pentachloride as described by Laurence[18] but can now be obtained from several commercial sources. To introduce one to three dye residues into each protein molecule (molecular weight 50,000), the protein is treated with an amount of dye reagent equal to 1 to 2% of its weight. The protein may be dissolved in 0.1 N phosphate buffer pH 7.5 or in 1% sodium bicarbonate. Ten milliliters of such a solution (cooled to 0.3°) are then treated with 0.5 ml. of dye reagent dissolved in 0.5 ml. of acetone. Stirring is used initially to form an emulsion of the reagent. The reaction mixture is kept at 0 to 3° for several hours to permit the reaction to go to completion. At the end of this time excess amounts of unhydrolyzed reagent, hydrolyzed sulfonic acid, and some of the sulfonamide due to traces of ammonia must be removed by centrifugation, dialyses, and treatment with ion exchange resins and other adsorbents. Before using the protein conjugate, it is necessary to make certain that it is free of traces of adsorbed dye as an impurity since such material could

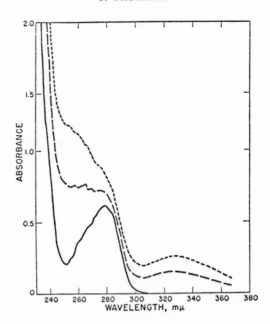

FIG. 12. Absorption spectra of bovine plasma albumin and of 1-dimethylamino-naphthalene-5-sulfonamido derivatives of the protein; 1 mg./ml. protein, pH 7.5 (after Shore and Pardee[57]). —, untreated plasma albumin; – – –, 2.1 moles of dye per mole of protein; ---, 4.0 moles of dye per mole of protein.

introduce large errors in subsequent studies. Chromatography and electro-phoresis on paper, in appropriate media, are the best criteria of purity. The procedure was originally devised by Weber[58] and has been presented in detail in the review by Laurence.[18] The reagents combine mainly with free amino groups in the protein in sulfonamide linkage.

The 1-dimethylaminonaphthalene-5-sulfonamide derivative of bovine serum albumin has an absorption maximum in the region 320–340 mμ (Fig. 12). It is possible to determine the number of conjugated groups by measuring the absorption at the wavelength maximum of the dye in a solution of known protein content. The extinction coefficient of the dye at its absorption maximum is 4.3×10^6 cm.^{2}/g. mole.[18] The emitted fluorescence has a maximum in the region 540–590 mμ.[57] The absorption in the region 320–340 mμ of the 1-dimethylaminonaphthalene-5-sulfonamide conjugates is, of course, ideal for studies designed to demonstrate transfer of energy from tryptophan containing proteins since the protein fluorescence almost overlaps the dye absorption, an ideal situation for energy transfer. The studies of Shore and Pardee[57] demonstrated the usefulness of this dye for such studies. As for polarization studies, 1-dimethylamino-

TABLE IV

RANGE OF FLUORESCENCE DEPOLARIZATION METHOD WITH VARIOUS DYES[a]

Dye	$\tau \times 10^{-8}$ second of conjugates	Molecular weight range[b]
3-Phenyl-7-isocyanatocoumarin	0.25	1000–60,000
1-Dimethylaminonaphthalene-5-sulfonic acid	1.2	10,000–200,000
Anthracene isocyanate	4.0	100,000–300,000
3-Hydroxypyrene 5:8:10 trisulfonic acid	9.0	40,000–2,500,000

[a] After Chadwick et al.[56] and Laurence.[18]
[b] For spherical molecules.

naphthalene-5-sulfonic acid is an excellent dye for conjugation to proteins of molecular weight 10,000–200,000 (spherical molecules). Although this region covers many of the important proteins, it is obvious that dyes with shorter and longer fluorescence lifetimes, τ, are required for investigations of peptides, proteins, and other macromolecules outside this range. In Table IV are shown τ values and practical molecular weight ranges for several dyes. The structures of the dyes are shown in Fig. 13.

Conjugation with dye has been used to study the interaction of F-actin with myosin by fluorescence polarization.[59] It was also found that addition of adenosinetriphosphate to the F-actomyosin conjugate produces an abrupt increase in polarization which decreases to control values over a

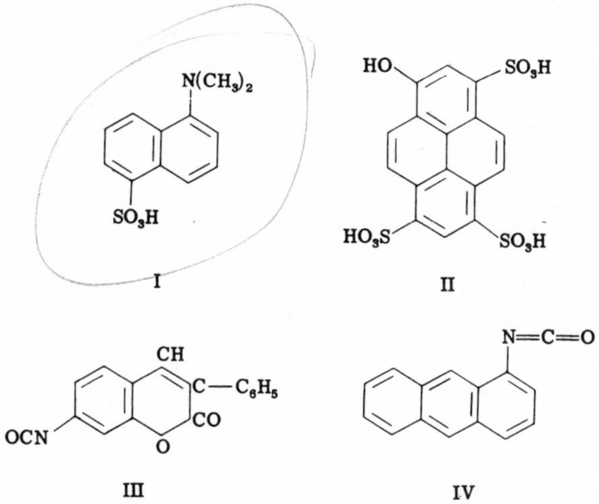

FIG. 13. Some dyes used for fluorescence labeling of proteins. I, 1-dimethylamino-naphthalene-5-sulfonic acid; II, 3-hydroxypyrene-5:8:10 trisulfonic acid; III, 3-phenyl-7-isocyanato-coumarin; IV, anthracene isocyanate.

period of minutes and that this process can be repeated each time that fresh adenosinetriphosphate is added. Massey[60] has used fluorescence polarization studies with dye conjugates of the enzyme fumarase and has applied the combination of techniques to determine the interaction between enzyme and substrate and enzyme and inhibitors. Hartley and Massey[61] have also applied these techniques in studies on the active center of the enzyme chymotrypsin.

B. As Protein Tracers and Immunofluorescence

There are many other important applications of fluorescent dye-protein conjugates in biology. The tracing of a protein when it is administered *in vivo* is one such use. Absorption of proteins from the intestinal tract or clearance by the kidney can be readily followed by this method. Measurement of plasma volume by dilution of fluorescent labeled plasma proteins is another useful application. De Barbieri and Scevola[62] have used fluorescent labeled cytochrome c to study the penetration of the protein through cell membranes *in vivo*. While their findings suggest that the intact protein can enter heart, liver, and kidney cells and combine with motochondria, the data is not conclusive. Nevertheless, this is an interesting type of study.

The widest applications of the fluorescent-dye method have been as a label for antibodies. The method was originally devised by Coons *et al.*[63] for the purpose of detecting and identifying antigenic material in tissue cells. Since then there have been many other applications some of which will be discussed later in this section. The use of labeled antibodies to detect antigen-antibody complexes in tissues is not a new idea. Initially colored dyes were used but these were not sufficiently sensitive or specific for histochemical purposes. The use of radioactively labeled antibodies solved the problem of sensitivity and specificity. However, it required the preparation of highly radioactive antibodies and autoradiography for detection. The large amounts of radioactivity not only present a hazard but introduce other problems as well. The use of simple labels with short half-lives, such as I^{131}, requires high initial radioactivity and frequent preparations due to the rapid decay of the isotope. Carbon-14 and tritium may be used but these must be incorporated with high specific activity into organic reagents before antibody labeling. The time required for detection by autoradiography may then be several days or weeks and the resolution of the resulting autoradiograms will frequently not approach that of fluorescence microscopy.

The general procedure after which most such studies are now patterned is the one published by Coons and Kaplan[64] in 1950. In principle, an antibody solution of high titer is conjugated with a fluorescent dye by treatment with a dye reagent. The resulting fluorescent antibody solution is freed of unreacted dye reagent and other dye products and is used as a specific histochemical stain on tissue sections; fluorescent proteins which are unreacted can be washed away during the preparation of the section. The section is then examined under a fluorescence microscope with appropriate filters to limit the exciting light. With fluorescein conjugated homologous antibody antigen is located by the deposition of the yellow-green fluorescence of the deposited fluorescein antibody.

Many of the papers which have been published in this field have been concerned with the preparation of suitable dye reagents and their purification and storage. Coons and Kaplan[64] in their initial studies used fluorescein isocyanate which they prepared from 4-nitrophthalic acid and resorcinol. The fluorescein isocyanate formed in this synthesis was not isolated and was used for protein conjugation without purification. Although the problems of synthesis, stability, and purity with fluorescein isocyanate were many, the usefulness of the method was so great that many applied it with and without modifications. More stable reagents of fluorescein and other dyes have now been introduced and many can be obtained in pure form from a number of commercial sources. Fluorescein isothiocyanate, one of the commercially available dye reagents, is stable and also considered to be superior to fluorescein isocyanate in yielding results at a higher dilution.[65] Conjugation of antibody protein with fluorescein isothiocyanate or with other isothiocyanates may be carried out in the following way.[65]

To an Ehrlenmeyer flask, fitted with a mechanical stirrer, are added 10 ml. of 0.85% NaCl, 3 ml. of 0.5 M carbonate-bicarbonate buffer, pH 9.0, and 3 ml. of acetone. The mixture is cooled in an acetone-dry ice bath until ice crystals appear and 10 ml. of a globulin fraction of known protein concentration is then added. The final solution should contain about 10 mg. of protein per ml. The mixture is again cooled until ice crystals appear and then 1.5 ml. of acetone, containing the fluorescein isothiocyanate, is added slowly with continuous stirring. The amount of dye reagent used should be 0.05 mg./mg. of protein. Following the addition of the dye reagent the mixture is maintained, with stirring, at about 4° for 18 hours; preferably in a cold room. To purify the protein conjugate, the mixture is dialyzed against 0.01 M phosphate buffer, ph 7.2. Dialysis is continued until the dialysate shows no fluorescence. Further purification may then be achieved by passing the conjugate through a Millipore filter. According to Coons and Kaplan,[64] gamma globulin mixtures purified by dialysis, precipitation, and treatment with chemical adsorbents still stain normal tissue in a nonspecific manner probably due to adsorbed unconjugated dye. They found that acetone-dried lyophilized liver preparations would remove most of this interfering material when shaken with the antibody conjugate in amounts of 100 mg. of tissue powder per milliliter. Two such treatments followed by high-speed centrifugation or filtration through Seitz filters are required. Chadwick et al.[66] claim that the tissue powder technique does not remove unreacted fluorescent material completely. They report that with both fluorescein isocyanate and fluorescein isothiocyanate, shaking of the protein conjugate with powdered activated charcoal for 1 hour removes essentially all nonprotein fluorescence. Of course, wherever possible the presence of adsorbed dye should be checked by chromatography and electrophoresis on paper, as described by Laurence.[18]

Condensation with dyes in the form of their amines can also be carried out by diazotization and coupling. The use of aminorosamine-B to prepare diazorosamine-B—antibody conjugates has been described by Borek and Silverstein.[67]

Details concerning the applications of the fluorescent labeled antibody technique to histological studies are outside the scope of this book. An excellent review of the subject was written by Coons.[68] The article includes a discussion of the instrumentation required for fluorescence microscopy. Attachments for fluorescence microscopy are now available as accessories to most microscopes and information concerning light sources and filters can be obtained from the manufacturers.

Applications of the fluorescent antibody technique are so numerous that only a few examples can be given.

(i) *Localization of antigen in tissue:* The procedure has been applied to the localization of many varieties of antigens including pneumococcal polysaccharides,[69] mumps virus, rickettsia,[70] herpes simplex virus,[71] streptococcal hyaluronidase,[72] and neurotropic influenza-A virus.[73]

(ii) *Staining and classification:* The procedure has been used to devise specific staining procedures for many microorganisms. The direct and indirect staining methods of Coons and Kaplan[64] and Weller and Coons[74] have been applied to other microorganisms. Carter and Leise[75] have compared the two types of staining techniques for fluorescence labeling of bacteria. These are summarized in Fig. 14.

Among the first to use fluorescent antibody for bacterial staining were Moody et al.[76] who applied it to the detection of *Malleomyces pseudomallei* in infected animals. Since then the procedure has been applied to many types of bacteria. A summary of recent applications to bacterial staining may be found in a paper by Poetschke et al.[77] The fluorescent antibody procedure has also been extended to provide specific staining procedures and to permit classification among *Endamoeba*[78] and other protozoan parasites[79] and yeasts.[80] Practically all studies carried out in this field have

Direct staining technique

Bacteria + fluorescent immune globulin → fluorescent bacteria

Indirect staining technique

Bacteria + immune globulin → bacteria immune globulin complex

Bacteria immune globulin complex + fluorescent antiglobulin → fluorescent bacteria

FIG. 14. Direct and indirect staining techniques.

been of a qualitative nature. However, Goldman[81] has used a microfluorometer to obtain more exact comparisons on the uptake of specific fluorescent antibody by intact amebae. In subsequent studies he has also attempted to determine the exact amount of antibody which reacts with an individual amoeba. The attempt at quantification is a most interesting development in this field and will be most important in the application of the fluorescent antibody procedure to serology and studies on antigen-antibody interaction, *in vitro*. It should be possible to carry out titrations of antigen or hapten with dye labeled antibody, using changes in fluorescence intensity, polarization, and spectra as a means of detecting and quantitating complex formation. These would be similar to the techniques reported by Velick *et al.*[54] for unlabeled antibody.

VIII. REVERSIBLE COMPLEXES OF MACROMOLECULES AND FLUORESCENT DYES

Fluorescent dyes can be used to investigate proteins and macromolecules without the necessity of conjugation. Some substances such as the *N*-benzyl derivatives of 3-chloro-6-methoxy-9 aminoacridine and aminonaphthalene sulfonic acids do not fluoresce in solution but are highly fluorescent when adsorbed onto proteins, alumina, paper, etc. According to Weber[17] and to Förster,[5] these molecules exist in a nonplanar form in solution in which form the radiative transition is forbidden. The adsorbed dye, on the other hand, exists in a planar configuration, in which form the transition is allowed. The applications of such a dye, which fluoresces only on adsorption to protein, to the estimation of albumin in serum has been reported by Rees *et al.*[82] The dye 1-anilinonaphthalene-8-sulfonic acid* is nonfluores-

---

* The sodium salt of 1-anilinonaphthalene-8-sulfonic acid can be obtained from The Eastman Kodak Company. It can be purified by precipitation with ammonium sulfate. The ammonium salt comes down as an oil which crystallizes after several days .The golden yellow crystals can be washed with a small amount of water and filtered.

cent in aqueous solution, but is highly fluorescent when absorbed onto plasma albumin. The dye binding in neutral aqueous solution is a specific property of native plasma albumin which is destroyed by heating. This specific binding in neutral solution involves 3 molecules of dye per molecule of protein. By contrast denatured protein molecules in acid solution bind over 20 molecules of dye each.

A. Assay of Plasma Albumin

To measure albumin content, 0.05 ml. of plasma is added to 7 ml. of solution containing the dye (4 mg. of 1-anilinonaphthalene-8-sulfonic acid dissolved in an equivalent amount of 0.1 N NaOH and diluted to 500 ml. with water). Solutions of human plasma albumin ranging from 1 to 6 g./100 ml. are treated in the same manner (0.05 ml. plus 7 ml. of dye solution) to obtain a standard curve (Fig. 15). All diluted samples and the reagent blank are measured in a fluorometer or fluorescence spectrometer. Values obtained with this unusually simple method compared extremely well with four other procedures of albumin assay. Comparisons with an electrophoretic method are shown in Fig. 16. Such agreement was found not only in normals but in patients with all types of disorders, except for jaundice (obstructive and hepatic). In the latter conditions low values were usually

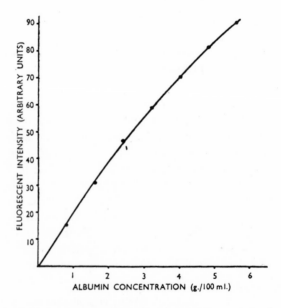

FIG. 15. A standard curve relating intensity of fluorescence of the dye protein mixture with the albumin concentration (after Rees et al.[82]).

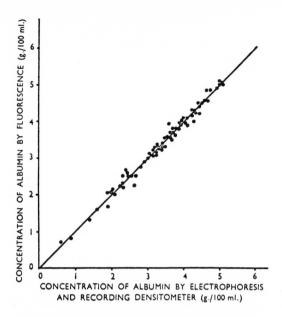

FIG. 16. Comparison of the albumin concentration determined fluorometrically and by electrophoresis and recording densitometer (after Rees et al.[82]).

obtained by the fluorescent dye method. According to Laurence,[83] bile pigments lower the fluorescence yield of adsorbed dye probably as a result of energy transfer between dye and pigment molecules on the protein surface. However, in all other instances the dye-fluorescence method is not only as accurate as other methods but is simpler, less time consuming, and requires extremely small amounts of plasma. More recently, Betheil[84] utilized the fluorescent dye, vasoflavine* (a sulfonated, methylated benzothiazole derivative) to measure plasma proteins by the dye inter-action procedure. The findings were similar to those of Rees et al.[82] in that the method was extremely simple and agreed with other methods (Table V) except in patients with biliary disturbances. However, Betheil[84] points out that the discrepancy occurs in only a small number of sera in which the bilirubin concentration is 7 mg./100 ml. or higher. The method has been employed routinely in the Clinical Chemistry Laboratory of the Bronx Municipal Hospital Center; over 12,500 determinations having been carried out. With vasoflavine, as little as 5 μliters of serum or plasma is sufficient for an analysis.

Kaufman and Singleterry have used fluorescent dyes to study micelle

* Vasoflavine can be obtained from the National Aniline Division, Allied Chemical Corporation, New York, New York.

TABLE V

COMPARISON OF VASOFLAVINE AND METHANOL FRACTIONATION METHODS AS APPLIED
TO PATHOLOGICAL HUMAN SERA[a]

No.	Albumin-globulin ratio	Albumin, gram %		
		Vasoflavine	Methanol	Difference
1	0.58	2.4	2.2	+0.2
2	0.60	2.6	2.4	+0.2
3	0.80	2.8	2.8	0
4	1.03	2.7	2.8	−0.1
5	0.63	2.8	2.5	+0.3
6	0.27	2.1	1.7	+0.4
7	—	2.7	2.9	−0.2
8	—	2.4	2.5	−0.1
9	0.30	2.0	1.7	+0.3
10	0.72	2.7	2.5	+0.2

[a] Values are averages from duplicate analyses (after Betheil[84]).

formation of fatty acids[85] and arylalkyl sulfonates[86, 87] by fluorescence depolarization methods.

B. Detection of Serum Lipoproteins

Kosaki et al.[88] showed that protoporphyrin has a marked affinity for tissue lipids and used protoporphyrin as a stain for fluorescence microscopy. Subsequently, Sulya and Smith[89] investigated protoporphyrin for the detection of lipids on filter paper. They showed that under appropriate conditions the porphyrin endows many lipids with a brilliant red fluorescence. More recently, Searcy and Bergquist[90] have shown that protoporphyrin combines with serum lipoproteins and thereby provides a means for detecting and possibly estimating serum lipoproteins on paper electrophorograms.

Ten microliters of serum are separated electrophoretically on paper. Following electrophoresis the strips are dried in an oven at 120° to 130° for about 30 minutes and are then immersed for 5 minutes in an acid protoporphyrin solution (25 mg. of protoporphyrin in 500 ml. of 0.05 N HCl). The paper strips are placed in distilled water for 5 minutes to remove the excess protoporphyrin reagent and while still damp they are viewed under ultraviolet light (long wavelengths; see Chapter 8 for the fluorescence characteristics of protoporphyrin). The principle protein bands are visible as bright red fluorescent bands against a light blue background; the albumin, a_2- and β-globulin fractions are most highly fluorescent. Lipid-bound globulin, migrating according to the procedure of Downs et al.,[91] is stained most intensely with the protoporphyrin reagent.

Weber[17] has pointed out the possibility of using fluorescent dyes of known size and lifetime of the excited state to measure microviscosities of solutions. Of course, such methods are meaningful only in the absence of dye binding.

GENERAL REFERENCES

Cherry, W. B., Goldman, M., Carski, T. R., and Moody, M. D., Fluorescent Antibody Techniques (in the diagnosis of communicable diseases). Public Health Service Publ. No. 729. U. S. Govt. Printing Office, Washington, D. C. (1960).

Coons, A. H., Immunofluorescence, *Public Health Repts. U. S.* **75,** 937 (1960).

Energy transfer with special reference to biological systems, *Discussions Faraday Soc. No.* **27** (1959).

Karreman, G., and Steele, R. H., On the possibility of long distance energy transfer by resonance, in biology, *Biochim. et Biophys. Acta.* **25,** 280 (1957).

Richards, O. W., Fluorescence microscopy, *in* "Medical Physics" (O. Glasser, ed.), Vol. III, p. 375. Year Book Publ., Chicago, Illinois, 1960.

Stevens, B., and Boudart, M., Vibrational energy transfer from collisional deactivation of simple molecules and fluorescence stabilization of complex molecules, *Ann. N. Y. Acad. Sci.* **67,** 570 (1957).

Theorell, H., and Nygaard, A. P., Kinetics and equilibria in flavoprotein systems, *Acta Chem. Scand.* **8** 877 (1954).

Uehleke, H., New methods of preparing fluorescence-labeled proteins, *Z. Naturforsch.* **13b,** 722 (1958).

REFERENCES

1. Duggan, D. E., and Udenfriend, S., *J. Biol. Chem.* **223,** 313 (1956).
2. Waalkes, T. P., and Udenfriend, S., *J. Lab. Clin. Med.* **50,** 733 (1957).
3. Shore, P. A., Burkhalter, A., and Cohn, V. H., Jr., *J. Pharm. Exptl. Therap.* **127,** 182 (1959).
4. Conn, R. B., Jr., and Davis, R. B., *Nature* **183,** 1053 (1959).
5. Förster, T., "Fluoreszenz Organischer Verbindungen," Vandenhoeck and Ruprecht, Göttingen, 1951.
6. Förster, T., *Discussions Faraday Soc. No.* **27,** 7 (1959).
7. Porter, G., and Weber, G., *Nature* **184,** 688 (1959).
8. *Discussions Faraday Soc. No.* **27** (1959).
9. Gemmill, C. L., *Anal. Chem.* **28,** 1061 (1956).
10. Lavorel, J., *J. Phys. Chem.* **61,** 1600 (1957).
11a. Pringsheim, P., "Fluorescence and Phosphorescence," pp. 342–343. Interscience, New York, 1949.
11b. Kallmann, H., and Furst, M., *in* "Liquid Scintillation Counting" (C. G. Bell, Jr., and F. N. Hayes, eds.). Pergamon Press, London, 1958.
12. Weber, G., *Nature* **180,** 1409 (1957).
13. Shore, V. G., and Pardee, A. B., *Arch. Biochem. Biophys.* **60,** 100 (1956).
14. Weber, G., *Biochem. J.* **47,** 114 (1950).
15. Whitby, L. G., *Biochem. J.,* **54,** 437 (1953).

228   6. PROTEINS

16. Duggan, D. E., Udenfriend, S., Bowman, R. L., and Brodie, B. B., *Arch. Biochem. Biophys.* **68**, 1 (1957).
17. Weber, G., *Advances in Protein Chem.* **8**, 415 (1953).
18. Laurence, D. J. R., *in* "Methods in Enzymology" (S. P. Colowick and N. O. Kaplan, eds.), Vol. IV, p. 174. Academic Press, New York, 1957.
19. Perrin, F., *J. phys. radium* **7**, 390 (1926).
20. Steiner, R. F., and McAlister, A. J., *J. Polymer Sci.* **24**, 105 (1957).
21. Steiner, R. F., *Arch. Biochem. Biophys.* **49**, 71 (1954).
22. Velick, S. F., *J. Biol. Chem.* **233**, 1455 (1958).
23. Vles, F., *Arch. phys. biol.* **16**, 137 (1946).
24. Teale, F. W. J., and Weber, G., *Biochem. J.* **65**, 476 (1957).
25a. Konev, S. V., *Doklady Akad. Nauk S.S.S.R.* **116**, 594 (1957).
25b. Vladimirov, Yu. A., *Doklady Akad. Nauk S.S.S.R.* **116**, 780 (1957).
26. Teale, F. W. J., and Weber, G., *Biochem. J.* **72**, 15P (1959).
27. Weber, G., *Biochem. J.* **75**, 345 (1960).
28. Teale, F. W. J., *Biochem. J.* **76**, 381 (1960).
29. Weber, G., *Biochem J.* **75**, 335 (1960).
30. Boroff, D. A., and Fitzgerald, J. E., *Nature* **181**, 751 (1958).
31. Boroff, D. A., *Intern. Arch. Allergy Appl. Immunol.* **15**, 74 (1959).
32a. Schantz, E. J., Stefanye, D., and Spero, L., *J. Biol. Chem.* **235**, 3489 (1960).
32b. Konev, S. V., and Kozunin, I. I., *Dairy Sci. Abstr.* **23**, 103 (1961).
33. Weber, G., and Teale, F. J. W., *Discussions Faraday Soc. No.* **27**, 1 (1959).
34. Theorell, H., and Bonnichsen, F. K., *Acta. Chem. Scand.* **5**, 1105 (1951).
35. Boyer, P. D., and Theorell, H., *Acta. Chem. Scand.* **10**, 447 (1956).
36. Fisher, H. F., and McGregor, L. L., *Biochim. et Biophys. Acta.* **38**, 562 (1960).
37. Shifrin, S., and Kaplan, N. O., *Advances in Enzymol.* **22**, 371 (1960).
38. Shifrin, S., Kaplan, N. O., and Ciotti, M. M., *J. Biol. Chem.* **234**, 1555 (1959).
39. Duysens, L. N. M., and Kronenberg, G. H. M., *Biochim. et Biophys. Acta* **26**, 437 (1957).
40. Winer, A. D., Schwert, G. W., and Millar, D. B. S., *J. Biol. Chem.* **234**, 1149 (1959).
41. Fisher, H. F., *Federation Proc.* **19**, 61C (1960).
42. Pfleiderer, G., and Hohnholz, E., *Biochem. Z.* **331**, 245 (1959).
43. Theorell, H., and Langan, T. A., *Acta Chem. Scand.* **14**, 933 (1960).
44. Langan, T. A., *Acta Chem. Scand.* **14**, 936 (1960).
45. Maxwell, E. S., de Robichon-Szulmajster, H., and Kalckar, H. M., *Arch. Biochem. Biophys.* **68**, 407 (1958).
46. Winer, A. D., and Schwert, G. W., *J. Biol. Chem.* **234**, 1155 (1959).
47. Maxwell, E. S., and de Robichon-Szulmajster, H., *J. Biol. Chem.* **235**, 308 (1960).
48. Straub, F. B., *Biochem. J.* **33**, 187 (1939).
49. Keilin, D., and Hartree, E., *Nature* **157**, 801 (1946).
50. Morell, D. B., *Biochem. J.* **51**, 657 (1952).
51. Walaas, E., and Walaas, O., *Acta Chem. Scand.* **10**, 122 (1956).
52. Hagins, W. A., and Jennings, W. H., *Discussions Faraday Soc. No.* **27**, 180 (1959).
53. Loomeijer, F. J., *Acta Physiol. et Pharmacol. Neerl.* **8**, 518 (1959).
54. Velick, S. F., Parker, C. W., and Eisen, H. N., *Proc. Natl. Acad. Sci. U. S.* **46**, 1470 (1960).
55. Creech, H. J. and Jones, R. N., *J. Am. Chem. Soc.* **62**, 1970 (1940); **63**, 1660 (1941).
56. Chadwick, C. S., Johnson, P., and Richards, E. G., *Nature* **186**, 239 (1960).
57. Shore, V. G., and Pardee, A. B., *Arch. Biochem. Biophys.* **62**, 355 (1956).
58. Weber, G., *Biochem. J.* **51**, 155 (1952).

59. Tsao, T. C., *Biochim. et Biophys. Acta* **11**, 227, 236 (1953).
60. Massey, V., *Biochem. J.* **53**, 67, 72 (1953).
61. Hartley, B. S., and Massey, V., *Biochim. et Biophys. Acta* **21**, 58 (1956).
62. De Barbieri, A., and Scevola, M. E., *Ital. J. Biochem.* **8**, 46 (1959).
63. Coons, A. H., Creech, H. J., Jones, R. N., and Berliner, E., *J. Immunol.* **45**, 159 (1942).
64. Coons, A. H., and Kaplan, M. H., *J. Exptl. Med.* **91**, 1 (1950).
65. Riggs, J. L., Seiwald, R. J., Burckhalter, J. H., Downs, C. M., and Metcalf, T. G., *Am. J. Pathol.* **34**, 1081 (1958).
66. Chadwick, C. S., Nairn, R. C., and McEntegart, M. G., *Biochem. J.* **73**, P41 (1959).
67. Borek, F., and Silverstein, A. M., *Arch. Biochem. Biophys.* **87**, 293 (1960).
68. Coons, A. H., *Intern. Rev. Cytol.* **5** 1 (1956).
69. Kaplan, M. H., Coons, A. H., and Deane, H. W., *J. Exptl. Med.* **91**, 15 (1951).
70. Coons, A. H., Snyder, J. C., Cheerer, F. S., and Murray, E. S., *J. Exptl. Med.* **9**, 31 (1950).
71. Biegeleisen, J. Z., Jr., Scott, L. V., and Lewis, V., Jr., *Science* **129**, 640 (1959).
72. Emmart, E. W., and Turner, W. A., Jr., *J. Histochem. and Cytochem.* **8**, 273 (1960.
73. Fraser, K. B., Nairn, R. C., McEntegart, M. G., and Chadwick, C. S., *J. Pathol. Bacteriol.* **78**, 423 (1959).
74. Weller, T. H., and Coons, A. H., *Proc. Soc. Exptl. Biol. Med.* **86**, 789 (1954).
75. Carter, C. H. and Leise, J. M., *J. Bacteriol.* **76**, 152 (1958).
76. Moody, M., Goldman, M., and Thomason, B., *J. Bacteriol.* **72**, 357 (1956).
77. Poetschke, Von. G., Uehleke, H. and Killisch, L., *Schweiz. Z. Pathol. u Bakteriol.* **22**, 758 (1959).
78. Goldman, M., *Am. J. Hygiene* **58**, 319 (1953).
79. Carver, R. K., and Goldman, M., *Am. J. Clin. Pathol.* **32**, 159 (1959).
80. Gordon, M. A., *Proc. Soc. Exptl. Biol. Med.* **97**, 694 (1958).
81. Goldman, M., *Proc. Soc. Exptl. Biol. Med.* **102**, 189 (1959); personal communication, 1960.
82. Rees, V. H., Fildes, J. E., and Laurence, D. J. R., *J. Clin. Pathol.* **7**, 336 (1954).
83. Laurence, D. J. R., personal communication, 1960.
84. Betheil, J. J., *Anal. Chem.* **32**, 560 (1960).
85. Kaufman, S., and Singleterry, C. R., *J. Colloid Sci.* **7**, 453 (1952).
86. Kaufman, S., and Singleterry, C. R., *J. Colloid Sci.* **10**, 139 (1955).
87. Kaufman, S., and Singleterry, C. R., *J. Colloid Sci.* **12**, 465 (1957).
88. Kosaki, T., Kotani, Y., Nakagawa, S., and Saka, T., *Mie Med. J.* **7**, 35 (1957).
89. Sulya, L. L., and Smith, R. R., *Biochem. Biophys. Research Communs.* **2**, 59 (1960).
90. Searcy, R. L., and Bergquist, L. M., *Clin. Chim. Acta* **5**, 941 (1960).
91. Downs, J. J., Geller, E., Lunan, K. D., and Mann, L. T., Jr., *J. Lab. Clin. Med.* **51**, 317 (1958).

7

VITAMINS, COENZYMES, AND THEIR METABOLITES

I. INTRODUCTORY REMARKS

THIS section comprises many classes of organic compounds. It is of interest that the vitamins were probably the first group of biologically important compounds to be assayed fluorometrically. This was so because many of them are aromatic structures containing substituent groups which endow the molecule with fluorescence in the visible region (riboflavin) or can be converted to such structures by simple procedures (thiochrome). In addition, many of the vitamins are present in tissues in microgram quantities requiring methods with the sensitivity of fluorescence for their assay.

We now know that vitamins are organic substances which must be supplied in the diet of mammals and other living organism. All of them undergo metabolism; the intermediate metabolic products in most cases are important as coenzymes. The final products of their metabolism are excreted in the urine and stools. Vitamins and coenzymes are of interest to scientists of many different disciplines and each group has different requirements as to sensitivity, specificity, and chemical form of the particular class of compound. Thus, the pharmaceutical chemist is interested in the assay of the vitamin itself in pure preparations or in vitamin mixtures, as a test of purity and stability. The nutritionist is interested in measuring the nutritionally active forms of the vitamin in foods before and after various processing procedures. The clinical nutritionist is interested in determining blood and tissue levels of the vitamin and its coenzyme forms in health and in various disease states. He is also interested in the urinary excretion of the metabolites in order to evaluate the over-all intake and disposition of a given vitamin. The enzymologist is mainly interested in assaying the coenzyme forms of the vitamin.

In these discussions we shall, in each case, summarize what is presently known about the fluorometric assay of the vitamin itself, of the forms in which it is active in tissues, and of the metabolic excretory products. The use of fluorescence to study the interaction of enzymes, coenzymes, and

230

substrates will be discussed in another section as will the utilization of coenzymes to measure enzyme activity (Chapter 9).

It must be emphasized that the vitamins and coenzymes represent a particularly active field of research and that for many of them there have been hundreds of papers published on analytical methods and their applications. In the ensuing section it is intended to present those procedures which have emerged over the course of many years as most generally applicable.

II. VITAMIN A

Vitamin A is one of many carotenoid substances found in plant and animal tissues. Although the vitamin is the form utilized in animals, many related structures which are found in plants yield vitamin A in animal tissues and are therefore pro-vitamins. Thus, β-carotene is cleaved at the central double bond to yield vitamin A alcohol. The alcohol, on the other hand, can be metabolized to vitamin A aldehyde, as in the visual cycle (Fig. 1). In addition to the free form, vitamin A also exists in ester form. Assays for vitamin A must therefore distinguish among these closely related structures. One international or USP unit of vitamin A is equivalent to 0.30 μg. of vitamin A alcohol, 0.344 μg. of the vitamin A acetate, or 0.6 μg. of carotene.

Vitamin A and its congeners absorb in the near ultraviolet because of their

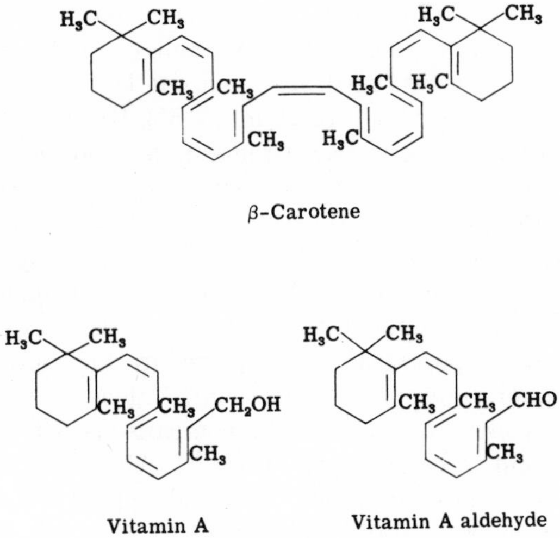

β-Carotene

Vitamin A   Vitamin A aldehyde

Fig. 1. Vitamin A and related metabolites.

conjugated polyene structure. The fluorescence characteristics of vitamin A in absolute ethanol have been reported as 325-mμ excitation and 470-mμ fluorescence (uncorrected).[1] More recently, Hagins and Jennings[2] reported corrected values of 327 mμ for excitation and 510 mμ for fluorescence. They also estimated the lifetime of the excited state to be approximately 0.73 × 10^{-9} second. The intensity of vitamin A fluorescence is sufficiently great to permit detection of quantities well below the microgram level.

Although it has long been known that vitamin A is a fluorescent compound, there have been few reports concerning the application of this fluorescence to the quantitative assay of the vitamin in tissue extracts. Sobotka et al.[3, 4] reported a procedure for fluorometric differentiation of vitamin A esters from vitamin A and applied it to their assay in fish liver oils. Vitamin A acetate, laurate, myristate, palmitate, and oleate were found to increase similarly in fluorescence upon ultraviolet irradiation. Under the same conditions, free vitamin A alcohol fluorescence decreased. By making measurements at varying time intervals it was possible to determine the percentage of free and esterified vitamin A in various preparations. In this instance fluorescence was used only for determination of the relative proportions of the two forms of vitamin A and not for measurement of absolute amounts.

In 1951 Fujita and Aoyama[5] reported a fluorometric procedure for the quantitative assay of total vitamin A (free and esterified) in various oils. A sample of oil containing 0.3–3 μg. of vitamin A (total) is dissolved in a stoppered centrifuge tube in 5 ml. of 5% KOH in absolute ethanol. Hydrogen or carbon dioxide is bubbled through the mixture (to remove oxygen) and the sample is heated at 75° for 30 minutes, then cooled rapidly. Following the saponification 10 ml. of benzene and 10 ml. of water are added and the tube is shaken and centrifuged. The benzene layer is transferred to another tube and is washed first with 20 ml. of 5% KOH in 60% aqueous methanol, then with 20 ml. of water. If the sample is not clear and transparent at this point, washing of the benzene is continued. An aliquot of the benzene is diluted to contain 0.1–1.0 μg. of vitamin A/ml. and compared with standards of the pure vitamin. Fujita and Aoyama[5] used a visual comparator for measurement of fluorescence. Under such conditions, carotene and vitamin D were found to interfere with the assay.

More recently, De[6] reported a similar procedure for vitamin A assay but used a fluorophotometer for measurement. Values were reported for animal livers, egg yolks, and plasma and comparisons with colorimetric procedures were found to be excellent (Table I). Although the specificity of the procedure was not established, it would appear that with some modifications as to instrumentation and details of isolation, it should be possible to devise a relatively specific assay for vitamin A. Certainly with

TABLE I

COMPARISON OF VITAMIN A VALUES ESTIMATED BY DIFFERENT METHODS[a]

| Material | Vitamin A content (I.U. per ml. or g.) | |
	Fluorescence	Colorimetric SbCl₃ method
Sheep plasma	0.80	0.84
Sheep liver	356	336
Rat liver	79	86
Human plasma	1.2	—
Egg yolk	4.5	5.1

[a] After De.[6]

the recent introduction of fluorescence spectrometry the application of fluorescence to vitamin A assay should be re-evaluated.

The application of fluorescence microscopy to the demonstration of vitamin A in tissues has been used since 1932. A review of its applications from 1932 to 1944 was presented by Popper.[7] Although the histologic procedure calls the green, fading fluorescence in tissues vitamin A, it was pointed out by Popper that not all such fluorescence can be claimed to be due to vitamin A. However, many interesting observations relating nutritional and pathologic states to tissue vitamin A levels have been made by use of fluorescence microscopy. A more recent application of the method, giving details of more recent instrumentation and other equipment, was presented by Blankenhorn and Braunstein.[8]

Hagins and Jennings[2] have carried out fluorescence studies on intact retinas. Illumination converts rhodopsin (retinene plus protein) to the orange-colored metarhodopsin which then fades to a yellow substance. The latter, when excited with light in the region of 400 mμ, emits an intense yellow fluorescence.

III. THIAMINE AND COCARBOXYLASE

Thiamine was one of the first metabolites to be assayed by fluorescence techniques. Although the molecule itself does not possess appreciable fluorescence, it is readily oxidized to the fluorescent product thiochrome (Fig. 2). This reaction has been made the basis for many assays of thiamine in blood, tissues, urine, food products, vitamin preparation, etc. It is one of the standard procedures for thiamine assay of the U. S. Pharmacopeia[9] and of the Association of Official Agricultural Chemists.[10]

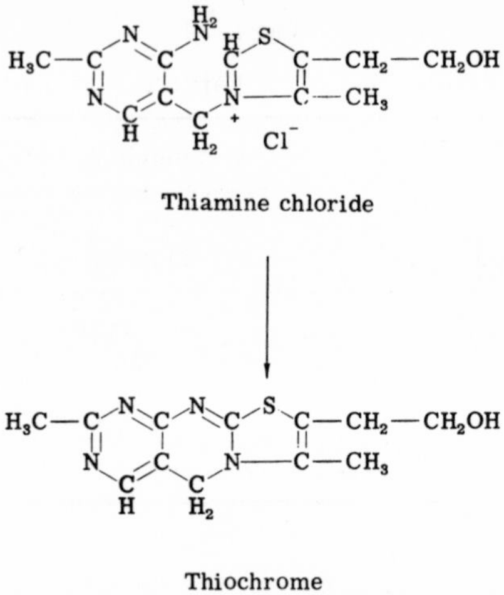

Thiamine chloride

Thiochrome

FIG. 2. Conversion of thiamine to thiochrome involves a dehydrogenation and the appearance of a double bond in conjugation with the pyrimidine ring.

Both thiamine and cocarboxylase (thiamine pyrophosphate) are present in animal and plant tissues. Other thiamine phosphates may also be present. Burch et al.[11] have described a microfluorometric procedure for the determination of free thiamine and total thiamine* in samples of fingertip blood. Their most recent procedure for total thiamine in whole blood is described.[12]

Sixty microliters of blood containing 0.0005–0.005 μg. of thiamine are transferred to a 6- × 50-mm. fermentation tube containing 275 μl. of 5% trichloroacetic acid and the contents are mixed with the aid of a vibrator.† After 30 to 60 minutes the tube is centrifuged and 250 μl. of the clear supernatant fluid are transferred to a 3-ml. test tube containing 30 μl. of 4 M potassium acetate. To the tube contents, which are now pH 4.5–4.8, are added 10 μl. of an acid phosphatase preparation.‡ The tube is covered with a rubber vial cap and hydrolysis is permitted to proceed for 4 to 16 hours at 25°. Following hydrolysis, 120 μl. of an alkaline potassium ferricyanide solution§ are added and the tube is vigorously mixed. After 15

---

* In human blood thiamine occurs chiefly as thiamine phosphates.

† Touching the tube against a vibrator, such as is used for massage purposes, brings about efficient mixing.

‡ A partially purified and highly active preparation obtained from human seminal fluid. For details see reference 12.

§ The oxidation reagent consists of 10 ml. of 7.5 N NaOH plus 0.65 ml. of 0.059 M potassium ferricyanide. This is made up within an hour of use.

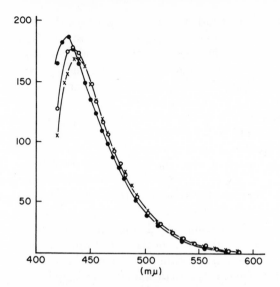

FIG. 3. Fluorescence spectra of thiochrome in isobutanol. The ordinate represents arbitrary units of fluorescence (after Yagi *et al.*[14]). ● — ●, reaction product in thiamine assay; ×—×, crystalline thiochrome in neutral or acid solvent; ○ — ○, crystalline thiochrome in alkaline solvent.

seconds, 120 μl. of a peroxide reducing agent* are added to destroy excess oxidant. Following this, 1.2 ml. of *n*-hexyl alcohol are added and the tube is shaken to extract the thiochrome. The tube is centrifuged (covered) and the clear alcohol layer is transferred to a cuvette.

Fluorescence is measured in a fluorometer or fluorescence spectrometer. According to Ohnesorge and Rogers,[13] excitation is at 365 mμ and fluorescence at 450 mμ. Although these are uncorrected values, the reported excitation maximum coincides with the absorption peak of thiochrome. Yagi *et al.*[14] reported a corrected fluorescence maximum of 430 to 435 mμ (Fig. 3). The appropriate filters for a fluorometer are those which are used for quinine assay. Standards and blanks are prepared by substituting 60 μl. of thiamine solution or water for the blood sample. Quinine solutions are used to standardize the instrument and to ascertain reproducibility of the chemical procedure.

Following these measurements, the sample, standards, and blanks are placed in a circular rack and are irradiated for 30 minutes with an AH-5 mercury lamp placed 3 inches from the samples.† A fan must be used to

* The reducing agent is prepared by adding 10 μl. of Superoxol to 10 ml. of 5.5 M NaH₂PO₄.

† Details of an appropriate circular rack for photochemical reactions are presented in Chapter 13, under chloroquin assay.

prevent heating. The irradiation process destroys the thiochrome and the residual fluorescence, some of it possibly due to riboflavin, is subtracted from the original reading leaving the fluorescence resulting from the thiamine. It may be that fluorescence spectrometers can distinguish the nonthiochrome fluorescence from other fluorescence and make it possible to dispense with the irradiation process.

Free thiamine may be measured by omitting the treatment with phosphatase. Although the thiamine phosphates yield the corresponding fluorescent thiochrome phosphates these are not extractable into the n-hexylalcohol and so do not interfere with the assay. Codecarboxylase and thiamine monophosphate may be calculated by subtracting the value for free thiamine from that of total thiamine. It should also be possible to measure the thiochrome phosphates directly and obtain more precise values.

The total thiamine in whole blood of humans and rats is about 5 μg./100 ml. In humans almost all of it is in the ester form, whereas in rats appreciable amounts of free thiamine are also found. Although the procedure was not applied to tissues. other than blood, it is stated by the author that such an adaptation should be relatively simple. For those who are not accustomed to dealing with miniature assays, it should also be possible, with not too much effort, to scale the assay up from the microliter to the milliliter range. It should also be pointed out that preliminary separation of the thiamine on base exchange resins can also be used to endow the method with greater specificity.

IV. RIBOFLAVIN AND THE FLAVIN COENZYMES

The riboflavin molecule is a condensed ring system, having many resonant forms and therefore exhibiting native fluorescence. This property of

Riboflavin

riboflavin and of its coenzyme derivatives, flavin mononucleotide (FMN) and flavin-adenine dinucleotide (FAD) has been known for a long time and there have been many procedures published for their assay. In general, these procedures either make use of the native fluorescence of the individual flavins or convert them to lumiflavin. Solvent extraction, chromatography, or paper electrophoresis have been used to separate the different flavins before assay.

Although there have been many published reports concerning the applications of riboflavin fluorescence to its quantitative assay, studies on the fluorescence itself are not many. The first detailed investigations of riboflavin and flavin adenine dinucleotide fluorescence were by Bessey et al.[15] and by Weber.[16] They reported that the fluorescence of FAD is much weaker than that of free riboflavin and that this is due to the quenching effect of the adenine part of the molecule. Various purines, when added to solutions of free riboflavin, are also effective quenching agents, indicating that purines form complexes with riboflavin (Fig. 4). These purine-flavin complexes differ in a number of ways from free riboflavin. They have appreciably different absorption spectra and the lifetimes of their excited states (τ) differ considerably (free riboflavin 1×10^{-8} second, caffeine or adenine complex 10^{-6} second). On the other hand, the lifetime of the excited state for FAD is the same as that for riboflavin. Quenching

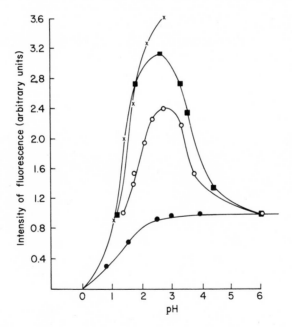

Fig. 4. Effect of pH on the fluorescence intensity of riboflavin and FAD in the presence of purine derivatives (after Weber[16]). $\times$—$\times$, riboflavin (5×10^{-6} M); ■ — ■, FAD (5×10^{-6} M); ○ — ○, riboflavin in 0.05 M adenosine; ● — ●, riboflavin in 0.05 M caffeine.

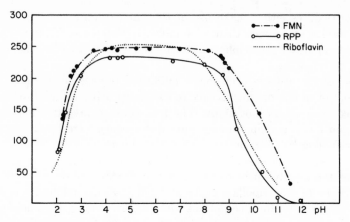

FIG. 5. Fluorescence of riboflavin, riboflavin-5-pyrophosphate (RPP), and flavin mononucleotide (FMN) as a function of pH. The ordinate represents arbitrary units of fluorescence (after Cerletti[18a]).

of riboflavin fluorescence by phenols was reported by Sakai.[17] Cerletti[18a] has reported that the fluorescence of riboflavin-5'-pyrophosphate parallels that of free riboflavin and FMN, both in intensity and in its variation with pH (Fig. 5). On the other hand, the effect of pH on the fluorescence of FAD is much different (Fig. 6). Much of this data is pertinent to enzyme-

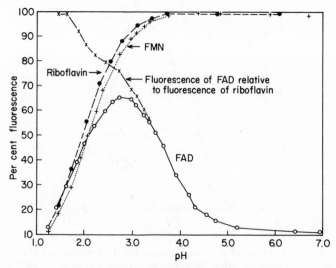

FIG. 6. Influence of pH on the fluorescence of riboflavin, flavin mononucleotide (FMN), and flavin-adenine dinucleotide (FAD) (after Bessey et al.[15]).

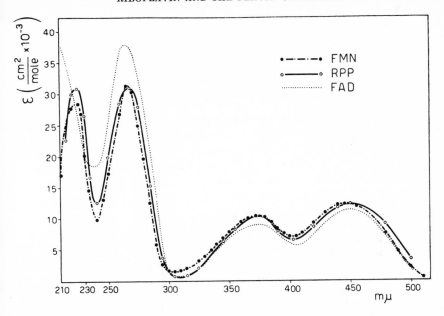

FIG. 7. Absorption spectra of riboflavin-5'-pyrophosphate (RPP), flavin mono-nucleotide (FMN), and flavin-adenine dinucleotide (FAD) in 0.1 M phosphate buffer pH 7.0 (after Cerletti[18a]).

coenzyme interaction and enzyme mechanisms and will be discussed in other sections (Chapters 6 and 9). What is important, from an analytical standpoint, is that the fluorescences of riboflavin and its nucleotide forms can be differentiated by their relative intensities, by the effects of quenching agents, by changes with pH, and by the degree of polarization in viscous solvents.[16] These findings should prove useful in distinguishing the related compounds analytically. Absorption (excitation) and fluorescence spectra of flavins are shown in Figs. 7 and 8.

One of the earlier methods for the measurement of riboflavin in blood was that of Najjar.[19] Based on earlier observations by Von Euler and Adler,[20] by Vivanco,[21] and by Ferrebee,[22] the method introduced a solvent extraction procedure to separate the riboflavin from interfering absorbing, fluorescing, and light scattering contaminants. In addition to solvent extraction, the procedure involves treatment with $KMnO_4$ to destroy interfering substances, measurement of fluorescence, and photodecomposition of the riboflavin to leave a residual blank. The procedure has sufficient sensitivity for urine. Although the method should be applicable to blood, the sensitivity is barely adequate. The most serious criticism of the method is that it does not distinguish between free riboflavin and its conjugates,

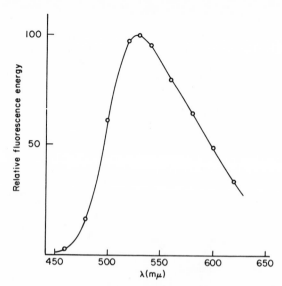

FIG. 8. Fluorescent spectra of riboflavin, FMN, and FAD in aqueous solution (after Yagi and Okuda[18b]).

FMN and FAD. Since these substances differ in their distributions between solvents and in their fluorescence intensities, application of the method to tissues containing all three would be difficult to interpret.

Several years later, Burch *et al.*[23] introduced a fluorometric procedure for assaying riboflavin and its natural derivatives in blood. This was a microprocedure sufficiently sensitive for application to as little as 0.05 to 0.2 ml. of serum. Application of this method to the assay of riboflavin and its coenzymes in tissues was later reported by Bessey *et al.*[15] The most recent modifications of these procedures were presented by Burch[24] in 1957.

TABLE II

PARTITION COEFFICIENTS OF RIBOFLAVIN, FMN, AND FAD

Compound	Partition coefficients[a]
Riboflavin	4.1
FAD	0.03
FMN	0.02

[a] Between benzyl alcohol and trichloracetic acid extracts neutralized to pH 6.8 (after Bessey *et al.*[15]).

The differential procedure is based on several factors:

(1) Riboflavin and FMN have the same molar fluorescence intensities and their fluorescence characteristics vary similarly as a function of pH whereas FAD differs considerably in this respect. At pH 6.8 FAD is 15% as fluorescent as free riboflavin.

(2) The partition coefficients of the three compounds between benzyl-alcohol and aqueous solutions at pH 6.8 are such that only riboflavin is extracted to an appreciable extent (Table II).

(3) FAD is completely hydrolyzed to FMN in trichloroacetic acid extracts at 38° overnight, to give the concomitant large increase in fluorescence.

A. Procedure for FAD and Total Flavins

All preparatory steps except hydrolysis should be carried out in the cold at 0 to 4°. Fresh tissues are homogenized with 10 ml. of water per gram of tissue. A 0.3-ml. aliquot of the homogenate is transferred to a test tube containing 3 ml. of 11% trichloroacetic acid, mixed, and after 15 minutes it is centrifuged. One aliquot (A_1) (0.2 ml.) is transferred to a fluorometer tube containing 1 ml. of 0.2 M K_2HPO_4 and mixed; the final pH should be 6.8. This is stored overnight in the dark. Another 0.2-ml. aliquot of the deproteinized homogenate (A_2) is transferred to a similar tube, but stored overnight at 38° in the dark, to permit hydrolysis of the FAD before the addition of 1 ml. of 1 M K_2HPO_4. Following neutralization, the flavins are highly sensitive to light and must be shielded from light until measurements are made.

Standards are prepared by substituting 0.3-ml. aliquots of riboflavin solutions for the initial tissue homogenate (range 0.03–2.0 μg.). Blanks utilize 0.3 ml. of distilled water in place of the homogenate. Fluorescence is measured with the excitation maximum at 450 mμ and the fluorescence maximum at 535 mμ, or appropriate filters. Following measurement of the fluorescence, the flavins are reduced by addition of 10 μl. of 10% sodium hydrosulfite in 5% $NaHCO_3$. Residual nonflavin fluorescence, which is normally fairly low, is subtracted from the original fluorescence reading. The FAD concentration in each tube is then equivalent to $A_2 - A_1/0.85$.[*] Comparison with the similarly treated riboflavin standards yields the FAD content in terms of riboflavin. A_2 is, of course, equivalent to total flavins.

---

[*] The factor 0.85 is used because FAD has a fluorescence equal to 15% of riboflavin.

B. Free Riboflavin

Except for blood, most animal tissues contain little free riboflavin. It may be separated from the flavin nucleotides by the following extraction procedure. A 2-ml. aliquot of the supernatant fluid, obtained after treatment with trichloroacetic acid, is transferred to a glass-stoppered centrifuge tube and neutralized by addition of 0.5 ml. of 4 M K_2HPO_4. An equal volume, 2.5 ml., of benzyl alcohol is added and the tube is shaken vigorously, then centrifuged. The riboflavin is returned to an aqueous solvent by transferring 2 ml. of the benzyl alcohol to another tube containing 30 ml. of toluene and 2 ml. of acetate buffer (1 volume of 0.1 N sodium acetate plus 1 volume 0.1 N acetic acid). Blanks and riboflavin standards are carried through the whole extraction procedure and fluorescence is again measured before and after reduction with sodium hydrosulfite.

TABLE III

FLUOROMETRIC AND ENZYMATIC ASSAY OF FAD IN RAT TISSUES[a]

Tissue	Fluorometric assay per g.	Enzymatic assay per g.
Liver 1	28.3	28.5
Liver 2	36.8	34.9
Kidney 1	19.2	21.8
Kidney 2	16.4	13.8
Heart 1	18.1	17.6
Heart 2	21.6	18.8

[a] The values are based on wet tissue weight (after Bessey et al.[15]).

According to Burch,[24] recovery of riboflavin, FMN, and FAD added to minced tissues averaged 95–100%. The fluorometric assay of FAD in crude tissues yields values comparable to those obtained by enzymatic assay with D-amino acid oxidase[25] (Table III). The application to purified preparations is just as good and far more sensitive than the enzyme procedure.

Kondo[26] has reported on the application of the procedures of Burch[24] to plant materials.

It should be pointed out that the manipulations used for assay and other procedures described by Bessey et al.[15] can also be used to ascertain the purity of the various coenzymes isolated from biological sources.

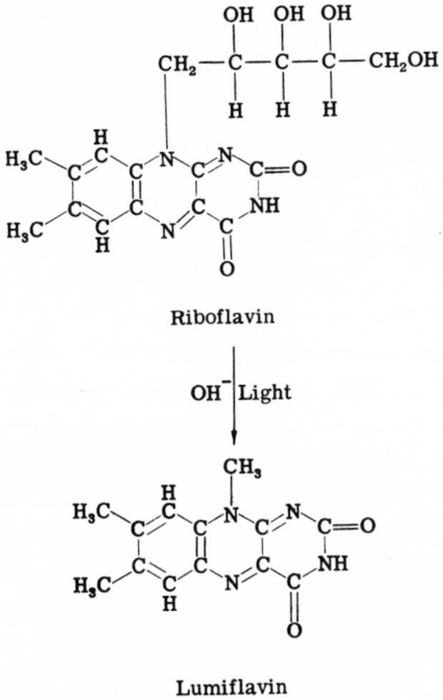

Riboflavin

OH⁻ Light

Lumiflavin

Fig. 9. Photochemical conversion of riboflavin to lumiflavin.

C. Lumiflavin Method

Riboflavin can also be converted to the more highly fluorescent derivative, lumiflavin, by photodecomposition in alkaline solution,[27] the formation of lumiflavin involving cleavage between the ribitol and flavin ring (Fig. 9). The same product is formed with FMN and FAD. One of the first applications of this reaction to flavin assay was by Kuhn et al.[28] Since then many other variations of the lumiflavin assay procedure have appeared. One of the most recent modifications for total flavins was first presented by Yagi[29] in 1956 and then a year later[30] in more detailed form.

PROCEDURE

The flavins are extracted from the tissue with hot water, proteins are removed by centrifugation, and an aliquot is made alkaline (pH 11) and irradiated under a fluorescent lamp until riboflavin decomposition is complete. The solution is then acidified and the lumiflavin is extracted into chloroform in which solvent fluorescence is measured. Since light absorbing and fluorescence quenching extraneous materials can interfere at both the photodecomposition and measurement stage, internal standards of riboflavin are run with each sample.

The tissue (0.1–1 g.) is minced, weighed, and transferred to a glass homogenizer containing several milliliters of water at 80°. After standing at this temperature for 3 to 5 minutes, homogenization is carried out and the homogenate is transferred to a graduated centrifuge tube and diluted to a definite volume (5–20 ml. depending upon the amount of tissue and the flavin content) and again heated at 80° for 15 minutes. After cooling and stirring the tube is centrifuged. One milliliter of the supernatant solution is transferred to a glass-stoppered centrifuge tube containing 1.0 ml. of water. To a second tube are added another 1-ml. aliquot and 1.0 ml. of riboflavin solution as an internal standard. This is followed by addition of 2.0 ml. of M NaOH to both tubes. Absolute standards and blanks are diluted in the same manner and all tubes are irradiated for 30 to 60 minutes in an apparatus comparable to the one shown in Fig. 10. The temperature should not rise above 30°. Following irradiation, 0.2 ml. of glacial acetic acid is added and the tubes are cooled to room temperature. Then 6.0 ml. of water-saturated chloroform are added and the tubes are shaken to extract the lumiflavin. The fluorescence of the chloroform extracts are measured and compared with standards and blanks carried through the entire procedure. The excitation spectrum of lumiflavin is the same as that of riboflavin and the fluorescent spectra are also similar.[31]

The lumiflavin procedure is suitable for determining total flavins in tissues. It should also be possible to use preliminary extractions, as described by Burch,[24] to differentiate FAD from riboflavin. Preliminary separation of the flavins can also be achieved by various chromatography procedures, as described below. Some tissues may contain fluorescent substances which are extractable into chloroform. Such material may be removed by extracting with chloroform before the irradiation. Ishiguro[32] has pointed out that, in urine, anthranilic acid and its glucuronide may interfere with the lumiflavin method. Treatment with permanganate, as described by Najjar,[19] to oxidize the anthranilic acid, and a preliminary

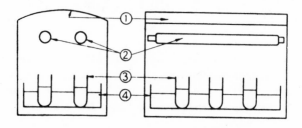

Fig. 10. Apparatus designed by Yagi[29] for the photochemical conversion of flavins to lumiflavin. ① mirror; ② fluorescent lamp (Mazda FL 20D); ③ centrifuge tube; ④ tube holder.

chloroform extraction from alkaline solution after photolysis, can effectively remove most interferences.

Yagi et al.[33] have described the use of chromatography on a Dowex 1 resin for separation of flavins before fluorescence assay. This procedure appears to have some advantages over previous ones. Cerletti and Siliprandi[34] reported the use of paper electrophoresis to effect separation of FAD, FMN, and riboflavin before fluorometric assay. Although separations were excellent, recoveries of the smaller amounts from tissue extracts were not quantitative. A similar electrophoretic procedure was described by Yagi.[30] A procedure for measuring riboflavin fluorescence directly on paper by a specially designed paper strip fluorometer was reported by Brown and Marsh.[35] More recently Yagi and Okuda[18b] and Yagi[30] reported a careful study of the fluorescence of riboflavin and its nucleotides on paper chromatograms. They noted that whereas in solution FAD is only 10% as fluorescent as riboflavin, on paper the nucleotide fluorescence is almost 50% that of the free flavin. The fluorescence spectra of riboflavin and its nucleotides were also measured on paper and found to be identical. However, free riboflavin and FMN were found to decompose rapidly on the paper when exposed to the ultraviolet light used to measure the fluorescence whereas FAD was relatively stable.

Fluorometric methods for assay of riboflavin in foods, preparations, and vitamin concentrates have been approved by the A.O.A.C.[36] and the U. S. Pharmacopeia.[37]

V. NICOTINAMIDE, PYRIDINE NUCLEOTIDES, AND THEIR METABOLITES

Neither niacin nor nicotinamide possess native fluorescence. However, the latter can be converted to a fluorescent derivative as can many of its congeners and there have been developed fluorescence methods for their assay. Such procedures are widely used for the assay of the coenzyme forms of niacin, diphosphopyridine nucleotide (DPN) and, triphosphopyridine nucleotide (TPN). A major metabolite, N'-methylnicotinamide, can also be assayed fluorometrically. Since the procedure for nicotinic acid involves chemical treatment followed by the method used for N'-methylnicotinamide this procedure will be presented last.

A. Native Fluorescence of Reduced Coenzymes

The pyridine nucleotides exist in oxidized and reduced forms (Fig. 11). In the latter form they absorb light in the ultraviolet at 340 mμ. Spectrophotometric assay of reduced pyridine nucleotides has played a key role

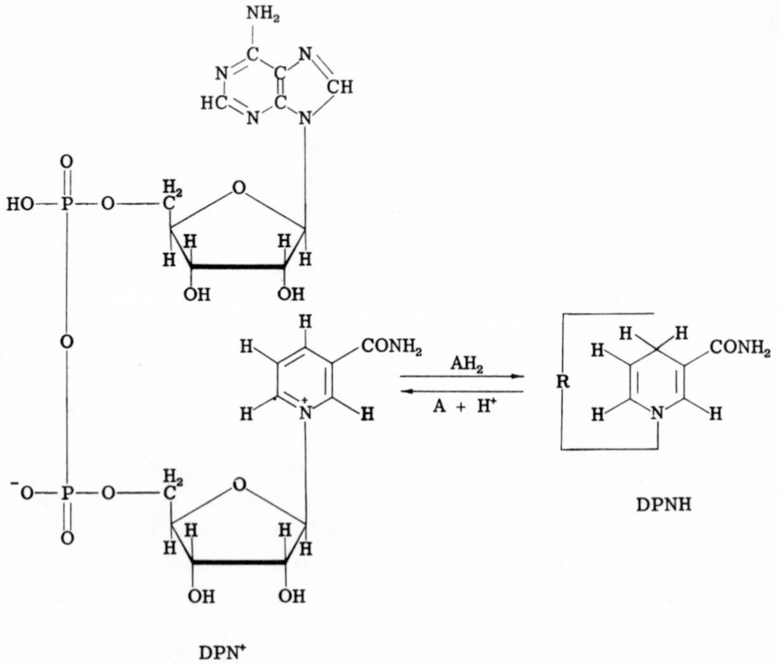

FIG. 11. Oxidized and reduced forms of diphosphopyridine nucleotide.

in modern enzymology. The reduced pyridine nucleotides when excited by ultraviolet radiation fluoresce, their native fluorescence being sufficiently intense to provide extremely sensitive procedures for following the activities of pyridine nucleotide enzymes. Furthermore, the fluorescence is maximal at neutral pH permitting direct measurement during enzyme incubation. Theorell and Bonnichsen[38] were among the first to make use of this fluorescence to study the kinetics and reaction mechanisms of this class of enzymes. Duysens and Amesz[39] and Chance and Baltscheffsky[40] have utilized the method for studying metabolism in whole cells and in cell particles. Winer and Schwert,[41] Velick,[42] Weber,[43] and others have utilized the native fluorescence of reduced pyridine nucleotides to study enzyme-coenzyme interaction and to demonstrate energy transfer from enzyme to coenzyme. The use of fluorescence as a tool in enzyme studies will be discussed in a later section of this book (Chapter 9).

Many uncorrected values have been reported for the fluorescence characteristics of DPNH. Values reported by Weber,[43] apparently corrected for instrumental artifacts, are 260 and 340 mμ as excitation maxima and 457 mμ as the fluorescence maximum, at 18°, in aqueous solution, at pH 8.3.

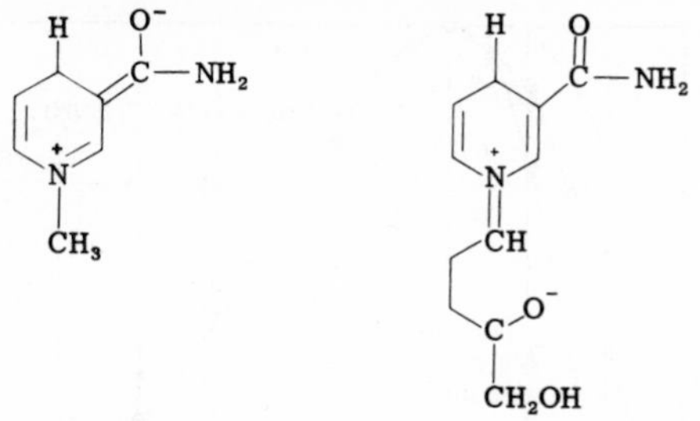

HNM   DPNH

FIG. 12. Postulated fluorescent structures of 4-hydro-N-methylnicotinamide (HNM) and DPNH according to Weber.[44]

HNM   DPNH

360· · ·excitation maximum, mμ · · ·340
468· · ·fluorescence maximum, mμ· · ·456

Several factors were shown to be involved in DPNH fluorescence. The excitation at 260 mμ was shown to be due to transfer of energy from the purine portion of the molecule since following treatment with nucleotide pyrophosphatase this excitation disappears (see Chapter 6, Fig. 3). Excitation at 260 mμ is also abolished in solvents such as propylene glycol. Apparently these changes have little effect on the 340-mμ excitation. Weber[44] also compared the fluorescence properties of 4-hydro-N-methyl-nicotinamide with those of DPNH and concluded that the fluorescent structures in each case were as shown in Fig. 12. Thus, the N-ribose linkage makes a specific contribution to both the excitation (absorption) and fluorescence spectra of DPNH.

Lowry et al.[45] have described procedures for using the native fluorescence of DPNH and TPNH for their assay in tissues and have discussed the effects of solvents, pH, and trace metals on this fluorescence. The results,

TABLE IV

EFFECT OF VARIOUS SOLVENTS ON NATIVE FLUORESCENCE OF DPNH AND TPNH

Solvent	DPNH	TPNH
H_2O^a	100	100
0.003 M NaOH	102	103
0.01 M NH$_4$OH	102	102
Ethanol	167	92
n-Propanol	297	74
Methyl cellosolve	209	173

^{a} The values in water are arbitrarily taken as 100 (after Lowry et al.[45]).

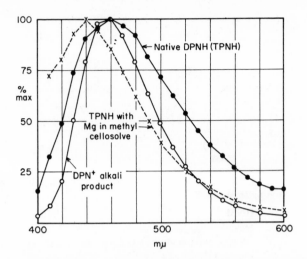

FIG. 13. Fluorescence spectra of pyridine nucleotides. Values are plotted as per cent of the maximum fluorescence. On an absolute scale DPNH fluorescence would have to be reduced by a factor of about 8 (after Lowry et al.[45]).

shown in Table IV, indicate that in water the intensity of fluorescence is independent of pH from 7 to 12, but pH changes in organic solvents influence fluorescence. Mg^{++} in NaOH or ethanol enhances both DPNH and TPNH fluorescence, although not to the same extent. Marked enhancement of TPNH fluorescence is brought about by divalent metals in alkalinized organic solvents. The metals also change the fluorescence spectrum of TPNH fluorescence (Fig. 13). Ethylenediamine tetraacetate can reverse both effects of the divalent metals. Lowry et al.[45] conclude that although the native fluorescences of DPNH and TPNH serve well for measurement in studies with purified enzymes they are not as sensitive as other fluorometric procedures for measuring these compounds.

Greengard[46] has pointed out that the native fluorescence of the reduced pyridine nucleotides serves as a better index of their true chemical structure and biological activity than does their light absorption. Attempts have been made to reduce DPN by various treatments including X-irradiation in the presence of ethanol and other reagents[47] and by electrolytic methods.[48] According to ultraviolet absorption data, material with the characteristic 340-mμ absorption of reduced pyridine nucleotides is formed by both these methods. However, they are enzymatically inactive and do not show the characteristic fluorescence of DPNH. On the other hand, reduction of DPN by sodium hyposulfite yields a product which not only has the characteristic 340-mμ absorption but also shows the enzymatic activity and the characteristic fluorescence of DPNH prepared by enzymatic reduction. Thus, the presence or absence of fluorescence appears to parallel the biological activity of the reduced coenzyme. The latter is, of course, a most delicate and specific indicator of chemical structure.

Two methods are available for converting the oxidized forms of the pyridine nucleotides to highly fluorescent derivatives.

B. DPN and TPN Fluorescence in Alkali

Kaplan et al.[49] found that the oxidized pyridine nucleotides are converted to highly fluorescent products when heated in alkali. Lowry et al.[45] were able to modify their procedure so as to stabilize the fluorophor and to measure its fluorescence spectrum (Fig. 13).

To determine oxidized DPN and TPN a sample is first made 0.2 N with respect to HCl for at least 30 seconds. This step destroys DPNH and TPNH which would otherwise interfere. Strong NaOH is then added to give a concentration of 6 N. The reaction is complete after 1 hour at room temperature, 30 minutes at 38°, or 5 minutes in a boiling water bath. The sample is then diluted at least fivefold with water and its fluorescence is measured (excitation, 360 mμ; fluorescence, 460 mμ). The final nucleotide concentration should be from 10^{-8} to 2×10^{-5} M and standards and blanks should be carried through the procedure. If desired, quinine in concentrations of 0.005 to 5 μg./ml. may be used as reference standards.

Following the reactions in alkali, the samples are stable for hours unless exposed to ultraviolet light. Stability to light is greater the lower the alkalinity. As for specificity, nicotinamide riboside and nicotinamide mononucleotide form the same fluorescent products as DPN and TPN. N'-methylnicotinamide gives a product having less than 2% the fluorescence of DPN. DPN and TPN can be determined in the presence of the other fluorescent materials by first measuring fluorescence on one aliquot and then destroying the coenzymes in another aliquot with Neurospora DPNase.[50] The enzyme specifically cleaves the riboside bond of the coenzymes so that the products no longer give alkaline fluorescence. Glucose and carbonyl compounds such as pyruvate, oxalacetate, and ketoglutarate, in concentrations above 1 mM, diminish the resulting fluorescence. To correct for such effects and for any light absorbing substances, internal standards should be carried through the procedure.

C. DPNH and TPNH by the Alkaline Method

In a mixture of oxidized and reduced pyridine nucleotides it is possible to destroy the oxidized coenzymes by heating in dilute alkali. The reduced coenzymes are then oxidized and treated with strong alkali to yield the typical alkali fluorescence products of DPN and TPN. To 0.1 ml. of sample, containing 2×10^{-10} to 10^{-7} mole of DPNH or TPNH is added 1 ml. of 0.04 N NaOH. After heating at 60° for 15 minutes, a 0.1-ml. aliquot is transferred to a cuvette containing 0.2 ml. of 0.015% H_2O_2 in 9 N NaOH (freshly prepared). After 1 hour, 2 ml. of water are added and the fluorescence is measured at the same wavelengths as described previously and compared to standards and blanks carried through the procedure.

D. Procedures Involving Condensation with Ketones

It was shown by Huff[51] that N'-methylnicotinamide condenses with acetone in alkaline solution to yield highly fluorescent products. In other studies Huff and Perlzweig[52] showed that DPN and TPN react in a similar manner. These observations were incorporated into analytical procedures for estimating N'-methylnicotinamide and the coenzymes in blood[52] and of the metabolite in urine.[53] According to Huff, a variety of carbonyl reagents can condense with structures of the type shown below to yield the

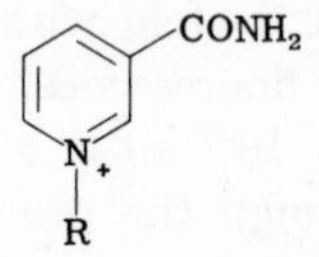

fluorescent products. The mechanisms suggested for the condensation of N'-methylnicotinamide with acetone are shown in Fig. 14. Huff was able to isolate and characterize product VI. However, it was apparent that in

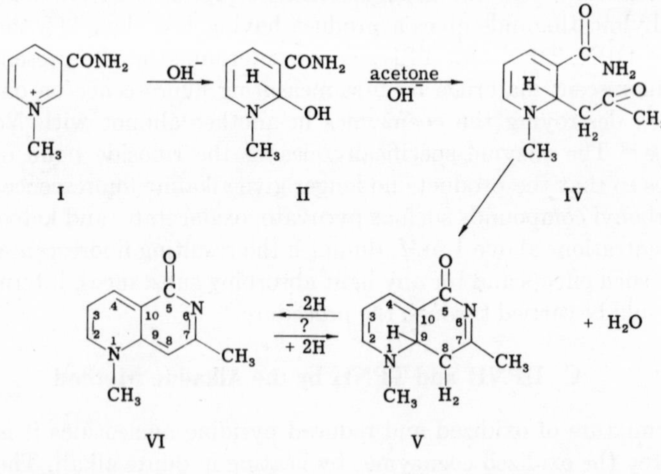

FIG. 14. Mechanism for the condensation of acetone with N'-methylnicotinamide according to Huff.[51] Compounds V and VI are responsible for the fluorescence.

the method of assay a mixture of two fluorophors is obtained, probably V and VI. Similar mechanisms can be formulated for condensation with other carbonyl compounds.

Asami[54] has reported a careful study of the acetone condensation procedure and its application to the measurement of N'-methylnicotinamide in urine. Carpenter and Kodicek[55] found that of several carbonyl compounds tested methyl ethyl ketone gave the highest fluorescence on condensation with N'-methylnicotinamide or with the coenzymes, and gave the lowest reagent blanks. It was also preferable to other ketones because of its low volatility. Based on these observations they developed procedures for the measurement of N'-methylnicotinamide in the presence of DPN and other related structures. Applications to the assay of human and rat urine were shown to be precise and specific. Later, Burch et al.[56] modified the methyl ethyl ketone procedure still further and applied it to the assay of pyridine nucleotides and N'-methylnicotinamide in blood and serum.

To determine DPN in whole blood, 25 μl. of blood are collected directly into a constriction pipette and transferred to a tube containing 1 ml. of 5% trichloroacetic acid. Hemolysis and clumping must be avoided. After 20 minutes the sample is centrifuged and 25 μl. aliquots of the supernatant fluid are transferred to fluorometer tubes. To each tube 25 μl. of methyl ethyl ketone are added followed by vigorous mixing. Before the phases have separated, 100 μl. of 1 N NaOH are added and the tubes are shaken again. After 6 minutes, 1 ml. of 0.17 N HCl is added (to about pH 1.8) and the samples are heated in boiling water for 3 minutes. After cooling to room temperature the fluorescence is measured and compared to standards and blanks carried through the procedure. With a filter instrument the same filters are used as for the thiamine assay.

Most of the DPN in whole blood is present in the red cells. That it is mainly DPN can be shown by treatment of filtrates with *Neurospora* DPN

TABLE V

DISTRIBUTION OF PYRIDINE NUCLEOTIDES IN RABBIT LIVER AND BRAIN

Compound	Liver μmole/g.	Brain μmole/g.
DPN+	0.54	0.24
TPN+	0.07	0.01
DPNH	0.22	0.11
TPNH	0.06	0.01
Total pyridine[a] nucleotides	0.90	0.37

[a] Experimentally determined (after Lowry et al.[45]).

TABLE VI

PYRIDINE NUCLEOTIDE CONTENTS OF FROG SCIATIC NERVE[a]

Sample	PN+ [b]	PNH[c]	Total
Right nerve 1	0.097	0.017	0.114
Left nerve 1	0.099	0.021	0.120
Right nerve 2	0.089	0.026	0.115
Left nerve 2	0.088	0.022	0.110
Rat liver	0.492	0.442	0.934

[a] Micromoles per gram (after Greengard et al.[57]).
[b] PN+ = oxidized pyridine nucleotides.
[c] PHN = reduced pyridine nucleotides.

nucleotidase[50] before carrying out the methyl ethyl ketone reaction. When this is done 85–90% of the fluorescence disappears. The procedure, along with enzymes such as *Neurospora* DPN nucleotidase and others, can be applied to measure the pyridine nucleotides and N'-methylnicotinamide in blood, serum, and tissue homogenates. Many of the enzyme and substrate assays presented in subsequent sections make use of the methyl ethyl ketone procedure to determine DPN or TPN changes.

The amounts and relative proportions of the oxidized pyridine nucleotides in brain and liver of the rabbit are shown in Table V. These were obtained by the various procedures which were presented in this section.

Greengard et al.[57] applied fluorometric procedures to determine the proportions of oxidized and reduced pyridine nucleotides in milligram quantities of peripheral nerve tissue normally and after degeneration. Normal values are shown in Table VI.

Lowry et al.[57a] have developed a fluorometric method using enzymatic cycling which can measure 10^{-15} moles of pyridine nucleotide.

E. Nicotinamide

Nicotinamide is, itself, nonfluorescent, nor does it yield fluorescent products under the conditions used for assay of N'-methylnicotinamide or the pyridine nucleotide coenzymes. However, Scudi[58] has shown that methylation of nicotinamide with methyl iodide permits fluorescence assay as for N-methylnicotinamide. A better procedure, proposed first by Friedemann and Frazier,[59] makes use of aqueous cyanogen bromide to convert nicotinamide to a pyridinium product capable of yielding the fluorescent reactions of N'-methylnicotinamide. Although no details were given, it was stated that treatment with cyanogen bromide is according to Bandier and Hald,[60] then substituting for the color reagent alkali and isobutanol, as described by Najjar.[61] Kodicek[62] has modified this procedure so that it can be used in the presence of N'-methylnicotinamide and other vitamins. To 1 ml. of solution containing nicotinamide add 0.4 ml. of freshly prepared cyanogen bromide solution.* Heat for 4 minutes in a water bath at 55°, then cool. An aliquot of the solution is then taken for assay by the methyl ethyl ketone method.

Le Clerc and Douzou[63] have used similar procedures for measurement of nicotinamide and nicotinic acid.

VI. VITAMIN B₆ GROUP

It is now known that the coenzyme form of vitamin B₆ is pyridoxal-5-phorphoric acid (codecarboxylase) which is required for amino acid decarboxylations, transamination, and many other reactions. However, as shown in Fig. 15, many related metabolites are found in animal, plant, and

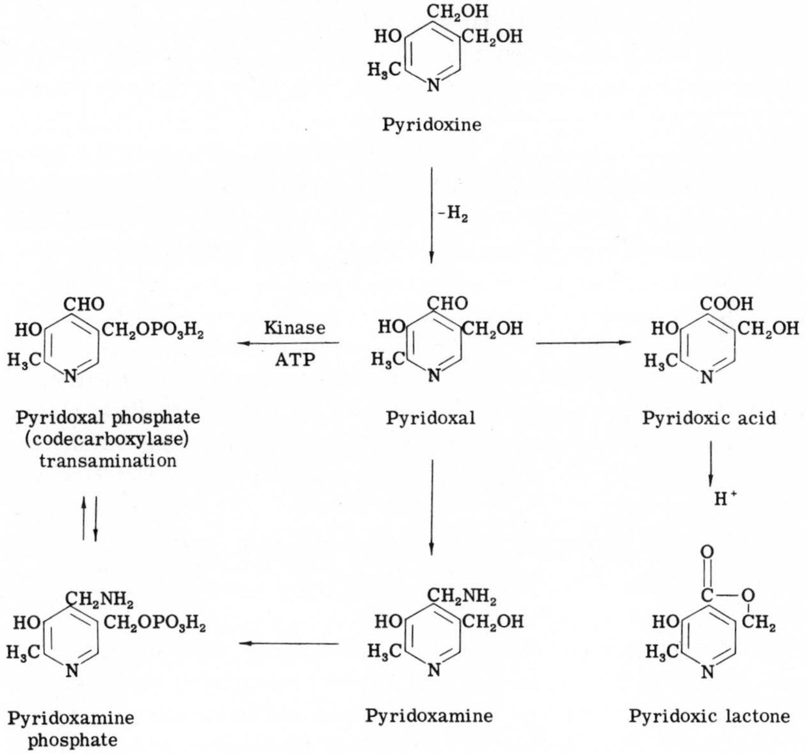

FIG. 15. Pyridoxine and its congeners.

---

* The cynogen bromide is prepared according to the procedure of Y. L. Wang and E. Kodicek, *Biochem. J.* **37**, 530 (1943).

bacterial sources which are either precursors or end products of metabolism. From the previous discussions concerning structure and fluorescence one would expect these phenolic pyridines to fluoresce. This was shown to be the case for the unphosphorylated metabolites.[1]

A. Pyrodixic Acid and Its Lactone

Although application of fluorescence assay to the biologically active forms of vitamin B_6 is relatively recent, the urinary metabolites, pyridoxic acid and its lactone, were assayed fluorometrically as early as 1944 by Huff and Perlzweig.[64] The fluorescence intensities of pyridoxic acid and its lactone are so great that it was possible to obtain sufficient excitation with the 365-mμ line of the mercury lamp in an all glass system although maximal excitation is at 330 mμ.[65] Huff and Perlzweig assayed pyridoxic acid by converting it to the corresponding lactone with acid treatment. The latter compound, at pH 9.0, fluoresces 25 times as much as does the parent acid. The increased fluorescence of the lactone is not due to a shift in the absorption and fluorescence spectra, but apparently to an increase in the quantum efficiency of fluorescence.

The procedure for pyridoxic acid assay, as originally described by Huff and Perlzweig,[64] utilized a commercial fluorometer with the same filters which are utilized for thiamine and for quinine assay. The urine is acidified, heated to form the lactone, and diluted from 100 to 1000-fold with sodium borate solution (final pH about 9.0) depending upon the amount of pyridoxic acid present. Obviously, without isolation procedures or fluorescence spectrometry the specificity of such a method is questionable. More recently, Reddy et al.[66] modified this procedure by introducing chromatographic isolation procedures to insure specificity. The resulting method permits quantitative recovery of pyridoxic acid added to urine and yields values which are much lower than those obtained with the Huff and Perlzweig procedure.

The following procedure for determining urinary pyridoxic acid is taken from the paper by Reddy et al.[66]

Materials: Dowex 1 (chloride) 10% cross-linkage, 200–400 mesh, and Dowex 50 (H$^+$), 12% cross-linkage, 200–400 mesh, are prepared as described by Price.[67] Synthetic 4-pyridoxic acid and its lactone can be prepared by the method of Heyl.[68] Twenty-four-hour urines are collected in amber bottles containing about 25 ml. of toluene, and after measurement of the volume, an aliquot of these samples is removed and stored at 0° until analyzed.

Preparation of analytical columns: A slurry of the resin is pipetted into the columns (1.2 cm. o.d.) to form a packed layer 3 cm. long. The Dowex 50 columns are washed successively with 50 ml. of 5 N HCl, 50 ml. of 2 N HCl, and 50 ml. of water. The Dowex 1 columns are washed with 50 ml. of 2 N HCl followed by 50 ml. of water. The columns are operated without added pressure, and are discarded after use.

Chromatography of 4-pyridoxic acid in urine: About 4% of the 24-hour urine is adjusted to a pH of 10.6 with saturated NaOH and filtered through Whatman No. 42 filter paper. Aliquots of this filtrate, representing 1% of the 24-hour excretion, are pipetted into duplicate 50-ml. graduated centrifuge tubes and 25 μg. of synthetic 4-pyridoxic acid is added to one tube. Then 1.5 ml. of 1.5 N NH₄OH is added to each centrifuge tube and the contents of both tubes are diluted to 40 ml., stirred well, and added to the Dowex 1 (Cl⁻) columns. The centrifuge tubes are rinsed with an additional 20 ml. of water and this water is used to wash the samples through the columns. The 4-pyridoxic acid is eluted from the Dowex 1 with 50 ml. of 0.05 N HCl and the effluent is passed directly onto the Dowex 50 columns. The latter are washed with 20 ml. of water and the 4-pyridoxic acid is eluted with 50 ml. of 2 N HCl.

The quantitative estimation consists of 4 steps: (1) delactonization, (2) lactonization, (3) adjustment of pH and dilution, and (4) reading in the fluorometer. The first two steps are to ensure complete conversion to a single entity, the pyridoxic acid lactone.

Delactonization: In the initial studies with this procedure, blank values were found to be unduly high. This was the result of considerable lactonization which occurred spontaneously at room temperature in the 2 N HCl used to elute the 4-pyridoxic acid from the Dowex 50 columns. A delactonization step was introduced to eliminate this problem. For delactonization, 4 ml. of the 2 N HCl eluate from the Dowex 50 is pipetted, in duplicate, into 50-ml. graduated centrifuge tubes. Two milliliters of 5 N NaOH are added to each tube which make the solution 0.33 N with respect to alkali and the tubes are heated in a boiling water bath for 5 minutes and then cooled. This period of time was found to be adequate for complete delactonization.

Lactonization: To neutralize the alkali, 2 ml. of 1 N HCl are added to the tube containing the sample to be lactonized. Then 2 ml. of 5 N HCl are added and the tube is heated in a boiling water bath for 15 minutes and cooled. It was found that 15 minutes of heating is adequate for lactonization. The second tube is not lactonized but is used to prepare the blank as described below.

Adjustment of pH and dilution: After lactonization the samples are made slightly alkaline by the addition of 11 ml. of 1 N NaOH and are then diluted to 30 ml. with water. From 0.5 to 5 ml. (usually 2 ml.) of this solution are immediately diluted to 10 ml. by the addition of 1% sodium borate solution. The pH of the final solution is 9.0 ± 0.3 which is optimal for the fluorescence measurements.

The blanks are handled differently because considerable lactonization takes place in acid solution at room temperature. To prepare the blank, 11 ml. of 1 N NaOH are added to the second of each pair of tubes which had been delactonized as described previously, followed by additions of 2 ml. of 1 N HCl, 2 ml. of 5 N HCl, and water to a volume of 30 ml. A 2-ml. aliquot of this mixture is immediately diluted to 10 ml. with 1% sodium borate for reading in the fluorometer. The fluorometric readings of the samples and blanks do not change for several hours in borate buffer.

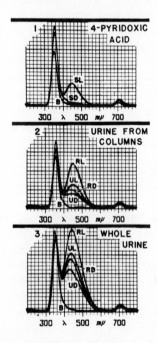

Fig. 16. Fluorescence spectra of pyridoxic acid and its lactone; excitation is at 350 mμ and the first peak in each instance represents light scattering (after Reddy *et al.*[66]).

1, lactonized standard (SL) and unlactonized standard (SD). *2*, material isolated from urine by chromatography; lactonized urine sample (UL), delactonized urine sample (UD); (RL) and (RD) are comparable samples to which known amounts of pyridoxic acid had been added as an internal standard. *3*, comparable to *2*, except urine was used directly without chromatographic purification.

Fluorescence measurements: Fluorescence measurements may be made with a filter photofluorometer or with a fluorescence spectrometer. The excitation maximum is at 350 mμ, and the fluorescence maximum at 450 mμ, uncorrected. The fluorescence characteristics of authentic pyridoxic acid and its lactone and those of the compounds obtained from urine are shown in Fig. 16. It is apparent that whole urine contains much interfering material wheras the material from the column extracts has spectral properties similar to those obtained with authentic pyridoxic acid.

The method is applicable to the measurement of pyridoxic acid in human urine when the compound is present in amounts above 2 μmoles/day. Individuals on ordinary diets excrete about 3–5 μmoles of pyridoxic acid a day and increases in the dietary intake of vitamin B₆ are reflected by increases in urinary excretion of the acid metabolite.

Fluorometric procedures for the assay of individual forms of vitamin B₆ and metabolites in animal and plant tissues have been reported by Fujita and colleagues in a series of four papers.[69-72] All the procedures involve conversion to the intensely fluorescent pyridoxic acid lactone and depend on isolation procedures for their specificity.

B. Pyridoxine

Pyridoxine[69] is extracted from the tissues and interfering anions are removed on a column. The compound is then adsorbed onto permutit, eluted, oxidized with $KMnO_4$ to pyridoxic acid, and heated in acid to form the lactone.

To 5 ml. of urine add 1 ml. of 1 N acetic acid, sufficient 1 N NaOH to adjust to pH 4.5 and make up to a 10-ml. volume. The urine sample is then passed through a column of Amberlite IRA-410 (acetate form—1 ml./minute) and the eluate is collected. Two 5-ml. portions of water are passed through the columns and are added to the eluates. To the combined eluates are added 3 ml. of 6 N H_2SO_4 and the samples are heated at 100° for 15 minutes. After cooling, 2.7 g. of barium hydroxide are added, the sample is adjusted to pH 5 with 1 N NaOH, then made up to a volume of 30 ml. After centrifugation, a 20-ml. aliquot is passed through a permutit column. The column is washed 3 times with 10-ml. portions of water and the washings are discarded. The pyridoxine is eluted with two 10-ml. portions of boiling 0.1 N H_2SO_4. The eluates are then diluted to 40 ml., the final solution containing 10% alcohol and being 0.1 N with respect to H_2SO_4. To 10 ml. of this solution are added 1 ml. of water (or 1 ml. of standard solution for an internal standard) and the sample is cooled in ice. To the chilled sample are added 2 ml. of 2% $KMnO_4$ (also chilled). The sample is mixed well and after exactly 50 seconds 2 ml. of 3% H_2O_2

are added to destroy the excess $KMnO_4$. At this point 3 ml. of 6 N HCl are added and the sample is heated at 100° for 8 minutes to form the lactone of pyridoxic acid. It is then made alkaline by addition of 2 ml. of 3 N ammonia, adjusted to volume and its fluorescence measured.

C. Pyridoxamine

The method for pyridoxamine[70] involves conversion to pyridoxine with nitrous acid which is then assayed as described previously. The eluates from the Amberlite IRA-410 columns are autoclaved at 130° for 1 hour, cooled, and centrifuged. To 20 ml. of the supernatant fluid are added 2 ml. of 6 N H_2SO_4 and 2 ml. of 24% $NaNO_2$ and the mixture is heated at 100° for 10 minutes. The solution is cooled and, after adding 0.5 ml. of 1 N acetic acid, is treated as for pyridoxine. Pyridoxine and pyridoxamine do not interfere with each other in the assay. Pyridoxal and pyridoxic acid have been removed by the column procedures and so do not interfere with the assay of either amine.

D. Pyridoxal

The third report[71] pertains to the assay of pyridoxal and pyridoxic acid. The procedure for pyridoxal is presented in detail.

A standard solution of pyridoxal hydrochloride is prepared by dissolving 20 mg. in 100 ml. of water containing 2 drops of concentrated HCl. An Amberlite IR-112 (strongly acidic cation exchanger) is used for adsorbing pyridoxal. Columns are prepared by placing 4.5 ml. of the Na^+ form of the resin in a tube 7 $\times$ 15 cm. Thirty milliliters of 1.14 NaOH are allowed to pass through the column at a rate of 1,2 ml./minute. After washing with water to remove alkali, 30 ml. of 1 N H_2SO_4 are allowed to pass through and again followed by water to remove free acid. Although the acid form of the column is used, treatment with NaOH is necessary to remove interfering substances from the resin. An Amberlite IRA-410 column (acetate form) is used to remove pyridoxic acid.

Procedure: Animal or plant tissue (1–10 g.) is homogenized in 5 to 50 volumes of dilute H_2SO_4 so that the final concentration of acid is about 0.1 N. The homogenate is autoclaved at 130° for 1 hour, cooled, and the volume is readjusted, if necessary. Following filtration, 1.5 ml. of 1 N NaOH is added to 5 ml. of the filtrate and the volume is adjusted to 10 ml. The sample is then heated at 75° for 5 minutes (to delactonize any pyridoxic lactone), adjusted to pH 4.5 with 1 N acetic acid, and made up to a volume of 20 ml. This solution is first passed through an Amberlite IRA-410 column and the effluent is further passed through the Amberlite

IR-112 column at a rate of 1 ml./minute. After the entire solution has passed through, the second column is washed twice with 10 ml. of water and the pyridoxal is eluted with 30 ml. of 1 N NaOH at a rate of 1 to 2 ml./minute. The effluent is made up to 50 ml. and 10-ml. portions are placed in 3 test tubes. One milliliter of water is added to two of the tubes and 1 ml. of a diluted pyridoxal standard solution is added to the third (usually 1 μg.). To each of the three tubes are added 1 ml. of 0.5 N NH₃ and 1 ml. of 0.2 M AgNO₃, and the tubes are heated at 100° for 40 minutes to oxidize the pyridoxal to pyridoxic acid. Following oxidation, 3 ml. of 6 N HCl are added to each tube and they are heated at 100° for 15 minutes, then cooled. Pyridoxic acid is thus converted to the lactone. One of the tubes is made alkaline (add 5 ml. of 3 N NaOH) and heated at 75° for 10 minutes. This converts the lactone back to the acid and thus acts as an internal blank. To all the tubes are added 2 ml. of 6 N HCl and 3 ml. of 6 N NH₃. To the two tubes still in the lactone form are also added 5 ml. of 3 N NaOH (in order to simulate the composition of the endogenous blank). The pH should now be about 9.5. If not, they must be brought to a pH between 9 and 10. The tubes are then made up to volume, 25 ml., centrifuged, and the supernatant solution assayed for pyridoxic lactone by the fluorometric procedure described previously. Proportionality is obtained in the range 0.2–4.0 μg. carried through the entire procedure. Recoveries added to tissue were reported to be quantitative, 100 ± 5%. Pyridoxine and pyridoxamine do not interfere with this assy, nor do pyridoxic acid and its lactone when they are removed by the Amberlite IRA-410 column.

In the final paper of the series, Fujita and Fujino[72] report the application of these fluorometric methods to the measurement of the four vitamin

TABLE VII

VARIOUS VITAMIN B₆ COMPONENTS AND METABOLITES IN HUMAN BLOOD AND URINE[a]

Sample	Pyridoxine μg./ml.	Pyridoxamine μg./ml.	Pyridoxal μg./ml.	Pyridoxic acid μg./ml.
Urine 1_____	0.03	0.01	0.10	1.4
Urine 2_____	0.05	0	0.13	1.2
Urine 3_____	0.04	0	0.08	1.4
Urine 4_____	0.14	0	0.17	1.8
Urine 5_____	0.12	0.02	0.15	1.8
Blood 1_____	0	0.20	0.02	0.11
Blood 2_____	0	0.15	0.05	0.10
Blood 3_____	0	0.19	0.07	0.13

[a] After Fujita and Fujino.[72]

TABLE VIII

FLUORESCENCE CHARACTERISTICS OF COMPOUNDS OF THE VITAMIN B_6 GROUP UNDER
VARIOUS PHYSICAL AND CHEMICAL CONDITIONS[a]

	Compound		
Characteristic	Pyridoxine	Pyridoxamine	Pyridoxal
Excitation maximum (mμ)	340	335	330
Fluorescence maximum (mμ)	400	400	385
Relative intensity	44	56	35
Effect of H_2O_2 (30%) (0.01 ml.)	No effect	No effect	Complete destruction
Ultraviolet irradiation:			
1 minute	Partial destruction in all cases		
5 minutes	Complete destruction in all cases		

[a] Each compound was dissolved in phosphate buffer, pH 6.75, at a concentration of 10^{-6} M (after Coursin and Brown[65]).

B_6 components in blood and urine. Some of their findings are shown in Table VII.

More recently, Coursin and Brown[65] have made use of the native fluorescence of pyridoxine, pyridoxal, and pyridoxamine to identify and estimate these substances in blood with a spectrophotofluorometer. All three substances fluoresce maximally in 0.1 M phosphate buffer, pH 6.75. The procedure used to distinguish the amine from the aldehyde is outlined in Table VIII. The spectral data in the table are uncorrected instrumental data. The reported absorption maxima of these compounds at pH 6.75 are: pyridoxine, 330 mμ; pyridoxamine, 315 mμ; and pyridoxal, 330 mμ.[73]

Apparently normal blood contains little pyridoxine. The following method for measuring pyridoxal and pyridoxamine in whole blood is described in the article by Coursin and Brown.[65]

A 1-ml. sample of fresh whole blood is extracted with 25 ml. of reagent grade acetone in an amber glass container. The mixture is shaken for 30 minutes, centrifuged, and the supernatant acetone layer is evaporated to dryness. The residue is taken up in 5 ml. of 0.1 M phosphate buffer of pH 6.75. The sample is then cleared either by high-speed centrifugation (20,000 r.p.m. for 30 minutes) or by filtration through a Morton sintered glass filter. A 1-ml. aliquot is placed in the spectrophotofluorometer and its fluorescence spectrum is recorded, with excitation set at 335 mμ. By using a micropipet, 0.01 ml. of a 30% solution of hydrogen peroxide is

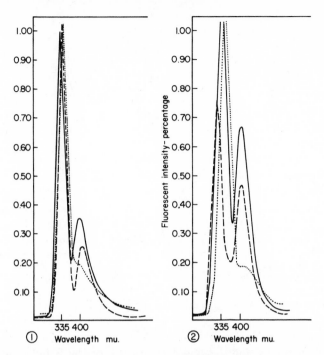

FIG. 17. Fluorescence spectra obtained on a whole blood sample from a patient before ① and after ② ingestion of 30 mg. of pyridoxine (after Coursin and Brown[65]). Peaks at 335 mμ represent light scattering. —, total B₆ content; – – –, residual pyridoxamine fluorescence following the destruction of pyridoxal with 30% hydrogen peroxide; ..., residual blank after irradiation with ultraviolet light.

added to the sample in the cuvette to destroy the pyridoxal and the fluorescence spectrum is measured once more with excitation at 335 mμ. The sample is then exposed to the ultraviolet radiation of a quartz mercury vapor lamp at 3 cm. This process destroys all the vitamin B₆ congeners and the residual fluorescence is taken as the blank. Standards of pyridoxal and pyridoxine are carried through the entire procedure. When these compounds are added to plasma they are recovered quantitatively.

As for specificity, the authors claim that neither pyridoxic acid nor pyridoxine could be detected in blood. Pyridoxic acid fluoresces at 430 mμ,[65] and it is clear from the fluorescence spectra shown in Fig. 17 that the material in blood fluoresces maximally between 385 and 400 mμ, consistent with the values expected for a mixture of pyridoxal and pyridoxamine. Furthermore, following treatment with hydrogen peroxide the fluorescence peak shifted to 400 mμ which is characteristic of pyridoxamine.

Fasella et al.[74a] have reported that metal complexes of the Schiff bases

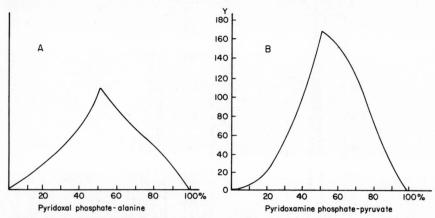

FIG. 18. Continuous titration of aluminum with (A) pyridoxal phosphate-alanine and (B) pyridoxamine phosphate-pyruvate. The ordinate represents corrected fluorescence intensity (Y) and the abscissa is mole fraction of the coenzyme-amino acid or keto acid derivative (after Fasella et al.[74a]).

formed between pyridoxal and alanine or pyridoxamine and pyruvate fluoresce in the visible region. The fluorescence can be seen on paper chromatograms and can be measured in solution. They also reported similar fluorescence for the metal complexes of the Schiff-base forms of

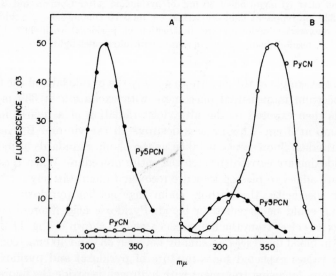

FIG. 19. Excitation spectra of the cyanohydrin derivatives of pyridoxal-5-phosphate (Py5PCN) and pyridoxal (PyCN). A, pH 3.8 and fluorescence emission at 420 mμ. B, pH 7.55 and fluorescence emission at 430 mμ (after Bonavita[74b]).

pyridoxal phosphate and pyridoxamine phosphate. Figure 18 shows the effect on fluorescence intensity (Y) when Al^{3+} is titrated against solutions containing equivalent amounts of pyridoxal-phosphate and alanine or of pyridoxamine-phosphate and pyruvate. Fluorescence is maximal when equivalent amounts of the reactants are present. This fluorescence is of interest since the phosphorylated compounds themselves do not fluoresce appreciably.

Bonavita[74b] has shown that the reaction of pyridoxal-5-phosphate and pyridoxal with cyanide in alkaline solution produces characteristic changes in the excitation and fluorescence spectra of both compounds. The pyridoxal-5-phosphate cyanohydrin derivative exhibits maximal fluorescence at pH 3.8. The excitation maximum is at 313 to 315 mμ and the fluorescence maximum at 420 mμ (uncorrected). The pyridoxal derivative exhibits maximal fluorescence at pH 9 to 10, with maximal excitation at 358 mμ and fluorescence at 430 mμ (uncorrected). With the proper choice of pH and excitation and fluorescence wavelengths, each compound can be determined in the presence of the other. Excitation spectra of each compound at its optimal pH and fluorescence wavelength is shown in Fig. 19. Bonavita considers the fluorometric procedure for pyridoxal-5-phosphate to be as sensitive and almost as specific as the available manometric procedures and to be more accurate and far simpler. He has also shown that the enzyme, phosphorylase, exhibits the characteristic fluorescence of the pyridoxal-5-phosphate cyanohydrin complex when it is treated with cyanide in the presence of 8 M urea.

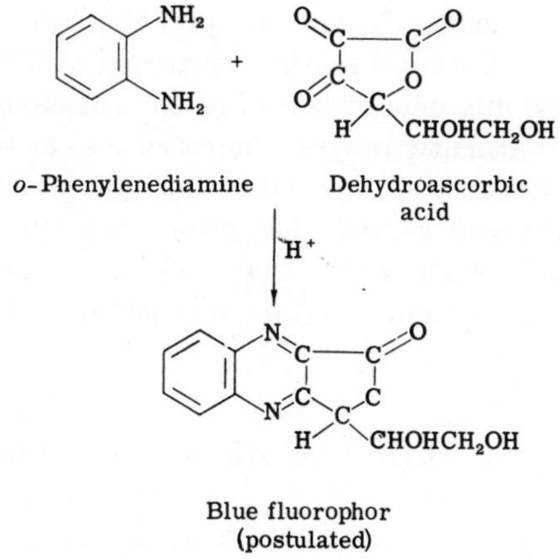

o-Phenylenediamine

Dehydroascorbic acid

H⁺

Blue fluorophor (postulated)

FIG. 20. Condensation of o-phenylenediamine with dehydroascorbic acid to yield a blue fluorescent product (according to Archibald[75]).

VII. ASCORBIC ACID

Archibald[75] reported a number of procedures for the determination of alloxan in tissues. One of these involved condensation of the drug with o-phenylenediamine (in acid solution) to yield a green fluorescent product (Chapter 13). In this report Archibald stated that under comparable conditions ascorbid acid also reacts to give a highly fluorescent product (blue). No details were reported but it was stated that dehydroascorbic acid was an intermediate in the reaction which probably proceeds as shown in Fig. 20. Although this reaction has not yet been applied to ascorbic acid assay, it seems highly promising and worthy of further study.

VIII. VITAMIN D

The application of fluorometric assay to the D vitamins is very recent (Jones et al.[76]). These investigators observed that both vitamin D_2 and vitamin D_3 yield highly fluorescent products when treated with trichloro-

Vitamin D₂   Vitamin D₃

acetic acid and other acids. The spectral properties of both products are qualitatively similar, excitation maxima occurring at 390 mμ and fluorescence maxima at 480 mμ, uncorrected (Fig. 21). Microgram quantities of the vitamins yield sufficiently intense fluorescence so as to obtain accurate measurements. Furthermore, a combination of procedures involving treatment with other acids as well* may even permit the differentiation of vitamins D_2 and D_3 by fluorometry. Such a distinction is necessary since, although both forms are found in nature, a large number of animal species can utilize only vitamin D_3.

IX. FOLIC ACID AND RELATED PTERINS

Pteridine is a nitrogen-containing, conjugated, bicyclic system. As with other conjugated ring systems substitution by phenolic and amino groups

---

* L. Friedman, personal communication.

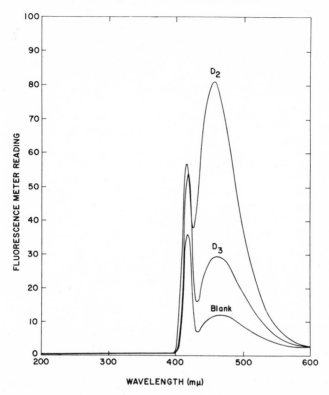

FIG. 21. Fluorescence spectra of vitamins D_2 and D_3 following treatment with trichloroacetic acid for 180 minutes according to Jones et al.[76] The peak at 410 mμ is due to scatter of the exciting radiation.

yields highly fluorescent derivatives. As shown in Fig. 22 folic acid and folinic acid are 2-amino-4-hydroxy-derivatives of pteridine; xanthopterin, a metabolite of the vitamins, is a 2-amino-4-6-dihydroxy derivative. It should be pointed out that folic and folinic acids contain a second fluorophoric system, p-aminobenzoic acid.

According to Duggan et al.,[1] folic and folinic acids both exhibit maximum fluorescence at 450 mμ when excited at 365 mμ (uncorrected). Both compounds also show considerable variation in fluorescence intensity as a function of pH and it is possible to make some distinction between them in this manner. However, when excitation is at 290 mμ, folinic acid gives an entirely different fluorescence spectrum with a maximum at 370 mμ. Folic acid, on the other hand, merely shows a diminution of the same 450 mμ fluorescence. Thus, as shown in Fig. 23, it is possible to distinguish these two related compounds when they are present in a mixture. It should be

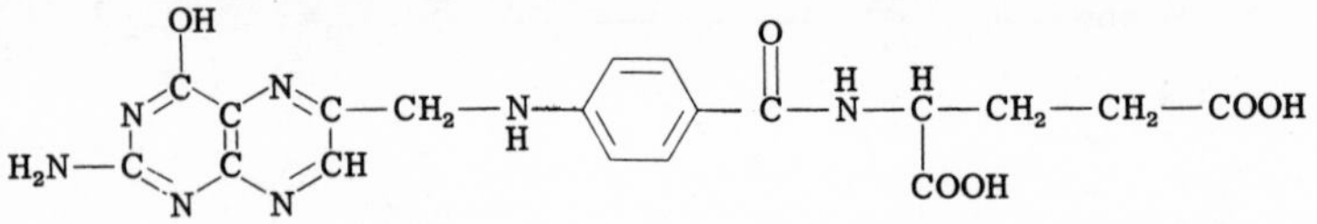

Folic acid

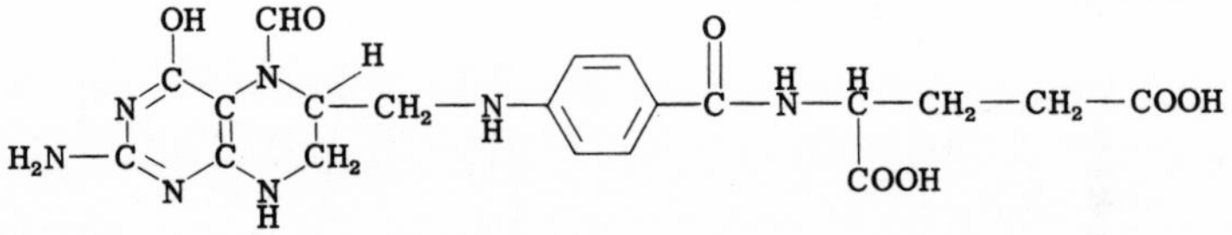

Folinic acid

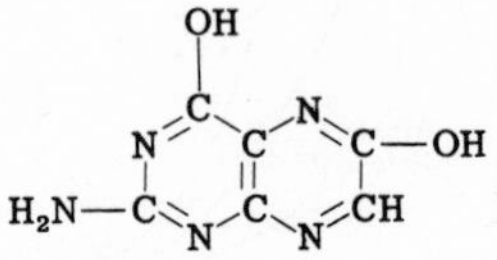

Xanthopterin 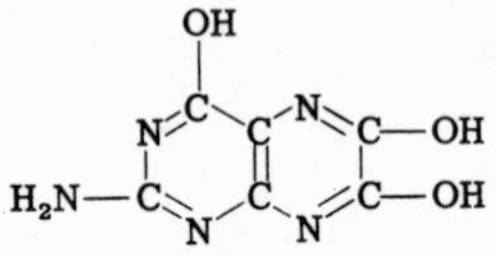   Leucopterin

FIG. 22. Folic acid and related pterins.

possible to incorporate this differential fluorescence into specific assays for these two agents.

Although fluorometric assay has not yet been used widely for folic and folinic acids, it has long been used in studies on the many pteridine derivatives found in nature. Simpson[77] has discussed the possibility of using fluorescence spectra as a means of identifying pterins and has presented fluorescence maxima for xanthopterin, leucopterin, and a number of pterins isolated from fish. On using the 365-mμ mercury line for excitation, xanthopterin and leucopterin exhibited an intense fluorescence band at 436 mμ and a weak band at 546 mμ. An interesting application was the use of fluorescence spectra to identify as a pterin the product formed by heating uric acid with water in sealed tubes at 200°. Hopkins[78] originally suggested that this may be xanthopterin. However, Simpson[77] showed that while the fluorescence characteristics of the uric acid product resembled that emitted by pterins, it was not exactly the same as that of xanthopterin, leucopterin, or other pterins which were available. Crowe and Walker[79] also used fluorescence analysis to identify a xanthopterin-like

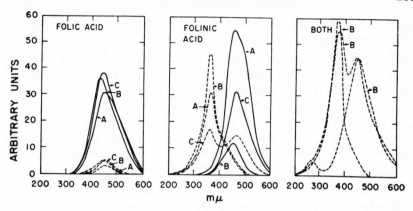

FIG. 23. Fluorescence spectra of folic and folinic acids. Solid lines represent spectra obtained by exciting at 370 mμ; broken lines by exciting at 280 mμ. Curve A, pH 1.0; curve B, pH 7.0; curve C, pH 10. Pure samples are 1.0 μg./ml.; mixtures contain 1.33 μg. folic acid and 10 μg. folinic acid per ml. (Duggan et al.[1]).

pigment which they isolated from human tubercle bacilli. Decker[80] devised a procedure for the quantitative fluorometric assay of leucopterin in the presence of xanthopterin and other pterins. He found that whereas in neutral solution leucopterin has little fluorescence and xanthopterin is highly fluorescent, with increasing alkalinity leucopterin fluorescence increases whereas xanthopterin fluorescence disappears. In 5 M NaOH essentially all of the fluorescence in a mixture of the two is due to leucopterin. By combining this with isolation procedures such as those described by Koschara,[81] he developed a procedure for leucopterin assay in

TABLE IX

FLUORESCENCE OF XANTHOPTERIN AND LEUCOPTERIN[a]

Compound	pH	Fluorescence maxima mμ
Xanthopterin	13	*463*, 516, 592
Xanthopterin	7	*455*, 516, 592, 620
Xanthopterin	1	460, *523*, 592, 620
Leucopterin	13 and above	463, *516*
		No fluorescence in neutral or acid solutions

[a] Excitation was with the 365-mμ mercury line. The values in italics represent the most intensely fluorescing bands (after Jacobson and Simpson[82]).

urine and other natural sources. With this procedure he found less than 100 μg. of leucopterin per liter of human urine.

Jacobson and Simpson[82] using photographic methods determined the fluorescence spectra of a number of pterins and related compounds. The fluorescence maxima of xanthopterin and leucopterin are shown in Table IX. Jacobson and Simpson[83] also used fluorometric procedures to investigate the pterins in a variety of tissue extracts in studies on antipernicious anemia factors. They reported the presence of xanthopterin in argentaffin tumors. Pirie and Simpson[84] identified a pigment in the eye of Squalus acanthias as xanthopterin by the characteristic fluorescence spectra at various pH values.

Kalckar and Klenow[85] made use of the fluorescence of xanthopterin to measure its disappearance in the presence of pterine oxidase at concentrations of 0.01 μg./ml. The oxidized product, leucopterin, is not fluorescent but can be converted to a fluorescent substance by addition of concentrated NaOH. Although no details were given, they reported that the method can be utilized for assaying xanthopterin in urine. An aliquot of a Norit eluate of urine is measured fluorometrically (excitation and fluorescence maxima as given before). A second aliquot is treated with milk xanthine oxidase (pterine oxidase). The fluorescence which is destroyed by the enzyme is due mainly to xanthopterin. With modern instrumentation and chromatography it should be possible to convert this to a specific and simple assay for xanthopterin in urine.

An interesting application of fluorescence has been in the determination of the acid-base properties and tautomerism of xanthopterin. Lowry et al.[86] showed that irradiation of folic acid with ultraviolet light leads first to cleavage of the side chain to yield in succession, 2-amino-4-hydroxy-6-formylpteridine, 2-amino-4-hydroxy-6-carboxypteridine, and 2-amino-4-hydroxypteridine. The intermediates in the photochemical reaction were detected by changes in intensity of fluorescence and by the effects of pH and quenching agents on the fluorescence intensity. Schou[87] extended these studies and indicated that xanthopterin exists in solution in a keto-

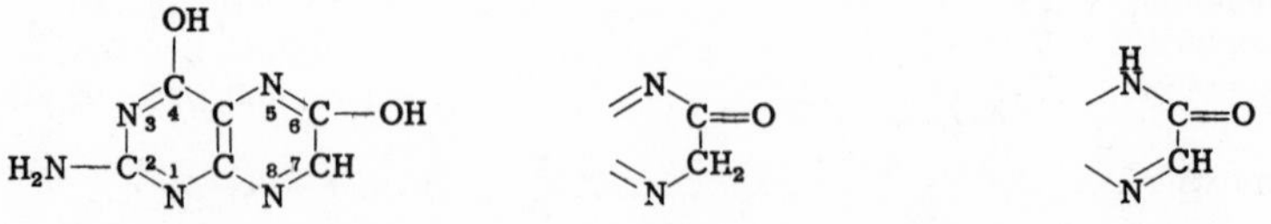

I
(enol)

II
(keto)

III
(lactam)

FIG. 24. Tautomeric forms of xanthopterin. According to Schou,[87] it is the enol form (I) which is the fluorescent form of the compound.

enol equilibrium catalyzed by acid-base (Fig. 24). Only the enol form is fluorescent. A third form (lactam) was ruled out as a possibility on the basis of ultraviolet absorption spectra.

Allfrey et al.[88] reported a fluorometric method for the determination of folic acid in plant and animal tissues. They used permanganate to convert the folic acid to 2-amino-4-hydroxypteridine-6-carboxylic acid, which is apparently more intensely fluorescent. The filters used by Allfrey et al. suggest that the acid product has spectral characteristics qualitatively similar to those of folic acid (see above). The use of chromatography and appropriate pH changes endow the method with some degree of specificity.

X. p-AMINOBENZOIC ACID

This substance, considered a growth factor in various organisms, is itself fluorescent. Duggan et al.[1] reported an excitation maximum of 294 mμ and a fluorescence maximum at 345 mμ (uncorrected). Fluorescence is maximal at pH 11, and is comparable in intensity to that of some of the most highly fluorescent compounds. Surprisingly, this fluorescence does not appear to have been utilized for assay purposes. This may be so because p-aminobenzoic acid, being an aromatic amine, reacts with numerous reagents which yield intense colors. Should increased sensitivity be required for p-aminobenzoic acid (below the microgram level) fluorescence assay offers great promise.

The folic acid molecule contains a p-aminobenzoic acid residue (see above). However, this portion of the molecule does not appear to contribute to the fluorescence since there is no excitation at the maximum for p-aminobenzoic (295 mμ). Upon reduction to folinic acid there does develop an excitation peak at 290 mμ.[1] However, the fluorescence is at 370 mμ rather than at 345 mμ, as for p-aminobenzoic. A more careful evaluation of the contribution of p-aminobenzoic acid to the fluorescence of folic and folinic acids would be desirable.

XI. VITAMIN B$_{12}$

Since vitamin B$_{12}$ is a porphyrin (Fig. 25) it should possess the characteristic visible fluorescence of this group of compounds (see Chapter 8). However, Duggan et al.[1] observed only one fluorescent band in the far ultraviolet with an excitation maximum at 275 mμ and a fluorescence maximum at 305 mμ, uncorrected. There have been no reports of the use of fluorescence for vitamin B$_{12}$ assay. There has, however been an interesting application of fluorescence to the vitamin B$_{12}$ coenzymes. Recently,

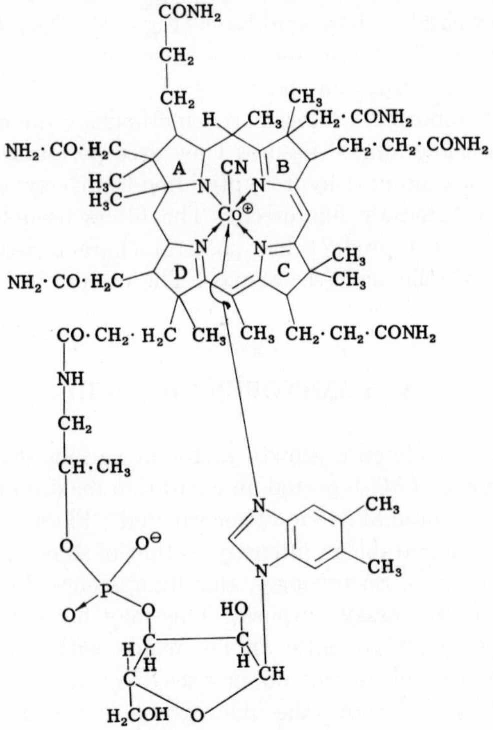

FIG. 25. Vitamin B_{12} (cyanocobalamin).

Barker and his associates[39, 90] have found vitamin B_{12} derivatives in bacteria which can function as actual coenzymes in specific enzyme reactions. Similar compounds were later found in animal tissues. Although some details of these vitamin B_{12} coenzymes remain to be elucidated, it is known that two organic bases are present in each coenzyme molecule. An adenine residue is present in all of them. However, the second base can be varied, adenine, benzimidazole, and dimethylbenzimidazole being found most frequently. The latter two compounds are highly fluorescent and can be extracted into organic solvents. These properties have been utilized to measure benzimidazole and 5,6-dimethylbenzimidazole in the coenzyme isolated from various natural sources.[91] In a typical assay, a sample con-

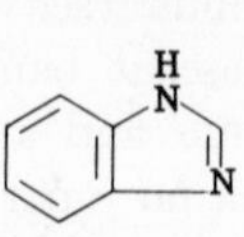

Benzimidazole 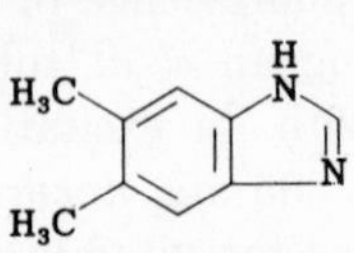   5,6-Dimethylbenzimidazole

TABLE X

FLUORESCENCE CHARACTERISTICS OF BENZIMIDAZOLE AND 5,6-DIMETHYLBENZIMIDAZOLE

Base	Excitation[a] maximum mμ	Fluorescence[a] maximum mμ
Benzimidazole	272	365
5,6-Dimethylbenzimidazole	285	380

[a] Uncorrected values obtained with the Aminco-Bowman spectrophotofluorometer (after Weissbach et al.[90]).

taining 0.17 μmole of benzimidazole-cobalamin coenzyme is dissolved in 0.6 ml. of 6 N HCl and heated in an evacuated and sealed tube for 18 hours at 150°. The solution is transferred to a 10-ml. beaker and dried under a stream of warm air. Some water is added and the solution is again evaporated to remove HCl. The residue is dissolved in 1 ml. of water, made alkaline with a drop of 10 N KOH, and extracted into 12 ml. of chloroform. The organic phase is transferred to a beaker and the chloroform is removed by a stream of warm air. The residue is dissolved in 0.1 N acetic acid and fluorescence is assayed in a fluorescence spectrometer at the wavelengths shown in Table X. Blanks and standards should be carried through the entire extraction procedure.

It may be of further interest that both benzimidazole bases fluoresce blue on paper chromatograms.

XII. TOCOPHEROLS (VITAMIN E)

The three major substances with vitamin E activity which have been isolated from natural sources are α-, β-, and γ- tocopherol. These may be considered as methyl derivatives of the parent substance, tocol.

α-tocopherol — 5,7,8-trimethyl tocol
β-tocopherol —   5,8-dimethyl tocol
γ-tocopherol —   7,8-dimethyl tocol

Although it is now known that the tocopherols fluoresce in the native state,[1] this occurs in the ultraviolet region and could not be detected until the recent advent of fluorescence spectrometers. Before this, Kofler[92, 93]

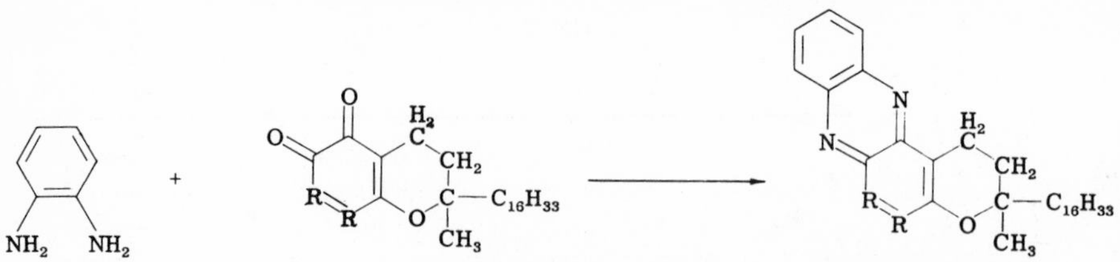

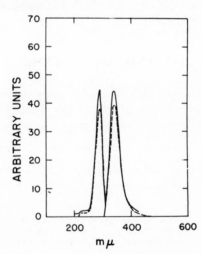

FIG. 26. Condensation of o-phenylenediamine with the oxidation product of tocopherol to yield a fluorescent phenazine derivative (after Kofler[93]).

devised a procedure for converting the tocopherols to visible fluorophors and applied it to the assay of the vitamins. The tocopherols are oxidized with nitric acid in alcoholic solution to yield, first, the orthoquinones (tocopherol red). Following solvent extraction, the oxidation products are treated with o-phenylenediamine in glacial acetic acid to yield the corresponding derivatives (Fig. 26). With α-tocopherol a single product is apparently formed with absorption maxima at 270 and 370 mμ and green fluorescence. With β- and γ-tocopherol a mixture of products are formed which can be distinguished from the α-tocopherol product. Since α-tocopherol has much higher biological activity than the other forms of the vitamin, this procedure offers the possibility of a chemical assay which is consistent with biological activity. Serum values (total) of 4 to 14 μg./ml. were found in man and 2–2.5 μg./ml. in rats. Further studies with this

FIG. 27. Excitation (295 mμ) and fluorescence (340 mμ) of α-tocopherol in ethanol (after Duggan[94]). —, α-tocopherol; – – –, α-tocopheryl acetate after reduction with lithium aluminum hydride.

procedure, including the use of newer instrumentation, would be desirable. The observation that the tocopherols fluoresce in their native state was made quite recently.[1] As shown in Fig. 27, d-α-tocopherol in ethanol fluoresces maximally at 340 mμ when excited at 295 mμ (uncorrected). Solutions of α-, β-, and γ-tocopherols in absolute ethanol yield, respectively, fluorescence intensities of 38.5, 44, and 45.5, the ratios corresponding almost exactly with ratios of the molecular extinctions at 295 mμ. All the tocopherols exhibit the same excitation and fluorescent spectra, but only the free tocopherols exhibit detectable fluorescence. α-Tocophenylquinone, the oxidized form of the vitamin, is nonfluorescent. Based on these observations, Duggan[94] developed the following procedure for measurement of free and conjugated tocopherols in blood and tissues: Purity of the reagents is essential for the assay. Matheson "hexane" is repeatedly washed with concentrated sulfuric acid until the acid layer is essentially colorless. It is then washed with dilute alkali and with water, dried, passed over a column of silica gel of sufficient size to retain the dark yellow band, and distilled. Absolute ethyl alcohol is distilled from potassium hydroxide. To prepare the lithium aluminum hydride reagent, 20 ml. of cold, freshly distilled ether is shaken with 200 to 300 mμ of finely powdered LiAlH$_4$ and filtered through Pyrex wool.

To a 2.0-ml. aliquot of plasma in a stoppered centrifuge tube are added 2.0 ml. water, 4.0 ml. of alcohol, and, after thorough mixing, 8.0 ml. of hexane. To a second aliquot is added a small volume of an alcoholic stock solution containing equimolar quantities of α-tocopherol and tocopheryl acetate (0.05 μmole constitutes a convenient internal standard for normal plasma). After thorough shaking to ensure equilibration of the added material with the sample, water, alcohol, and hexane are added as before. Thereafter, sample and internal standard are treated identically. The centrifuge tubes are securely stoppered, using a minimum quantity of silicone lubricant on the joint, agitated for 10 minutes on a mechanical shaker, and centrifuged at low speed. The individual components are then determined as follows.

A. Tocopherol

A 1.0-ml. aliquot of the clear organic phase is removed and diluted with 3 volumes of alcohol. Fluorescence is measured at 340 mμ with excitation at 295 mμ. A commercial fluorescence spectrometer was used in the original studies.

B. Tocopheryl Acetate

A 4.0-ml. aliquot of the organic phase is withdrawn and evaporated to dryness under nitrogen. A stream of nitrogen is then drawn through the apparently dry tube for several minutes more to ensure removal of the last traces of alcohol. The residue is redissolved in 4.0 ml. of hexane, cooled in ice, and treated with 1.0 ml. of the $LiAlH_4$ reagent. After intermittent shaking for 5 minutes, 1.0 ml. of cold alcohol is added carefully, then 1 volume of 0.1 N sulfuric acid. The tubes are stoppered and shaken as before; then an aliquot of the clear organic phase is removed and diluted with 3 volumes of alcohol, and its fluorescence measured as before.

A standard solution, containing an appropriate volume of the tocopherol and tocopheryl acetate alcoholic stock solution, diluted to 2.0 ml. with water may be carried through each of the foregoing procedures to serve as checks upon the internal standards. Samples of water (instead of plasma) must be carried through both procedures as reagent blanks.

The fluorescence of sample B will be equivalent to the total of free and esterified tocopherol originally present in the sample, while A will be due to free tocopherol only.

FOR FREE TOCOPHEROL

$$[X_a/(S_a - X_a)] \times (\mu M\ S/2.0) = \mu\text{moles free tocopherol/ml. plasma} \quad (1)$$

where X_a and S_a are the fluorescence intensities, corrected for blank, of the untreated tissue sample and internal standard, respectively, and $\mu M\ S$ is the number of micromoles of free tocopherol added to the internal standard.

FOR TOTAL TOCOPHEROL (TO FIND ACETATE BY DIFFERENCE)

$$[X_B/(S_B - X_B)] \times (\mu M\ S/2.0) = \mu\text{moles total tocopherols/ml. plasma} \quad (2)$$

where X_B and S_B are the respective fluorescence intensities of the $LiAlH_4$-treated tissue samples and internal standard, and $\mu M\ S$ is the total number of micromoles of tocopherol and tocopheryl acetate added to the internal standard.

The values of the $LiAlH_4$-treated and untreated aliquots of the external standards should be equal, respectively, to the denominators of (2) and (1), if the original external standard was equal to the added internal standard. These external standards tend to give slightly higher results, however, owing to the light absorption by the tissue sample at either the exciting or fluorescence wavelengths. Where a significant difference exists between results based upon internal and external standards, the former is to be preferred. The internal standard, furthermore, may be made to agree with the absolute or external standard by correcting for calculated percentage transmission over the fluorescent light path as described elsewhere (Chapter 4).[1]

For each compound, the fluorescence is proportional to concentration over a wide range of concentrations, extending through three orders of magnitude, from the lowest detectable concentration of approximately 0.01 μg./ml. through about 10 μg./ml. An essentially linear relationship still exists at levels as high as 40 μg./ml. Recovery of α-tocopheryl acetate from water, plasma, and homogenates of various tissues is essentially quantitative.

Plasma tocopherol levels of 12 normal human subjects were determined by both the fluorometric procedure and the modified Emmerie-Engel method,[95] triplicate determinations being made on each sample. The mean value for the fluorometric method was 10.8 ± 1.9 μg./ml., that for the latter method was 11.3 ± 1.9 μg./ml. Only negligible differences between values of normal plasma treated by spectrofluorometric procedures A and B were observed, indicating the presence of little or no esterified tocopherols.

As compared to the chemical methods extensively used for the estimation of tocopherols in biological materials, the fluorometric method affords several marked advantages and suffers from one relative disadvantage. The latter relates to the fact that the molar fluorescence intensities of the α-, β-, and γ-tocopherols vary with their respective molar extinction coefficients, thus posing a problem where precise determinations of mixtures of these compounds are to be made. Furthermore, the minor differences in absorption maxima of the pure compounds which have been reported have, for all practical purposes, no analogy in either excitation or fluorescence maxima, owing probably to the relatively poor resolution of the fluorescence maxima, thus rendering a differential analysis on this basis impossible. On the other hand, the fluorometric method offers several distinct advantages in terms of sensitivity, specificity, and the relative mildness of conditions employed. Based upon a fluorescence intensity of 10 times the lowest achievable blank value as an arbitrary lower limit for reproducibility, concentrations of 0.005 μg./ml. can be estimated with a fair degree of precision. Less stringent purification of solvents results in a several fold decrease in ultimate sensitivity, but still permits accurate determination of concentrations of tocopherol at least 1 order of magnitude lower than those assessable with the most sensitive colorimetric methods.

Vitamin A and other fat-soluble reducing substances which constitute the major source of interference in the commonly used iron-dipyridyl and phosphomolybdic acid methods[96] show no fluorescence similar to that of the tocopherols, thus obviating the necessity of hydrogenation or chromatographic separation. Larger quantities of vitamin A will interfere, however, by absorption of tocopherol fluorescence in the 320 to 330-mμ region, but this difficulty is readily circumvented through correction obtainable by either internal standards or with absorption data as previously described.[1] The phenolic estrogens are the only normal constituents of mammalian tissues which might conceivably interfere under the conditions of extraction and fluorescence measurement employed in this method.

XIII. VITAMIN K

The reduced forms of vitamins K$_1$ and K$_2$ are phenolic naphthalenes and as such would be expected to fluoresce. However, Duggan et al.[1] were unable to detect significant fluorescence in a variety of solvents including

Compound	R
Vitamin K$_1$	Phytyl
Vitamin K$_2$	Difarnesyl
Menadione	H

aqueous acid, alkali, and alcohol. On the other hand, reduced menadione, a synthetic compound which is structurally quite similar to vitamins K$_1$ and K$_2$, was found to be highly fluorescent. In 95% ethanol, concentrations well below a microgram per milliliter could be detected; the excitation maximum was at 335 mμ and the fluorescence maximum at 480 mμ (uncorrected). Just why vitamins K$_1$ and K$_2$ do not fluoresce is not clear. However, studies by Beyer and Kennison[97] and by others, indicate that vitamin K is readily destroyed by ultraviolet light. It may be that this represents another instance where photodecomposition occurs more readily than does fluorescence. Careful studies of the photochemical decomposition of the K vitamins are certainly warranted. It may be that, under appropriate conditions (absence of oxygen or otherwise), they may be stable to ultraviolet light and also fluorescent. Kofler[98] has reported a fluorometric procedure for menadione based on the condensation with o-phenylenediamine to yield a phenazine derivative in a manner similar to that described for tocopherol (see above). The procedure has been applied to the assay of menadione in plasma and urine in experimental

Ubiquinone (coenzyme Q)

studies. The natural K vitamins cannot condense in a similar manner due to the substitution of the phytyl group in the 3 position.

It is of further interest that ubiquinone, or coenzyme Q, is, in its reduced form, also a structure which should be capable of fluorescing. Thus far no studies on the fluorescence of ubiquinones have appeared.

REFERENCES

1. Duggan, D. E., Udenfriend, S. L., Bowman, R., and Brodie, B. B., *Arch. Biochem. Biophys.* **68**, 1 (1957).
2. Hagins, W. A., and Jennings, W. H., *Discussions Faraday Soc. No.* **27**, 180 (1959).
3. Sobotka, H., Kann, S., and Loewenstein, E., *J. Am. Chem. Soc.* **65**, 1959 (1943).
4. Sobotka, H., Kann, S., and Winternitz, W., *J. Biol. Chem.* **152**, 635 (1944).
5. Fujita, A., and Aoyama, M., *J. Biochem. (Tokyo)* **38**, 271 (1951).
6. De, N. K., *Indian J. Med. Research* **43**, 3 (1955).
7. Popper, H., *Physiol. Revs.* **24**, 205 (1944).
8. Blankenhorn, D. H., and Braunstein, H., *J. Clin. Invest.* **37**, 160 (1958).
9. "U. S. Pharmacopeia" 16th Revision, p. 909. Mack Publ. Co., Easton, Pennsylvania, 1960.
10. "Official Methods of Analysis," Assoc. Official Agr. Chemists, Washington, D. C., 1960, 9th ed., p. 655.
11. Burch, H. B., Bessey, O. A., Love, R. H., and Lowry, O. H., *J. Biol. Chem.* **198**, 477 (1952).
12. Burch, H. B., *in* "Methods in Enzymology" (S. P. Colowick and N. O. Kaplan, eds.) Vol. III, p. 946. Academic Press, New York, 1957.
13. Ohnesorge, W. E., and Rogers, L. B., *Anal. Chem.* **28**, 1017 (1956).
14. Yagi, K., Tabata, T., Kotaki, E., and Arakawa, T., *Vitamins (Kyoto)* **9**, 391 (1955).
15. Bessey, O. A., Lowry, O. H., and Love, R. H., *J. Biol. Chem.* **180**, 755 (1949).
16. Weber, G., *Biochem. J.* **47**, 114 (1950).
17. Sakai, K., *Nagoya J. Med. Sci.* **18**, 245 (1956).
18a. Cerletti, P., *Anal. Chim. Acta* **20**, 243 (1959).
18b. Yagi, K., and Okuda, J., *Chem. & Pharm. Bull. (Tokyo)* **6**, 659 (1958).
19. Najjar, V. A., *J. Biol. Chem.* **141**, 355 (1941).
20. Von Euler, H., and Adler, E., *Z. physiol. Chem. Hoppe-Seyler's* **223**, 105 (1934).
21. Vivanco, F. B., *Naturwissenschaften* **23**, 306 (1935).
22. Ferrebee, J. W., *J. Clin. Invest.* **19**, 251 (1940).
23. Burch, H. B., Bessey, O. A., and Lowry, O. H., *J. Biol. Chem.* **175**, 457 (1948).
24. Burch, H. B. *in* "Methods in Enzymology" (S. P. Colowick and N. O. Kaplan, eds.), Vol. III, p. 960. Academic Press, New York, 1957.
25. Warburg, O., and Christian, W., *Biochem. Z.* **298**, 150 (1938).
26. Kondo, H., *J. Agr. Chem. Soc. Japan* **30**, 690 (1956).
27. Warburg, O., and Christian, W., *Biochem. Z.* **266**, 377 (1933).
28. Kuhn, R., Wagner-Jauregg, T., and Kattschmidt, M., *Ber.* **67**, 1452 (1934).
29. Yagi, K., *J. Biochem. (Tokyo)* **43**, 635 (1956).
30. Yagi, K., *Bull. soc. chim. France* p. 1543 (1957).
31. Yagi, K., Tabata, T., Kotaki, E., and Arakawa, T., *Vitamins (Kyoto)* **8**, 61 (1955).
32. Ishiguro, I., *J. Vitaminol. (Osaka)* **2**, 264 (1956).
33. Yagi, K., Okuda, J., and Matsuoka, Y., *Nature* **175**, 555 (1955).
34. Cerletti, P., and Siliprandi, N., *Biochem. J.* **61**, 324 (1955).

35. Brown, J. A., and Marsh, M. M., *Anal. Chem.* **25**, 1865 (1953).
36. "Official Methods of Analysis," Assoc. Official Agr. Chemists, Washington, D. C., 1960, 9th ed., p. 658.
37. "U. S. Pharmacopeia" 16th Revision, p. 907. Mack Publ. Co., Easton, Pennsylvania, 1960.
38. Theorell, H., and Bonnichsen, F. K., *Acta. Chem. Scand.* **5**, 1105 (1951).
39. Duysens, L. N. M., and Amesz, J., *Biochim. et Biophys. Acta.* **24**, 19 (1957).
40. Chance, B., and Baltscheffsky, H., *J. Biol. Chem.* **233**, 736 (1958).
41. Winer, A. D., and Schwert, G. W., *J. Biol. Chem.* **234**, 1155 (1959).
42. Velick, S., *J. Biol. Chem.* **233**, 1455 (1958).
43. Weber, G., *Nature* **180**, 1409 (1957).
44. Weber, G., *J. chim. phys.* **55**, 878 (1958).
45. Lowry, O. H., Roberts, N. R., and Kapphahn, J. I., *J. Biol. Chem.* **224**, 1047 (1957).
46. Greengard, P., *Photoelec. Spectrometry Group Bull. No.* **11**, 292 (1958).
47. Swallow, A. J., *Biochem. J.* **61**, 197 (1955).
48. Bacon, K., *Arch. Biochem. Biophys.* **60**, 505 (1956).
49. Kaplan, N. O., Colowick, S. P., and Barnes, C. C., *J. Biol. Chem.* **191**, 461 (1951).
50. Ciotti, M. M., and Kaplan, N. O., *in* "Methods in Enzymology" (S. P. Colowick and N. O. Kaplan, eds.), Vol. III, p. 891. Academic Press, New York, 1957.
51. Huff, J. W., *J. Biol. Chem.* **167**, 151 (1947).
52. Huff, J. W., and Perlzweig, W. A., *J. Biol. Chem.* **167**, 157 (1947).
53. Levitas, N., Robinson, J., Rosen, F., Huff, J. W., and Perlzweig, W. A., *J. Biol. Chem.* **167**, 169 (1947).
54. Asami, T., *J. Vitaminol. (Osaka)* **3**, 189 (1957).
55. Carpenter, K. J., and Kodicek, E., *Biochem. J.* **46**, 421 (1950).
56. Burch, H. B., Storrick, C. A., Bicknell, L., Kung, H. C., Alejo, L. G., Everhart, W. A., Lowry, O. H., King, C. C., and Bessey, O. A., *J. Biol. Chem.* **212**, 897 (1955).
57. Greengard, P., Brink, F., Jr., and Colowick, S. P., *J. Cellular Comp. Physiol.* **44**, 395 (1954).
57a. Lowry, O. H., Passonneau, J. P., Schulz, D. W., and Rock, M. K. *J. Biol. Chem.* **236**, 2746 (1961).
58. Scudi, J. V., *Science* **103**, 567 (1946).
59. Friedemann, T. E., and Frazier, E. I., *Science* **102**, 97 (1945).
60. Bandier, E., and Hald, J., *Biochem. J.* **33**, 264 (1939).
61. Najjar, V. A., *Bull. Johns Hopkins Hosp.* **74**, 392 (1944).
62. Kodicek, E., *Analyst* **72**, 385 (1947).
63. Le Clerc, A. M., and Douzou, P., *Compt. rend. Acad. Sci.* **236**, 2006 (1953).
64. Huff, J. W., and Perlzweig, W. A., *J. Biol. Chem.* **155**, 345 (1944).
65. Coursin, D. B., and Brown, V. C., *Proc. Soc. Exptl. Biol. Med.* **98**, 315 (1958).
66. Reddy, S. K., Reynolds, M. S., and Price, J. M., *J. Biol. Chem.* **233**, 691 (1958).
67. Price, J. M., *J. Biol. Chem.* **211**, 117 (1954).
68. Heyl, D., *J. Am. Chem. Soc.* **70**, 3434 (1948).
69. Fujita, A., Matsuura, K., and Fujino, K., *J. Vitaminol. (Osaka)* **1**, 267 (1955).
70. Fujita, A., Fujita, D., and Fujino, K., *J. Vitaminol. (Osaka)* **1**, 275 (1955).
71. Fujita, A., Fujita, D., and Fujino, K., *J. Vitaminol. (Osaka)* **1**, 279 (1955).
72. Fujita, A., and Fujino, K., *J. Vitaminol. (Osaka)* **1**, 290 (1955).
73. Melnick, D., Hochberg, M., Himes, H. W., and Oser, B. L., *J. Biol. Chem.* **160**, 1 (1945).

74a. Fasella, P., Lis, H., Siliprandi, N., and Baglioni, C., *Biochim. et Biophys. Acta* **23**, 417 (1957).

74b. Bonavita, V., *Arch. Biochem. Biophys.* **88**, 366 (1960).

75. Archibald, R. M., *J. Biol. Chem.* **158**, 347 (1945).

76. Jones, S. W., Wilkie, J. B., Morris, W. W., Jr., and Friedman, L., abstract presented at the 138th Meeting *Am. Chem. Soc., New York, September, 1960* p. 60C.

77. Simpson, D. M., *Analyst* **72**, 382 (1947).

78. Hopkins, F. G., *Phil. Trans. Roy. Soc. London Ser. B* **186**, 661 (1895).

79. Crowe, M. O., and Walker, A., *Phys. Rev.* **86**, 817 (1952).

80. Decker, P., *Z. physiol. Chem. Hoppe-Seyler's* **274**, 223 (1942).

81. Koschara, W. S. S., *Z. physiol. Chem. Hoppe-Seyler's* **240**, 127 (1936).

82. Jacobson, W., and Simpson, D. M., *Biochem. J.* **40**, 3 (1946).

83. Jacobson, W., and Simpson, D. M., *Biochem. J.* **40**, 9 (1946).

84. Pirie, A., and Simpson, D. M., *Biochem. J.* **40**, 14 (1946).

85. Kalckar, H. M., and Klenow, H., *J. Biol. Chem.* **172**, 349 (1948).

86. Lowry, O. H., Bessey, O. A., and Crawford, E. J., *J. Biol. Chem.* **180**, 389 (1949).

87. Schou, M. A., *Arch. Biochem.* **28**, 10 (1950).

88. Allfrey, V., Teply, L. J., Geffen, C., and King, C. G., *J. Biol. Chem.* **178**, 465 (1949).

89. Barker, H. A., Weissbach, H., and Smyth, R. D., *Proc. Natl. Acad. Sci. U. S.* **44**, 1093 (1958).

90. Weissbach, H., Toohey, J. I., and Barker, H. A., *Proc. Natl. Acad. Sci. U.S.* **45**, 521 (1959).

91. Barker, H. A., Smyth, R. D., Weissbach, H., Toohey, J. I., Ladd, J. N., and Volcani, B. E., *J. Biol. Chem.* **235**, 480 (1960).

92. Kofler, M., *Helv. Chim. Acta.* **25**, 1469 (1942); **26**, 2166 (1943).

93. Kofler, M., *Helv. Chim. Acta.* **28**, 26 (1945); **30**, 1053 (1947).

94. Duggan, D. E., *Arch. Biochem. Biophys.* **84**, 116 (1959).

95. Quaife, M. L., and Harris, P. L., *J. Biol. Chem.* **156**, 499 (1944).

96. Rosenkrantz, H., *J. Biol. Chem.* **224**, 165 (1957).

97. Beyer, R. E., and Kennison, B. D., *Arch. Biochem. Biophys.* **84**, 63 (1959).

98. Kofler, M., *Helv. Chim. Acta* **28**, 702 (1945).

8

METABOLITES, GENERAL

I. GENERAL REMARKS

THIS section is devoted to chemical methods for the estimation of metabolites other than those discussed in the previous chapters. It includes some observations and applications concerning carbohydrates, aldehydes, ketones, acids, purines, nucleotides, deoxyribonucleic acid, and porphyrins. These are not intended to be all inclusive but rather to serve as examples of biological assays which are of importance in themselves and also as models for similar assays of related metabolites.

II. CARBOHYDRATES

Carbohydrates condense with a variety of reagents to yield colored derivatives. With each reagent the extent of the reaction and the nature of the color depends on whether the substance is a hexose or a pentose and whether it is an aldose or a ketose. Some of the colored derivatives formed in this way are also fluorescent and can be assayed fluorometrically at concentrations below those detectable by colorimetry.

A. Detection on Paper Chromatograms

Phloroglucinol yields green fluorescence with certain ketoses but does not react with aldoses and uronic acids.[1] The reaction is used on paper chromatograms. Chargaff et al.[2] utilized another reagent, m-phenylenediamine, for the identification of reducing sugars on chromatograms by fluorescence. Chromatograms are developed according to the procedure described by Partridge[3] and are then dried in air or for a short time at 105°. They are then sprayed with a 0.2 M solution of m-phenylenediamine dihydrochloride in 76% ethanol, and heated for 5 minutes at 105°. Less than 10 μg. of the individual sugars yield well-defined fluorescent spots when viewed under a quartz lamp (360-mμ mercury line). The most intense fluorescence is observed with the pentoses, arabinose, xylose, and ribose, which yield orange-yellow fluorescence. Glucose, fructose, galactose, mannose, rhamnose, fucose, and sorbose yield yellow fluorescent derivatives. Little fluorescence is given with ascorbic acid or the poly-hydroxy alcohols. The intensity of fluorescence is also dependent upon the solvent systems used for development of the chromatograms. In situations where separations are adequate, it should be possible to quantify the fluorescence measurements by using chromatogram adapters with fluorom-eter devices (Chapter 3).

B. Blood Sugar; Ultramicro Method

A fluorometric assay which should have wide applications in clinical investigations is the microestimation of blood sugar devised by Momose and Ohkura.[4] These investigators observed that 5-hydroxy-1-tetralone condensed with hexoses, or polysaccharides containing hexoses, in strong acid to yield highly fluorescent derivatives. They were able to isolate the fluorophor formed with glucose and to identify it as benzonaphthene-dione.[5] The series of reactions which may take place are shown in Fig. 1. The 5-hydroxytetralone first condenses with the aldehyde group of glucose.

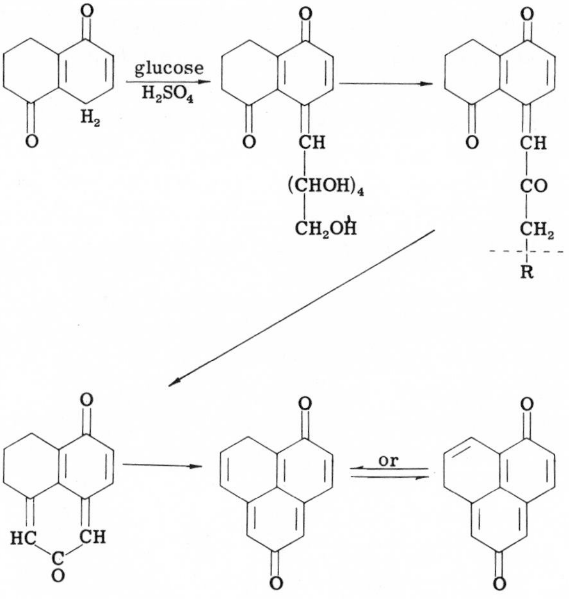

FIG. 1. Condensation of 5-hydroxyl-1-tetralone with glucose. Dehydration, cleavage, and condensation with the carbonyl group of the tetralone yield an intermediate product. This, upon further oxidation, yields benzonaphthenedione which can exist in the two resonant forms shown in the figure (after Momose and Ohkura[5]).

Dehydration occurs leaving an active methylene group in the 3 position of the sugar which combines with the carbonyl group of the tetralone. The elimination of three carbons of the sugar may occur either before or after the condensation. The intermediate, dihydrobenzonaphthenone, is readily oxidized to the final product, benzonaphthenedione. The same fluorescent compound is also obtained from the interaction of 5-hydroxytetralone with mannose, galactose, and fructose. Glyceraldehyde also yields this product. However, the extent of the reaction is not as great with the latter compound under the conditions devised for glucose.

A detailed study of the reaction of 5-hydroxytetralone with glucose has been presented by Momose and Ohkura.[6] In this they present data for the absorption and fluorescence characteristics of the fluorophor (Fig. 2). It should be pointed out that the absorption maximum is at 420 mμ whereas Momose and Ohkura utilize the 365-mμ mercury line for excitation. Further sensitivity or specificity might be obtained by exciting at the absorption maximum. The procedure as described yields fluorescence which is proportional to concentration up to about 40 μg./ml. Less than 1 μg./ml. can be determined.

The procedure for blood assay is carried out in the following manner.

The blood sample is measured with a micropipet and 0.02 ml. is introduced into 0.50 ml. of water in a microcentrifuge tube, with rinsing. After hemolysis has occurred, 0.5 ml. of 5% trichloroacetic acid is added and the tube is set aside for 5 minutes. At the end of this period 0.98 ml. of water is added to adjust the volume to 2.00 ml. and, after mixing, the tube is centrifuged. Two 0.5-ml. aliquots of the clear supernatant solution are transferred to glass-stoppered test tubes containing 2 ml. of cold tetralone reagent (100 mg. of 5-hydroxy-1-tetralone dissolved in 400 ml. of concen-

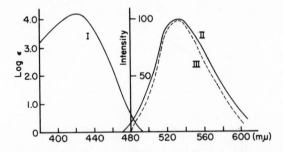

FIG. 2. Spectral characteristics of the glucose condensation product with 5-hydroxy-1-tetralone.

Curve I, absorption spectrum of benzonaphthenedione; curve II, fluorescence spectrum of benzonaphthenedione; curve III, fluorescence spectrum of the reaction mixture of 5-hydroxy-1-tetralone and glucose according to the procedure of Momose and Ohkura.[6]

TABLE I

COMPARISON OF BLOOD SUGAR VALUES OBTAINED WITH THE FLUOROMETRIC METHOD
AND THE HAGEDORN AND JENSEN METHOD[a,b]

Blood sample	1	2	3	4	5	6	7	8
Fluorometric assay	89	88	102	96	81	89	103	75
Hagedorn and Jensen procedure	91	89	100	98	82	92	102	73

[a] See reference 7.
[b] After Momose and Ohkura.[4]

trated sulfuric acid, stored in the refrigerator). The tubes are kept in an
ice bath during addition of the blood extracts. Following this, the tubes
are heated in a boiling water bath for 40 minutes, again cooled in ice water,
and diluted with 7.5 ml. of water. After mixing, fluorescence is measured
utilizing the excitation characteristics described previously. Standards are
prepared by substituting 0.5 ml. of glucose solution (10 μg./ml.) for the
original 0.5-ml. volume and omitting the blood sample. For reagent blanks
merely omit the blood.

Glucose, added to blood, is recovered quantitatively and the values obtained with this method correspond almost exactly with those obtained by
the Hagedorn and Jensen[7] procedure (Table I). It should be pointed out
that the final volume of the analysis is 10 ml. With present instruments
1-ml. volumes are used routinely. If only a single aliquot is utilized from
each blood sample, it should be possible to carry out this assay on less than
0.001 ml. of blood. Further modifications as to excitation wavelengths and
instrumentation can increase sensitivity even further. Undoubtedly, this
is the most sensitive procedure for blood sugar yet devised.

C. 2-Deoxysugars

2-Deoxysugars can be assayed by a method originally suggested by
Velluz et al.[8] Aldehydes of the type R—CH$_2$—CHO yield highly fluorescent products with 3,5-diaminobenzoic acid in concentrated mineral acids
at elevated temperatures. Applications of this method to the assay of
acetaldehyde will be described later in this section. Deoxyribose can also
be assayed in mixtures of other pentoses and hexoses.[9, 10] Utilization of the
reaction of 3,5-diaminobenzoic acid with deoxyribose in a microfluorometric procedure for deoxyribonucleic acid is also described later in this
section.

2-Deoxy-D-glucose is an analog of glucose which is widely used in studies

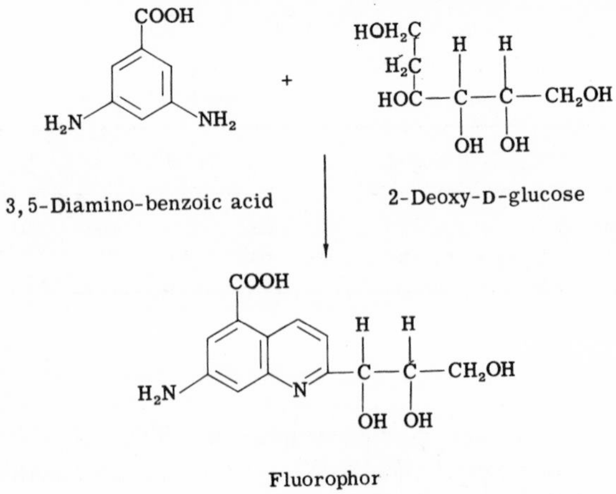

FIG. 3. Condensation of 2-deoxy-D-glucose with 3,5-diaminobenzoic acid to yield the fluorophor, 2-(1-glycerol), 5-carboxy-7-amino quinoline.

on glucose metabolism and transport. Blecher[11] has devised an extremely sensitive and specific method for assay of 2-deoxy-D-glucose based on the reaction with 3,5-diaminobenzoic acid. In this case the fluorophor is presumed to be 2-(1-glycerol), 5-carboxy-7-amino quinoline (Fig. 3).

Solutions containing from 0.02 to 1.0 μmole of 2-deoxyglucose in volumes up to 3.0 ml. are adjusted to 3-ml. volume with water. Following this, 3.0 ml. of a 0.01 M solution of 3,5-diaminobenzoic acid in 5.0 M phosphoric acid are added and the mixtures are heated in a boiling water bath for 15 minutes. After this they are permitted to stand at room temperature for 20 minutes (preferably in the dark to avoid photodecomposition). Any dilutions that may be necessary, for instrumental reasons or otherwise, may be made at this point utilizing 2.5 M phosphoric acid as the diluent. With a fluorescence spectrometer the quinoline fluorophor shows maximal excitation at 410 mμ and maximal fluorescence at about 495 mμ (uncorrected instrumental readings). The procedure, as described by Blecher, yields significant and proportional measurements in the fluorometer from 0.003 to 0.125 μmole. The assay can be made even more sensitive by using smaller volumes and more sensitive instrumentation than described by Blecher.[11] The method is highly precise and is highly specific. Glucose and related hexoses give no fluorescence. This is an important feature of the method since 2-deoxy-D-glucose is usually studied in solutions containing significant amounts of glucose. Only 2-deoxyribose yields fluorescence with this procedure, amounting to about 10% of the fluorescence intensity given by equivalent concentrations of 2-deoxy-D-glucose.

D. Reducing Oligosaccharides

A further application of fluorescence to carbohydrate assay is the procedure for reducing oligosaccharides reported by Wadman et al.[12] Oligosaccharides containing four or more monosaccharide units are difficult to separate by paper chromatography since they exhibit low R_f values in most solvents. Bayly and Bourne[13] solved this problem by treating the sugars with benzylamine to form the corresponding N-benzylglycosylamines. Wadman et al.[12] studied a number of fluorescent amines and found that N-(1-naphthyl)-ethylenediamine reacted with oligosaccharides to yield highly fluorescent glycosylamine derivatives which migrated with large R_f values on paper chromatograms. This reagent has been used to determine oligosaccharides in complex mixtures.

A 0.1-ml. aliquot of the mixture to be investigated, containing 1 mg. of mixed oligosaccharides, is added to an equal volume of a 10% solution of N-(1-naphthyl)-ethylenediamine dihydrochloride in triethylamine-ethanol-water (5:4:1). Five microliter aliquots of the mixture are applied to Whatman No. 1 paper and, before developing, the paper is heated at 100° for 30 minutes. The reaction between the amine and the sugars takes place on the paper during this period of heating. Following this the chromatogram is developed (descending) with n-butanol-ethanol-water-ammonia (40:12:16:1) for about 18 hours. After drying, the chromatogram is examined under an ultraviolet lamp (360 mμ). The unreacted amine is found as a fluorescent spot immediately behind the solvent front and the sugar derivatives are also seen as fluorescent spots arranged in order of their increasing size (Table II). The sugars may be estimated directly on the paper by using an adaptor for fluorescence assay on paper or in solution after elu-

TABLE II

MOVEMENT OF FLUORESCENT OLIGOSACCHARIDE AMINE DERIVATIVES ON A
PAPER CHROMATOGRAM

Compound	Distance	Ra[a]
Unreacted amine	39.4	1.00
Glucose derivative	35.2	0.89
Disaccharide derivative	26.0	0.66
Trisaccharide derivative	16.0	0.41
Tetrasaccharide derivative	10.2	0.26
Pentasaccharide derivative	7.5	0.19
Hexasaccharide derivative	3.4	0.09

[a] Ra values are distances traveled relative to the unreacted N-(1-naphthyl)-ethylenediamine (after Wadman et al.[12]).

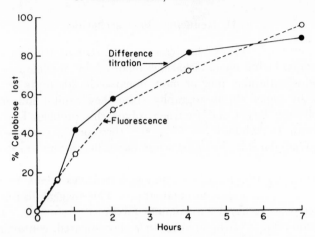

Fɪɢ. 4. Measurement of the hydrolysis of cellobiose by fluorometry and differential titration (after Wadman *et al.*[12]).

tion. The most precise determinations are made by cutting out the individual spots, eluting the fluorescent material with 5 ml. of 1% trisodium phosphate, and determining fluorescence in a fluorometer. Known amounts of the sugars should be run on the same sheet of paper for calibration standards. Other portions of the paper are used for blanks. The glycosylamines are maximally excited at about 355 mμ. The only information available as to the fluorescence maximum is that it is above 400 mμ. As little as 0.05 μg. of glycosylamine derivative can be detected. Wadman *et al.*[12] compared the appearance of acid and enzyme hydrolysis products from various polysaccharides using the fluorescent procedure and less direct macroprocedures. Excellent correlation was found in all instances. Measurement of the hydrolysis of cellobiose by emulsin, as followed by determining fluorometrically the disappearance of the disaccharide, is shown in Fig. 4. Comparison with differential titration of reducing activity is plotted for comparison

It should also be possible to apply a similar technique to analyse mixtures of monosaccharides. However, this apparently will require a different amine reagent.

III. ALDEHYDES AND KETONES

A. Acetaldehyde

Velluz *et al.*[14] have described an analytical method for acetaldehyde based on the reaction with 3,5-diaminobenzoic acid to yield the fluorescent indole derivative, 2-methyl, 5-carboxy, 7-amino quinoline. No applications to tissue extracts were given.

B. Succinic Semialdehyde

Salvador and Albers[15] used the same type of reaction to measure the γ-aminobutyric acid metabolite, succinic semialdehyde. Equal volumes of sample and 0.25 M 3,5-diaminobenzoic acid are mixed and heated for 1 hour in a water bath at 60°. An aliquot of the mixture is then diluted with water and the fluorescence is measured. The excitation and fluorescence characteristics of the reagent and of the succinic semialdehyde derivative are shown in Fig. 5. This method has been used in the assay of glutamic-γ-aminobutyric acid transaminase (see Chapter 9).

Camber[16] introduced salicylol hydrazide as a fluorometric reagent for carbonyl containing compounds. Chen[17a] prepared a large number of salicylol hydrazones and studied their fluorescence characteristics and other physical properties. He found that most of the hydrazones had the same excitation maximum (350 mμ) as the unreacted salicylol hydrazide and exhibited the same fluorescence maxima (425 mμ). However, the unreacted hydrazide can be separated from the hydrazones by solvent extraction procedures. Butyraldehyde, acetophenone, and various ketosteroids give highly fluorescent derivatives of this type. The acetaldehyde and benzaldehyde derivatives do not show appreciable fluorescence, indicating some degree of specificity of the reaction. Sawicki and Stanley[17b] have used

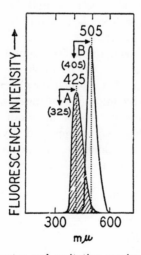

Fig. 5. Fluorescence spectra and excitation maxima (parentheses) of 3,5-diaminobenzoic acid (curve A) and of the condensation product with succinic semialdehyde (curve B) (after Salvador and Albers[15]).

the reagent in a spot test for glyoxal and other aldehydes on paper. Salicylol hydrazide has not yet been applied to a specific analytical problem in biology.

C. Acetol, Diacetyl, and Acetylmethylcarbinol

The compound acetol, CH_3COCH_2OH, is an intermediate in the metabolism of acetone, 1,2-propanediol, lactic acid, and pyruvic acid, (in some organisms), so that methods for its assay are of interest. Baudisch and Deuel[18] first described the condensation of o-aminobenzaldehyde with acetol to yield the highly fluorescent 2-methyl, 3-hydroxyquinoline (3-hydroxyquinaldine) (Fig. 6). Fluorometric methods for the determination of acetol in biological material have been developed by Forist and Speck[19] and by Werle and Stiess.[20] More recently, Huggins and Miller[21] reported the following procedure for acetol assay applicable to enzyme studies. Duplicate aliquots containing 1–25 μg. of acetol are transferred to test tubes graduated at the 10-ml. mark. The samples are diluted with water to 4 ml. and 0.5 ml. of ethylenediamine tetraacetate are added. To one tube is added 1 ml. of o-aminobenzaldehyde (0.02%) and to the other 1 ml. of water (blank). After mixing, 1.25 ml. of 0.52 N sodium hydroxide are added and the tubes are shaken on a mechanical shaker for 1 hour at room temperature. Following this, the tubes are acidified with 1.5 ml. of 1 N hydrochloric acid. After mixing, an excess of solid sodium bicarbonate is added to bring the pH to about 8. The tubes are then diluted to the 10-ml. mark with water and fluorescence is measured. The 365-mμ mercury line can be used for excitation; emission is at about 400 to 440 mμ. The exact excitation and fluorescence spectra of the product have not been reported.

A large number of related compounds were found to give no fluorescence under these conditions. When heated in alkali, glucose apparently yields acetol and therefore interferes.[20] A method for glucose assay based on conversion to acetol followed by fluorometric measurement may be practical. However, glucose yields no acetol in alkali at room temperature and does not interfere in the assay as reported. Pyruvic acid, dihydroxyacetone, glyceraldehyde, glycol aldehyde, and acetaldehyde, when present in 10 times the concentration of acetol, interfere. In what way they inter-

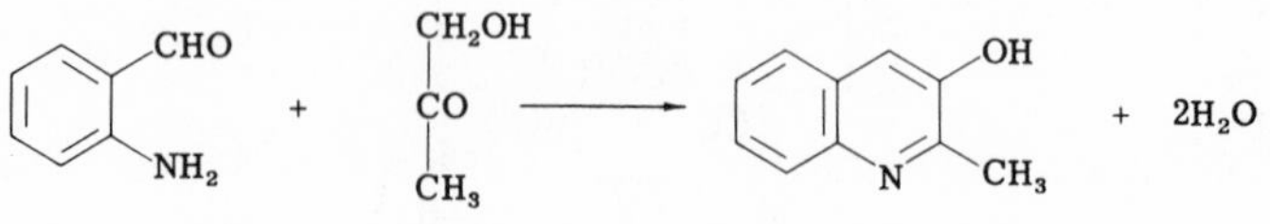

FIG. 6. The condensation of o-aminobenzaldehyde with acetol to yield the fluorophor, 3-hydroxyquinaldine (after Baudisch and Deuel[18]).

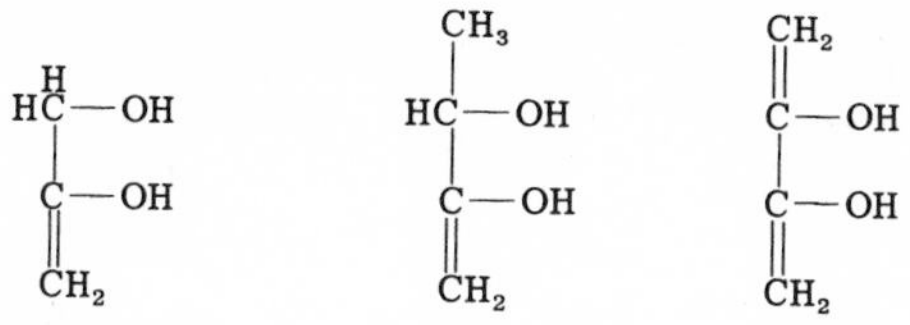

$$
\begin{array}{ccc}
& CH_3 & CH_2 \\
\overset{H}{H}\overset{|}{C}\!-\!OH & H\overset{|}{C}\!-\!OH & \overset{\|}{C}\!-\!OH \\
\overset{|}{C}\!-\!OH & \overset{|}{C}\!-\!OH & \overset{\|}{C}\!-\!OH \\
\overset{|}{C}H_2 & \overset{|}{C}H_2 & \overset{|}{C}H_2 \\
\text{Acetol} & \text{Acetylmethyl} & \text{Diacetyl} \\
& \text{carbinol} &
\end{array}
$$

Fig. 7. Comparison of the structures of acetol, acetylmethylcarbinol, and diacetyl in their enol forms.

fere is not stated. Diacetyl and acetylmethylcarbinol yield fluorescence equivalent to acetol on a molar basis. The structures of these three compounds are, of course, quite similar particularly when they are in their enol forms in alkaline solution (Fig. 7). Huggins and Miller[21] found that the condensation product with acetol did not fluoresce at pH 2.6 whereas those formed with diacetyl and acetylmethylcarbinol remained highly fluorescent even at the acid pH. Reaction with the latter compound suggests that o-aminobenzaldehyde may prove useful analytically in the field of valine, leucine, and isoleucine metabolism. Other intermediates in this area such as acetolactic acid, acetolhydroxybutyric acid, and acetylethylcarbinol are also structurally related to acetol. A procedure for thymine assay involving prior oxidation to acetol followed by fluorometric assay of the latter is presented later in this chapter.

The antioxidant, diacetyl ($CH_3CO \cdot COCH_3$), a relatively simple molecule, possesses native fluorescence both in the gas phase and in solution in water and organic solvents. Although the fluorescence efficiency is quite low, about 0.15, when excited at 405 or 436 mμ,[22, 23] it is of interest that the molecule fluoresces at all. Furthermore, the fluorescence is markedly quenched by oxygen. The small amount of green fluorescence which may be observed in acetone solutions is generally due to traces of diacetyl. The latter is apparently one of the products of the photodecomposition of acetone.

IV. POLYCARBOXYLIC ACIDS

A. General

In highly concentrated sulfuric acid, polycarboxylic acids react with resorcinal to yield fluorescent derivatives. The chemical mechanisms of many of these reactions are not yet fully understood. However, many of the fluorescent products have been identified.[24] Polycarboxylic acids with

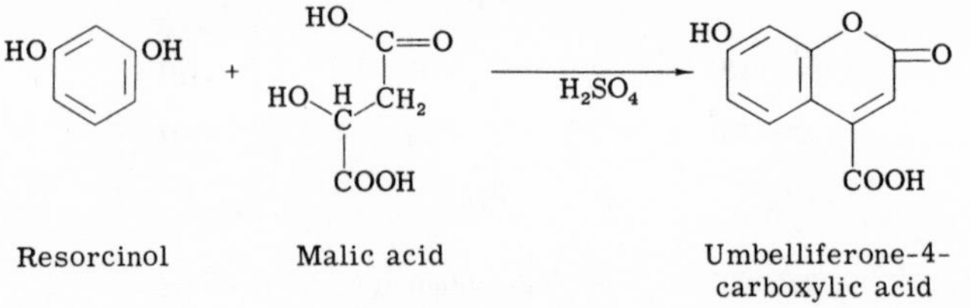

Resorcinol   Malic acid   Umbelliferone-4-
                                         carboxylic acid

FIG. 8. Condensation of malic acid with resorcinol to yield the fluorophor, umbelli-
ferone-4-carboxylic acid.

no other functional groups react to produce compounds similar to fluorescein. Those acids which are unsaturated or which contain hydroxyl groups
produce umbelliferone derivatives. The reaction of resorcinol with malic
acid is shown in Fig. 8. Keto acids also react with resorcinol to form fluorescent products. The composition of these are not known. The various polycarboxylic acid derivatives have different fluorescent characteristics which
have not yet been studied with fluorescence spectrometry. The products
formed with the Kreb's cycle intermediates all fluoresce in the visible
region when dissolved in alkali. The colors and intensities of fluorescence,
obtained with the exciting light at 436 mμ and a single cutoff filter, are
shown in Table III. These are obviously not the optimal conditions for
fluorescence assay for all the compounds.

Frohman and Orten[24] applied this procedure to determine the tricar-

TABLE III

COLOR AND INTENSITY OF FLUORESCENCE OF RESORCINOL-POLYCARBOXYLIC
ACID DERIVATIVES[a]

Acid	Color of fluorescence	Relative fluorescence
Fumaric	Blue-violet	24
α-Ketoglutaric	Blue-green	35
Oxalacetic	Blue-green	12
Succinic	Yellow-green	20
Aconitic	Light-blue	75
Malic	Blue-violet	22
Isocitric	Light-blue	58
Citric	Sky-blue	89

[a] In each case 40 μg. of acid was used and the final volume was 5 ml. A solution of
quinine, 1 μg./5 ml., was set to a reading of 39. A blue filter (405 mμ maximum) was used
for excitation and a filter cutting off below 475 mμ was used in front of the detector
(after Frohman and Orten[24]).

boxylic-acid cycle compounds following their chromatographic separation. In their procedure silica gel is used as the adsorbent and a mixture of tertiary amyl alcohol and chloroform as the eluent. To each fraction is added 0.1 ml. of 0.4% resorcinol in ethanol. The samples are then evaporated to dryness. A boiling salt water bath (108°) is required to vaporize the tertiary amyl alcohol. Each tube is aspirated for a minute or two to remove the last traces of the alcohol vapor. To the dry tubes are added 0.1 ml. of concentrated sulfuric acid and they are heated for 20 minutes in a boiling water bath. When the tubes are cool, 5 ml. of 1.5 N NaOH are added and they are mixed and permitted to come to room temperature. The various acids are recognized from the characteristic color of the fluorescence and the elution volume. Standard curves are prepared for each acid by carrying known amounts through the entire evaporation and condensation procedure. Reagent blanks are prepared in a similar manner. All of the acids can be recovered quantitatively except for pyruvic and lactic acids which are lost during the evaporation. Perhaps the addition of ammonia or an organic base to the solvent before evaporation would prevent losses of these two acids.

An interesting property of the resorcinol products is that they can all be extracted from acid solution into organic solvents. Frohman and Orten[24] utilized this to concentrate each derivative for radioactive studies. They evaporated the ether, which they used as solvent, onto planchettes for counting. With modern scintillation counters it should prove convenient to measure specific activity by first measuring fluorescence and then extracting into solvent, adding a scintillation fluorophor, and measuring radioactivity.

Barr[25] carried out a detailed study of the chemistry and fluorescence of two types of resorcinol condensations; with succinic acid and the fluorescein type product and with malic acid and the umbelliferone type product. He showed, for instance, that the fluorescence of the succinic acid derivative is greatest in acid solution, pH 1.5–2.0. At such pH values the malic acid derivative and other umbelliferones are not fluorescent. Their pH optimum is about 10.5. Furthermore, the succinic acid derivative fluoresces most intensely when excited between 420 and 520 mμ, whereas the malic acid derivative fluoresces maximally when the exciting light is between 320 and 420 mμ. By making use of these differences in fluorescence properties, Barr[25] developed an assay to determine both malic acid and succinic acid when present in the same mixture such as in plant extracts. It is apparent from these studies that with modern instrumentation it may be possible to distinguish many of the carboxylic acid-resorcinol condensation products without the requirement of prior isolation procedures. However, it appears that the distinction will be into groups rather than into individual acids;

fluorescein derivatives, umbelliferone derivatives, and keto acid derivatives.

Fluorometric assay following resorcinol condensation has been used to determine acetoacetic acid in urine[26] in studies on diabetes. Although it is now apparent that such a procedure is not specific for acetoacetic acid, the urinary excretion of the keto acid in diabetes is so great that significant increases can be observed even in the presence of interfering compounds.

B. Malic Acid

By using orcinol (3,5-dihydroxytoluene) instead of resorcinol, and a simple isolation procedure, Hummel[27] was able to develop a highly specific method for measuring malic acid even in the presence of large amounts of other tricarboxylic acid cycle intermediates, lactic acid, glucose, fructose, purines, and amino acids. The procedure for determining malic acid in blood or tissue homogenates is as follows:

The solution is deproteinized with trichloroacetic acid; the final acidity should be 1 N. After centrifugation, an aliquot of the supernatant fluid (containing 0.1–1 μg. of malic acid) is transferred to a conical centrifuge tube and diluted to 1 ml. with 1 N HCl. Following this, add 0.1 ml. of 0.1% 2,4-dinitrophenylhydrazine in 2 N HCl and 0.5 ml. of 10% calcium chloride. After 30 minutes add 0.3 ml. of 5 N NH$_4$OH and 6 ml. of absolute ethanol. Under these conditions calcium malate is highly insoluble and precipitates out. The dinitrophenylhydrazine reacts with keto acids and carbohydrates which may otherwise interfere and converts them to alcohol soluble derivatives. The tubes are left standing overnight to ensure complete precipitation and are then centrifuged. The supernatant fluid is carefully and completely poured off and the tube is dried in an oven at 105° for 15 minutes. Three milliliters of orcinolsulfuric acid solution* are then added. The tubes are mixed carefully heated to 100° for 10 minutes, then cooled under the tap and diluted to 10 ml. with concentrated sulfuric acid. The condensation product, 7-hydroxy-5-methyl coumarin, fluoresces an intense blue color in this strongly acidic solution. Details of excitation and fluorescence should be ascertained before the method is applied.

Fluorescence is proportional to concentration and recoveries from human blood were reported to be quantitative (95–102%). Hummel determined the malic acid content of human blood using 0.2-ml. volumes of finter tip blood. The average obtained on 13 individuals was 0.46 mg. %. Following

---

* The orcinol is purified by recrystallization from benzene and 80 mg. are dissolved in 100 ml. of 12.5% sulfuric acid. The acid concentration is fairly critical and should be checked carefully. The reagent is stable for months when kept in an amber bottle in the cold.

exercise, when the lactic acid had risen fourfold, the fluorometrically determined values for malic acid remained unchanged.

Many other compounds condense with polycarboxylic and keto acids in strong sulfuric acid. Undoubtedly, new applications will be discovered from time to time. Thus, β-naphthol has been used to determine malic acid in the presence of fumaric acid making it possible to assay the enzymes fumarase[28, 29] and malic dehydrogenase by simple and sensitive fluorometric methods.[30] The malic acid assay was apparently based on an unpublished method of Dr. John F. Speck.

Details of the assay as reported by Loewus et al.[30] are described.

An aliquot of tissue extract containing 0.003–0.010 μmole of malic acid is transferred to an 18- $\times$ 150-mm. test tube immersed in an ice bath. The volume is adjusted to 1.0 ml. with water and 2 ml. of β-naphthol-H_2SO_4 reagent* are added, slowly and with shaking. Then another 5 ml. of the reagent are added and, after mixing, the tube is heated for 10 minutes on a boiling water bath. It is then cooled rapidly in an ice-water bath. Before fluorometric assay the tube must be shaken owing to the high viscosity of the solution. With filter instruments fluorescence is highest when excitation is at the 365-mμ mercury line and the secondary filter cuts off radiation below 410 mμ. A reagent blank carried through the procedure gives an appreciable reading. Lowry et al.[29] used incubation mixtures without deproteinization and found no interference. On the other hand, Loewus et al.[30] found fairly large blanks in trichloroacetic acid filtrates of enzyme mixtures. For this reason they isolated malic acid by an extraction procedure. It may well be that trichloroacetic acid or an impurity in it is responsible for the blank fluorescence.

C. Lipofucsin

It has been known for some time that yellow fluorescent granules appear in various animal tissues. The material, called lipofucsin, is also found in heart where its concentration increases with age. It has, therefore, come to be termed the "age pigment." Strehler et al.[31] have reviewed much of the earlier studies in this area and have demonstrated that in extracts of tissues the yellow orange fluorescence is associated with the cephalin fraction.[32] They have also shown that authentic cephalin and lecithin acquire the same yellow-orange fluorescence when undergoing oxidative rancidity. Apparently, highly unsaturated fatty acids resulting from

---

* β-Naphthol is prepared as a 0.004 M solution in 0.004 M NaOH. To prepare the sulfuric acid reagent, add 100 ml. of ice-cold concentrated sulfuric acid to 1 ml. of the alkaline reagent. The alkaline solution can be stored but the acid solution must be prepared fresh before use.

oxidation are fluorescent although no direct reports of this have appeared. It appears that from the standpoint of lipid metabolism, aging is merely an *in vivo* appearance of rancidity. Obviously, the fluorescence of the lipids can be utilized to study the process of rancidity wherever it occurs, be it medicine or food processing. Hyden and Lindstrom[33] have reported the following spectral data for human neuronal lipofuscin: absorption peaks at 260 and 375 mμ and fluorescence peaks at about 450 and 545 mμ.

V. PURINES, PYRIMIDINES, AND NUCLEIC ACIDS

There have been numerous reports concerning the fluorescence of purines, pyrimidines, and nucleic acids in the visible region.[34] However, it is now generally recognized that with the exception of the guanines none of these compounds yields visible fluorescence in the native state. Impurities and light scattering may explain the earlier reports. However, purines and pyrimidines are resonant structures and have high absorption coefficients in the ultraviolet region (260 mμ). Such structures might be expected to emit fluorescence in the ultraviolet region.

A. Purines

Shore and Pardee[35] investigated a large number of free purines and their corresponding nucleosides and nucleotides and could detect no fluorescence in water. They also reported that adenylic acid and adenosine did not fluoresce in 0.1 N HCl or in 0.1 N NaOH. They pointed out, however, that the limitations of their instrumentation were such that had these purines fluoresced with a quantum efficiency of less than 1%, it would not have been detected. Shortly thereafter, Weber[36] reported that, in agreement with Shore and Pardee,[35] he could observe no fluorescence emission from solutions of adenine or adenylic acid although the method which was used was supposedly capable of detecting fluorescence even at quantum efficiencies of 0.1%. On the other hand, Duggan *et al.*[37] observed significant fluorescence emission in solutions of many purines and purine nucleosides and nucleotides. Excitation and fluorescence maxima and optimal pH values for fluorescence of a number of purine derivatives were reported in their paper. These values were obtained during a large scale study on over a hundred biologically important compounds. Because of the immensity of that study, it was not possible to determine detailed variations of fluorescence with pH for all the purine derivatives, nor to determine quantum yields. More recently, many of these measurements were repeated (Table IV) and quantum yields were estimated by the comparative method of Parker and Rees[38a] using tryptophan as the fluorescent reference of known quantum yield.[38b] For adenine and its derivatives, quantum yields

TABLE IV

FLUORESCENCE CHARACTERISTICS OF PURINES AND THEIR DERIVATIVES

Compound	Absorption peak mμ	Excitation maximum[a] mμ	Fluorescence maximum[a] mμ	Solvent
Adenine	262	272	380	0.1 N H$_2$SO$_4$
Adenosine	257	272	390	5 N H$_2$SO$_4$
Adenylic acid	257	272	390	5 N H$_2$SO$_4$
Adenosine diphosphate	257	272	390	5 N H$_2$SO$_4$
Adenosine triphosphate	257	272	390	5 N H$_2$SO$_4$
Guanine	272[b]	285	360	0.1 N H$_2$SO$_4$
Guanosine	280[b]	285	390	0.1 N H$_2$SO$_4$
Guanylic acid	280[b]	285	390	0.1 N H$_2$SO$_4$
Tryptophan[c]	278	285	365	0.1 N phosphate buffer, pH 7

[a] Uncorrected instrumental readings, Aminco-Bowman spectrophotofluorometer.
[b] Shoulder.
[c] Tryptophan data were obtained on the same instrument at the same time and are presented for comparison.

of about 0.2% were obtained. The value found for unsubstituted guanine in dilute solution at pH 11 was about 1.5 to 2.0%. By comparison, the quantum yield of tryptophan fluorescence is 20%.[38c] Such values would have been too low for detection by the technique of Shore and Pardee,[35] but they should have been detected by Weber.[36] However, it should be pointed out that pH and the presence of other ions are critical for purine fluorescence (Table V). It may well be that Weber utilized conditions of pH or

TABLE V

EFFECT OF pH AND ACIDITY ON FLUORESCENCE OF ADENOSINE AND ADENOSINE-3-PHOSPHATE

Medium	Relative fluorescence intensity	
	Adenosine	Adenosine-3-phosphate
5 N H$_2$SO$_4$[a]	100	100
3 N HCl	—	0
0.01 N H$_2$SO$_4$	30	23
pH 7 (phosphate)	0	0
1 N NH$_4$OH	0	0

[a] Value in 5 N H$_2$SO$_4$ taken as 100.

ionic composition in which the adenine derivatives do not fluoresce. Guanine derivatives are even more fluorescent than the corresponding adenine derivatives and they exhibit fluorescence in alkaline as well as acid media. A report of guanylic acid fluorescence on acidified paper chromatograms has also appeared.[39] The N-methylated guanines are even more intensely fluorescent than guanine itself.[38b]

What is the significance of fluorescence with such a low quantum yield? First, one must seriously consider impurities as being responsible for the fluorescence. However, this seems unlikely for many reasons. Purified compounds have been used from a variety of sources and these all have the same spectral characteristics.[38b] The fluorescence characteristics of adenosine, adenosine diphosphate, and adenosine triphosphate were not only similar, but the intensities of the emitted fluorescences showed some relationship to the molecular weights, or to the adenosine contents. It may be that the fluorescence in strong sulfuric acid is that of a degradation product. However, there is also significant fluorescence of a similar sort in 0.1 N H_2SO_4. It would appear, therefore, that adenine and guanine derivatives do possess native fluorescence in acid media. While the quantum yields are low, the fluorescence can be used for analytical purposes. A simple and specific method for the fluorometric assay of guanine in nucleic acid hydrolyzates has been developed.[38b] The distinction between free adenine and its nucleoside and nucleotides in strong sulfuric acid should be quite useful as an analytical procedure. The free base shows no fluorescence in 7 N sulfuric acid, whereas all its derivatives fluoresce almost maximally. It should also be possible to determine guanine and its derivatives in the presence of adenine and its derivatives since only the former fluoresce in alkaline media. The fluorescence intensities of xanthine, hypoxanthine, and uric acid are extremely low, and it may be that the previous reports concerning the fluorescence of xanthine and uric acid[37] were due to contamination of the samples. A careful study of the fluorescence properties of all the purines and their derivatives is certainly called for and should lead to many interesting findings of both practical and theoretical significance. It should be pointed out, however, that the extremely low fluorescence yields of the purines means that solutions used for fluorometric assay will absorb light to an appreciable extent. Under such conditions the relationship between fluorescence and concentration may not be linear (see Chapter 1).

B. Nucleotides

Rorem[40] has described a procedure for detecting phosphate esters on paper chromatograms by a fluorometric procedure. After development, the chromatogram is dipped into an alcoholic solution of quinine. The dried chromatogram is then viewed under ultraviolet light (either 360 or 254 mμ). Phosphate esters, including sugar phosphates and purine and pyrimidine nucleotides, show up as light areas against an intense gray-blue fluorescent background. Microgram quantities can be detected.

C. Homopurines

Fischer and Neumann[41] prepared a number of homopurines and found them to be highly fluorescent. They particularly studied oxyhomouric acid and oxyhomoxanthine:

Homouric acid   Homoxanthine

The oxyhomopurines have an additional resonating group compared to the purines; $O{=}C{-}C{=}C{-}C{=}O$, instead of $C{=}C{-}C{=}O$ so that their absorptions are shifted towards the visible. In alkaline media oxyhomouric acid absorbs maximally at 325 mμ and oxyhomoxanthine at 312 mμ. The uric acid derivative fluoresces an intense light blue and less than 0.1 μg./ml. can be measured even with exciting light at 360 mμ. The xanthine derivative fluorescence is blue violet and also less intense.

D. Fluorescence of Purine Complexes

An interesting observation concerning the ability of purines to quench or augment the fluorescence of hydrocarbons and other substances was pointed out by Weil-Malherbe.[42] Apparently, purines form complexes with many compounds including hydrocarbons. The complexes form under different conditions with different compounds and also endow the purines with solubilizing activity. The differential effects of purines on fluorescence may be of practical use in developing analytical methods.

E. Thymine, 5-Methylcytosine, and Their Derivatives

Udenfriend and Zaltzman[38b] surveyed many pyrimidines and their nucleosides and nucleotides. With the exception of free thymine the pyrimidine compounds were without fluorescence. Thymine fluoresced

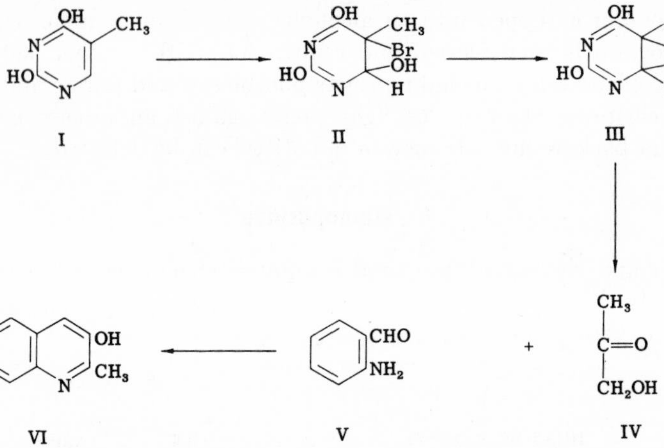

FIG. 9. The release of acetol (IV) from thymine (I) and the subsequent condensation with o-aminobenzaldehyde (V) to form the fluorophor, 3-hydroxyquinaldine (VI) (according to Roberts and Friedkin[43]).

maximally at pH 11 ($\varphi = 0.15\%$); excitation maximum was at 290 mμ (corrected), fluorescence maximum at 380 mμ (uncorrected). The native fluorescence does not appear to be suitable for measuring thymine in biological mixtures.

Chemical procedures for fluorometric assay have been developed for thymine and 5-methylcytosine.[43] The procedures can also be applied to nucleosides and nucleotides of these pyrimidines as well to deoxyribonucleic acid. The method depends on the oxidation of the pyrimidine with bromine water to acetol and the subsequent conversion of the acetol to 3-hydroxyquinaldine by condensation with o-aminobenzaldehyde (Fig. 9). The following micromodification is applicable to amounts of thymine from 0.002 to 2.0 μg.; the volumes can be modified to suit more macro procedures.

To 5.0 μl. of a solution containing 0.002–2.0 μg. of thymine in 0.01 M ethylenediamine tetraacetate solution are added 3.4 μl. of bromine water (12.5% saturated and freshly diluted). After 20 minutes at room temperature add 19 μl. of alkaline o-aminobenzaldehyde solution (5 volumes of 0.3 N NaOH in 0.01 M ethylenediamine tetraacetate plus 2 volumes of 1% o-aminobenzaldehyde). The mixture is warmed at 65° for 20 minutes and then diluted with 1.0 ml. of 0.1 M phosphate buffer, pH 6.9. Standards and blanks are treated in the same manner. It is not necessary to hydrolyze thymine containing compounds to carry out this reaction. The procedure can be used directly to determine the thymine content of thymidine, thymidine-5-phosphate, and even of intact deoxyribonucleic acids. Equimolar amounts of these derivatives when assayed yield about identi-

TABLE VI

FLUOROMETRIC ASSAY OF THYMINE CONTENT OF VARIOUS THYMINE DERIVATIVES[a,b]

Compound	Thymine content based on:	
	Weight mg./ml.	Fluorometric assay mg./ml.
Thymidine _____	0.20	0.20
Thymidine-5'-phosphate _____	0.20	0.24
Deoxyribonucleic acid _____	0.20	0.18

[a] Solutions of thymine, thymidine, thymidine-5'-phosphate, and deoxyribonucleic acid were prepared, each containing the equivalent of 0.20 mg. of thymine per ml. The thymine solution was used as the fluorometric standard.

[b] After Roberts and Friedkin.[43]

cal amounts of fluorescence as shown in Table VI. The only other pyrimidine to give this reaction is 5-methylcytosine. Other interferences are those already described under the procedure for acetol.

F. Deoxyribonucleic Acid

Since thymine is uniformly present in deoxyribonucleic acid and is the only base to give this acetol fluorescence (5-methylcytosine is a rare component), this assay can be used to estimate deoxyribonucleic acid in tissues. The procedure is excellent when applied to the nucleic acid isolated by standard procedures (Table VII). However, it can be used directly on unfractionated whole homogenates to provide a most simple and sensitive procedure for deoxyribonucleic acid assay. The release of acetol from saline homogenates of spleen, thymus, bone marrow, and lung can yield results that agree very well with values obtained by other procedures (also shown in Table VII). However, certain tissues yield high values. The advantage of the method as applied directly to homogenates is its simplicity and the fact that amounts of tissue containing as little as 0.05 μg. of deoxyribonucleic acid can be assayed.

Another unique component of deoxyribonucleic acid is the sugar deoxyribose and many colorimetric methods have been devised for estimating the nucleic acid based on deoxyribose assay of isolated material. Recently, Kissane and Robins[44] developed an extremely sensitive fluorometric method for deoxyribonucleic acid based on the formation of a fluorescent quinaldine by the reaction of deoxyribose with 3,5-diaminobenzoic acid, as first suggested by Velluz et al.[9]

Tissue homogenate containing 0.05–6.4 μg. of deoxyribonucleic acid is

TABLE VII

A COMPARISON OF METHODS USED FOR DETERMINATION OF DNA IN VARIOUS
RABBIT ORGANS[a,b]

| Tissue | DNA soluble in hot trichloroacetic acid | | | | DNA in saline homo-genates |
	Fluoro-metric method mg./g.	Diphenyl-amine method mg./g.	Phosphate method mg./g.	Ultra-violet absorption method mg./g.	Fluoro-metric micro-method mg./g.
Spleen	10.8	11.7	11.5	11.9	11.2
	10.2	10.5	10.6	10.8	
Thymus	24.3	23.7	24.5	24.9	25.2
	25.6	25.0	25.6	26.8	
Bone marrow	6.58	6.38	6.38	6.55	6.20
	7.04	6.92	6.88	7.11	
Kidney	3.30	3.61	3.40	3.58	4.61
	3.34	3.66	3.38	3.56	
Liver	1.63	2.03	1.61	1.75	2.68
	1.58	1.78	1.52	1.69	
Cerebral hemi-sphere	0.74	0.74	0.95	0.79	1.21
	0.70	0.70	0.84	0.82	
Lung	5.31	5.84	5.46	5.56	5.45
	5.73	6.52	5.90	6.08	

[a] All values in the table are based on wet weight of tissue.
[b] After Roberts and Friedkin.[43]

precipitated with cold trichloroacetic acid in a microcentrifuge tube and the precipitate is extracted with alcohol to remove lipids and other interfering organic extractable material. The remaining precipitate is taken to dryness. Deoxyribose standards and blanks are taken to dryness too. To each of the tubes are added 10 ml. of freshly prepared 2 M 3,5-diaminobenzoic acid. After vigorous mixing the tubes are capped and heated at 60° for 30 minutes. After this the solutions are diluted by adding 1.0 ml. of 0.6 N perchloric acid and fluorescence is assayed (excitation maximum, 420 mμ; fluorescence maximum, 520 mμ; uncorrected).

Deoxyribose is liberated only from the purine residues by the hydrolysis brought about during the heating with the strongly acidic solution of 3,5-diaminobenzoic acid. Accordingly, a factor of 2 must be used to correct for total deoxyribose. That the material assayed in this way was deoxyribose was shown by comparing excitation and fluorescent spectra obtained from tissue extracts with those from standards. These were found to be exactly the same. Deoxyribonucleic acid when added to homogenates was recovered quantitatively and a comparison of this fluorometric assay with macrocolorimetric procedures was excellent. In the paper by Kissane and Robins,[44] a micromodification is described which permits the measurement of as little as 0.002 μg. of deoxyribonucleic acid. This micromodification was applied as a histochemical method to determine the localization of the nucleic acid in brain tissues. Microgram quantities of tissue were used for these assays. Proportionality between fluorescence intensity and quantity of deoxyribonucleic acid or brain tissue is shown in Fig. 10.

Nucleic acids, both ribose and deoxyribose, interact with various dyes in a manner which can be utilized for assay, structure determination, and for studies on other interactions with nucleic acids. Thus, Auramine O (dimethylamino diphenylamine) exhibits little fluorescence in aqueous solution. However, in the presence of trace amounts of nucleic acid, it emits an intense yellow fluorescence.[45] Apparently, the dye-nucleic acid complex has a greater fluorescence efficiency than the free dye. It has also been shown that acriflavine, which is by itself highly fluorescent, binds to nucleic acids. However, in this case the fluorescence intensity decreases with complex formation. The dye acridine orange has been used as a stain to distinguish between deoxyribonucleic acid and ribonucleic acid in

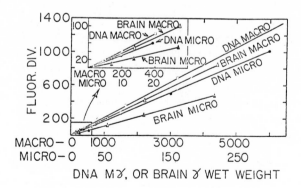

FIG. 10. Proportionality between fluorescence intensity and quantity of deoxyribonucleic acid and of whole brain (according to the procedure of Kissane and Robins[44]).

fixed tissues. Different colors of fluorescence are produced with the two nucleic acids. Schummelfeder et al.[46] and Beers et al.[47] have investigated this interaction and found that there are a number of complexes formed with each acid and that the state of aggregation is an important factor in determining the color of the fluorescence.

The standard Schiff reagent (basic fuchsin plus sulfurous acid) is widely used to localize aldehydes in tissues. It is perhaps best known for its use in the Feulgen reaction for histological localization of deoxyribonucleic acid. Kasten et al.[48a] have described a number of Schiff-type reagents which, upon reaction with aldehydes, give rise to intense fluorescence. Acridine-Y-SO_2 and acriflavine-SO_2 are particularly useful reagents as far as sensitivity and specificity of staining. Applications to polysaccharide staining are also discussed.

VI. PORPHYRINS

The naturally occurring porphyrins may be considered to be derivatives of the substance porphin, which is made up of four pyrrole rings joined into a

Porphin

ring system by four methene bridges (CH groups). Porphyrins are highly colored and the visible absorption bands of each porphyrin provide a characteristic means of identification. Solutions of porphyrins in organic solvents or in acid solution are intensely fluorescent. Excitation is usually in the visible, coinciding with some of the absorption bands. The emitted fluorescence occurs in the red and in the infrared. Fluorescence in this region of the spectrum requires detecting devices sensitive to red and infrared. Porphyrin fluorescence can be detected in standard instruments equipped with 1P21 and 1P28 photomultiplier tubes. However, the sensitivity is far from maximal and so much amplification is required that the scattered light becomes so large, by comparison, as to present a problem. With a red sensitive detector (RCA 7102 tube) sensitivity is increased markedly and scattered light is minimal. A comparison of the uncorrected

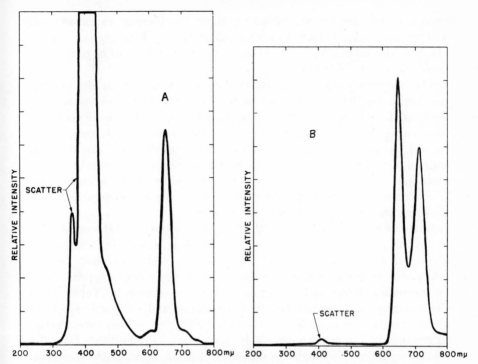

FIG. 11. Comparison of the uncorrected fluorescence spectra of tetraphenylporphin obtained with a 1P21 phototube and with an infrared sensitive 7102 tube. The solvent is mineral oil (from the American Instrument Co., Instruction Bull. No. 816[48b]). *A*, with 1P21 tube; *B*, with 7102 tube.

fluorescence spectra of tetraphenyl porphine obtained with a 1P21 tube and a 7102 tube[48b] is shown in Fig. 11. Not only is the 660-mμ peak increased enormously over the interfering light scatter with the red sensitive tube, but a second peak becomes apparent at about 710 mμ. Although most studies to date have not made use of red sensitive detectors, it is obvious that they should be of considerable use in the porphyrin field.

The naturally occurring porphyrins include hemoglobin, myoglobin, cytochromes, chlorophyll, and numerous functional and nonfunctional pigments. This chapter will be limited essentially to animal pigments. Chlorophyll will be discussed in the next chapter on fluorescence in plants (Chapter 11).

Dhéré[49] carried out the most intensive studies on the fluorescence of the porphyrins and much of our present knowledge concerning the fluorescence characteristics of the different porphyrins found in nature comes from studies by Dhéré and his colleagues. An excellent review of the physical properties of porphyrins, including their fluorescence characteristics, was written by Vannotti.[50] An earlier publication by Lemberg and Legge[51] also provides interesting information about the fluorescence of porphyrins and their metabolic products.

The biologically active forms of the porphyrins in animals are metalloporphyrin complexes of iron or copper. Such metal complexes are not fluorescent. In fact, it has been shown that metalloporphyrins in combination with proteins (hemoglobin) quench the protein fluorescence. They do this by an effective capture of the energy which would otherwise be emitted as ultraviolet fluorescence characteristic of the aromatic amino acids (see Chapter 8). On the other hand, chlorophyll, a magnesium porphyrin complex, is highly fluorescent. When the iron is detached from the heme porphyrins the fluorescence appears.

Dhéré described and classified the fluorescence spectra of the porphyrins in a variety of solvents. The same spectra (four bands) are observed in alcohol, dioxane, ammonium hydroxide solution, and in pyridine. In acid solution or in acidic or nonpolar solvents only three bands appear. Data for protophorphyrin and hematoporphyrin are shown in Table VIII. Coproporphyrin, uroporphyrin, mesoporphyrin, and etioporphyrin have almost exactly the same fluorescence characteristics as hematoporphyrin. In pyridine and in many organic solvents, protoporphyrin differs from the other porphyrins. Although the difference is small, fluorescence spectrom-

TABLE VIII

FLUORESCENCE SPECTRA OF PORPHYRINS[a]

	Pyridine Fluorescence bands, mμ			
Porphyrin	α	β	γ	Principal band
Protoporphyrin	701	683	663	634
Hematoporphyrin	691	673	654	625

	2 N Hydrochloric acid		
	Band I	Band II	Band III
Protoporphyrin	655	—	603
Hematoporphyrin	653	619	597

[a] According to Dhéré and Bois[52] as presented by Vannotti.[50]

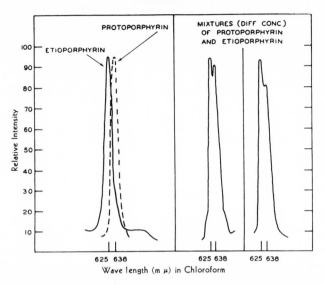

FIG. 12. Differentiation of etioporphyrin from protoporphyrin with a fluorescence spectrometer (from Goldstein et al.[53]).

eters can distinguish them. Goldstein et al.,[53] utilizing a fluorescence spectrometer of their own design, were able to distinguish etioporphyrin from protoporphyrin (Fig. 12). Although the separation is not large, it should be pointed out that ordinary absorption spectrophotometry fails to distinguish between these two compounds. Of further interest is the close agreement between the values of Dhéré and those of Goldstein et al.[53] for the principal fluorescence band; for protoporphyrin, 634 mμ vs. 638 mμ, and for etioporphyrin, 625 mμ vs. 625 mμ. In the case of the acid-type solvent, Howerton[48b] has obtained values of 605 and 655 mμ for hematoporphyrin (Fig. 13) as compared to 597 and 653 mμ reported by Dhéré. The intermediate band at 619 mμ is a minor one and may be obscured by the two major bands.

Fluorometric studies are of interest in studies on the biosynthesis and metabolism of the porphyrins in vitro and in vivo.[54] Tissues and enzyme preparations which are capable of carrying out the first stages of porphyrin synthesis usually show large increases in porphyrin fluorescence which can be used to follow enzyme activity. The most practical use of fluorometry in the field of porphyrins is as a diagnostic aid in various pathologic conditions. An excellent discussion concerning the clinical aspects of porphyrin metabolism has been presented by Schwartz.[55]

Protoporphyrin, the porphyrin portion of hemoglobin, is also found in the free form in erythrocytes and endows these cells with red fluorescence.

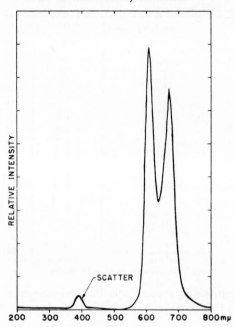

FIG. 13. Fluorescence spectrum of hematoporphyrin obtained with a commercial fluorescence spectrometer utilizing the 7102 (infrared sensitive) phototube.[48b]

Normally, this is very low, but in iron deficiency, lead poisoning, and other pathologic conditions the levels can be markedly increased (Table IX). Measurement of erythrocyte protoporphyrin and coproporphyrin, as described by Schwartz and Wikoff,[56] can be of diagnostic value in these disorders. Protoporphyrin is not found in urine.

TABLE IX

ERYTHROCYTE PROTOPORPHYRIN LEVELS OBSERVED IN VARIOUS DISEASES[a]

Subjects	μg. Protoporphyrin per 100 ml. of cells
Normals	20–50
Iron deficiency	200–800
Lead poisoning	200–2000
Increased erythropoiesis	40–150
Pernicious anemia	15–30
Infection	50–300

[a] After Schwartz.[55]

TABLE X

Urinary Coproporphyrin Values in Various Diseases[a]

Subjects	µg./day
Normal males	100–300
Normal females	75–275
Lead poisoning	500–3000
Poliomyelitis	300–900
Liver disease	300–800
Pernicious anemia	150–400
Aplastic anemia	250–500
Hemolytic anemia	200–400
Hodgkin's disease	200–1000
Azotemia	10–75
Acute alcoholism	250–500

[a] After Schwartz.[55]

A. Coproporphyrin in Urine and Feces

Coproporphyrin is normally the predominant porphyrin in urine and feces. About one-half of the urinary coproporphyrin in freshly voided urine is present in a nonfluorescent form which is converted to coproporphyrin under mild oxidative conditions; standing in air overnight, treatment with dilute solutions of iodine, or hydrogen peroxide. Total coproporphyrin methods utilize an oxidative step to include this fraction. Urinary coproporphyrin values observed in various pathologic conditions are shown in Table X. It is apparent that changes can be very large in certain states. The value of urinary coproporphyrin in detecting early stages of lead poisoning has been well established.

Several modifications of a procedure for urinary coproporphyrin have appeared from the same laboratory.[57, 58] A description of the procedure as described by Schwartz[55] is presented.

To 5 ml. of urine* in a separatory funnel add 5 ml. of a mixture of glacial acetic acid and saturated sodium acetate (4:1 v/v), 100 ml. of ethyl acetate, and 15–20 ml. of water. Shake well and then discard the aqueous phase. Wash the ethyl acetate once with 0.005% aqueous iodine (to convert the coproporphyrin precursor to coproporphyrin) and continue washing with 20- to 30-ml. portions of 1% sodium acetate until the washings show no red fluorescence when examined under ultraviolet light. The coproporphyrin is re-extracted from the ethyl acetate with 4 portions of

---

* Urines should be kept neutral or slightly alkaline since coproporphyrin destruction takes place under acid conditions.

about 5 ml. each of 1.5 N hydrochloric acid (until coproporphyrin fluorescence no longer appears in the acid extracts).† The combined hydrochloric acid extracts are diluted with 1.5 N hydrochloric acid to a convenient volume and the porphyrin concentration is measured fluorometrically. Standards and reagents blanks are prepared in a similar manner. It has been shown that only a small fraction of the porphyrin determined by this procedure is not coproporphyrin. In some instances, measurements of coproporphyrin in feces, erythrocytes, and liver may be of diagnostic value. Fluorescence spectrometry has also been used to identify coproporphyrin Type I in tubercle bacilli.[59]

B. Uroporphyrin

The bulk of the uroporphyrin excreted in the urine is an artifact resulting from the chemical conversion from porphobilinogens by the procedures used for isolation and estimation. Furthermore, less is known about the uroporphyrins and their significance. Nevertheless, a uroporphyrin assay is utilized in many laboratories as an adjunct to other diagnostic tests The procedure by Schwartz[55] converts porphobilinogen to uroporphyrin before assay.

A 5-ml. aliquot of urine is adjusted to pH 4 by addition of glacial acetic acid. The sample is then heated in a boiling water bath for 15 minutes. After cooling, add 1 ml. of saturated sodium acetate and 15 ml. of water and shake in a separatory funnel with 100 ml. of ethyl acetate. The aqueous phase is set aside and the ethyl acetate is washed once with 0.005% aqueous iodine and then with portions of 1% sodium acetate until all red fluorescing porphyrin is extracted. The original aqueous phase and all the washings are combined and made up to volume with hydrochloric acid, the final concentration being 1.5 N. Fluorescence may be assayed directly at this stage. However, this extract is not pure or homogenous and further purifications involving adsorption and elution from alumina or calcium phosphate can be used to endow the method with greater specificity. Electrophoresis on paper, prior to fluorometric assay, has also been used successfully.[60]

Talman[61] has described a modification which permits the determination of both coproporphyrin and uroporphyrin in a single aliquot of urine. The method is well presented and contains additional details and explanations of the procedure.

An interesting property of hematoporphyrin is its ability to localize in lymphatic and neoplastic tissues where it can be detected in situ by its

---

† It may be possible to achieve quantitative extraction with smaller volumes of acid by adding a less polar solvent such as heptane to the ethyl acetate, as in the procedure for serotonin assay (Chapter 5).

red fluorescence in ultraviolet light. Rasmussen-Taxdal *et al.*[62] have reported that as much as 1 g. of hematoporphyrin may be safely administered to patients. With such doses it was possible to visualize neoplasms by their fluorescence through the skin and through the bowel wall. They suggest that the method should prove useful to the surgeon in visualizing and delineating neoplastic tissue during operations.

C. Urobilinogen and Urobilin

The catabolism of hemoglobin involves a series of degradative steps. Major products are biliverdin and bilirubin which are open tetrapyrrole chains. These and related compounds pass from the liver into the bile and are referred to as "bile pigments." These pigments are secreted with the bile into the intestine where they undergo further metabolism. A major metabolite formed in the intestine is urobilinogen (stercobilinogen) much of which is excreted in the feces. Urobilinogen which is reabsorbed from the intestine is mostly removed by the liver. However, in some pathological states the liver does not do this as efficiently. The result is that urobilinogen is excreted in the urine. Measurement of urinary urinobilinogen can serve therefore, as an index of liver and biliary function.

Urobilinogen is excreted in urine in two isomeric forms. Since the methine bridges are totally reduced they are not fluorescent (Fig. 14). When urobilinogen solutions stand in air they are oxidized to urobilin. The latter can be determined fluorometrically. Adler[63] oxidized urobilinogen with an alcoholic solution of iodine in the presence of zinc acetate. This

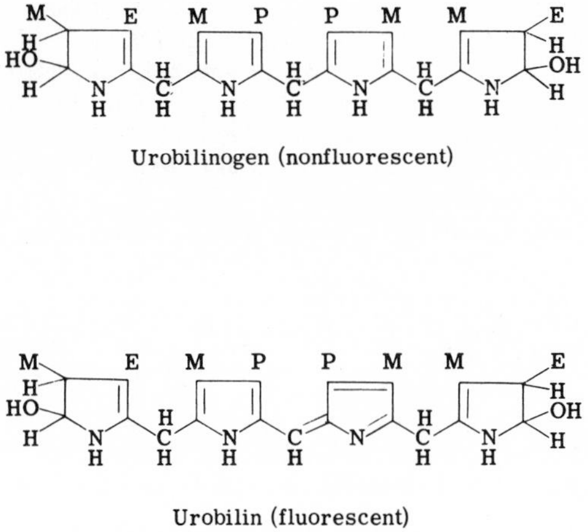

Urobilinogen (nonfluorescent)

Urobilin (fluorescent)

FIG. 14. Urobilinogen and urobilin.

yielded the strongly fluorescent (green) urobilin zinc complex which could be determined fluorometrically. According to Lemberg and Legge,[51] this is an extremely sensitive test for urobilin and urobilinogen, and with appropriate modifications as to chemistry and conditions for excitation and fluorescence detection it should be possible to utilize it as a precise and specific analytical procedure. Other bile pigments which contain oxidized methine bridges should also yield fluorescent products.

REFERENCES

1. Smith, I. "Chromatographic Techniques," p. 168. Interscience, New York, 1958.
2. Chargaff, E., Levine, C., and Green, C., *J. Biol. Chem.* **175,** 67 (1948).
3. Partridge, S. M., *Nature* **158,** 270 (1946).
4. Momose, T., and Ohkura, Y., *Talanta* **3,** 155 (1959).
5. Momose, T., and Ohkura, Y., *Chem. & Pharm. Bull. (Tokyo)* **6,** 412 (1958).
6. Momose, T., and Ohkura, Y., *Chem. & Pharm. Bull. (Tokyo)* **7,** 31 (1959).
7. Hagedorn, H. C., and Jensen, B. N., *Biochem. Z.* **135,** 46 (1923).
8. Velluz, L., Amiard, G., and Pesez, M., *Bull. soc. chim. France* **15,** 678 (1948).
9. Velluz, L., Pesez, M., and Amiard, G., *Bull. soc. chim. France* **15,** 680 (1948).
10. Pesez, M., *Bull. soc. chim. biol.* **32,** 701 (1950).
11. Blecher, M., *Anal. Biochem.* **2,** 30 (1961).
12. Wadman, W. H., Thomas, G. J., and Pardee, A. B., *Anal. Chem.* **26,** 1192 (1954).
13. Bayly, R. J., and Bourne, E. J., *Nature* **171,** 385 (1953).
14. Velluz, L., Pesez, M., and Herbain, M., *Bull. soc. chim. France* **15,** 681 (1948).
15. Salvador, R. A., and Albers, R. W., *J. Biol. Chem.* **234,** 922 (1959).
16. Camber, B., *Nature* **174,** 1107 (1954).
17a. Chen, P. S., Jr., *Anal. Chem.* **31,** 296 (1959).
17b. Sawicki, E., and Stanley, T. W., *Chemist Analyst* **49,** 107 (1960).
18. Baudisch, O., and Deuel, H. J., Jr., *J. Am. Chem. Soc.* **44,** 1586 (1922).
19. Forist, A. A., and Speck, J. C., Jr., *Anal. Chem.* **22,** 902 (1950).
20. Werle, E., and Stiess, P., *Biochem. Z.* **321,** 485 (1951).
21. Huggins, C. G., and Miller, N. O., *J. Biol. Chem.* **221,** 711 (1956).
22. Almy, G. M., Fuller, H. Q., and Kinzer, G. D., *J. Chem. Phys.* **8,** 37 (1940).
23. Almy, G. M., and Gillette, P. R., *J. Chem. Phys.* **11,** 188 (1943).
24. Frohman, C. E., and Orten, J. M., *J. Biol. Chem.* **205,** 717 (1953).
25. Barr, C. G., *Plant Physiol.* **23,** 443 (1948).
26. Leonhardi, G., and Glasenapp, I. V., *Z. physiol. Chem. Hoppe-Seyler's* **286,** 145 (1951).
27. Hummel, J. P., *J. Biol. Chem.* **180,** 1225 (1949).
28. Lowry, O. H., Roberts, N. R., Wu, M.-L., Hixon, W. S., and Crawford, E. J., *J. Biol. Chem.* **207,** 19 (1954).
29. Lowry, O. H., *in* "Methods in Enzymology" (S. P. Colowick and N. O. Kaplan, eds.), Vol. IV, p. 374. Academic Press, New York, 1957.
30. Loewus, F. A., Tchen, T. T., and Vennesland, B., *J. Biol. Chem.* **212,** 787 (1955).
31. Strehler, B. L., Mark, D. D., Mildvan, A. S., and Gee, M. V., *J. Gerontol.* **14,** 430 (1959).
32. Mildvan, A. S., and Strehler, B. L., *Federation Proc.* **19,** 231 (1960).
33. Hyden, H., and Lindstrom, B., *Discussions Faraday Soc. No.* **9,** 436 (1950).

34. West, W., *in* "Chemical Applications of Spectroscopy" (W. West, ed.), p. 732. Interscience, New York, 1956.

35. Shore, V. G., and Pardee, A. B., *Arch. Biochem. Biophys.* **60**, 100 (1956).

36. Weber, G., *Nature* **180**, 1409 (1957).

37. Duggan, D. E., Bowman, R. L., Brodie, B. B., and Udenfriend, S., *Arch. Biochem. Biophys.* **68**, 1 (1957).

38a. Parker, C. A., and Rees, W. T., *Analyst* **85**, 587 (1960).

38b. Udenfriend, S., and Zaltzman, P., *Anal. Biochem.* **3**, 49 (1962).

38c. Teale, F. W. J., and Weber, G., *Biochem. J.* **65**, 476 (1957).

39. Mathews, R. E. F., *Nature* **171**, 1065 (1953).

40. Rorem, E. S., *Nature* **183**, 1739 (1959).

41. Fischer, F. G., and Neumann, P., *Ann.* **572**, 230 (1951).

42. Weil-Malberbe, H., *Biochem. J.* **40**, 363 (1946).

43. Roberts, D., and Friedkin, M., *J. Biol. Chem.* **233**, 483 (1958).

44. Kissane, J. M., and Robins, E., *J. Biol. Chem.* **233**, 184 (1958).

45. Oster, G., *Compt. rend. acad. sci.* **232**, 1708 (1951).

46. Schummelfeder, N., Ebschner, K., and Krogh, E., *Naturwissenschaften* **44**, 467 (1957).

47. Beers, R. F., Hendley, D. D., and Steiner, R. F., *Nature* **182**, 242 (1958).

48a. Kasten, F. H., Burton, V., and Glover, P., *Nature* **184**, 1797 (1959).

48b. Howerton, H. K., American Instrument Co., Instruction Bull. No. 816 (October, 1960).

49. Dhéré, C., "La fluorescence en biochimie." Les Presses Universitaires, Paris, 1937.

50. Vannotti, A., "Porphyrins—Their Biological and Chemical Importance" (translated by C. Rimington), pp. 18–30. Hilger and Watts, London, 1954.

51. Lemberg, R., and Legge, J. W., "Hematin Compounds and Bile Pigments." Interscience, New York, 1949.

52. Dhéré, C., and Bois, E., *Compt. rena. acad. sci.* **183**, 321 (1926).

53. Goldstein, J. M., McNabb, W. M., and Hazel, J. F., *J. Chem. Educ.* **34**, 604 (1957).

54. Shemin, D., and Kumin, S., *J. Biol. Chem.* **198**, 827 (1952).

55. Schwartz, S., Clinical Aspects of Porphyrin Metabolism. Veterans Administration Tech. Bull. TB 10–94, Washington D. C., 1953.

56. Schwartz, S., and Wikoff, H. M., *J. Biol. Chem.* **194**, 563 (1952).

57. Schwartz, S., Zieve, L., and Watson, C. J., *J. Lab. Clin. Med.* **37**, 843 (1951).

58. Zieve, L., Hill, E., Schwartz, S., and Watson, C. J., *J. Lab. Clin. Med.* **41**, 663 (1953).

59. Crowe, M. O., and Walker, A., *J. Opt. Soc. Am.* **47**, 1044 (1957).

60. Sterling, R. E., and Redeker, A. G., *Scand. J. Clin. & Lab. Invest.* **9**, 407 (1957).

61. Talman, E. L., *in* "Standard Methods of Clinical Chemistry" (D. Seligson, ed.), Vol. 2, p. 137. Academic Press, New York, 1958.

62. Rasmussen-Taxdal, D. S., Ward, G. E., and Figge, F. H. J., *Cancer* **8**, 78 (1955).

63. Adler, A., *Deut. Arch. klin. Med.* **138**, 309 (1922).

9

FLUORESCENCE IN ENZYMOLOGY

I. GENERAL REMARKS

THE application of spectrophotometry to enzymology had a great influence on the rapid development of the field. In contrast to previous methods, spectrophotometric assay required small amounts of enzyme and substrates, was simple, and could be applied so as to measure changes in absorption directly in an incubation mixture. Many of the compounds which have been used so successfully in developing spectrophotometric assay procedures are also fluorescent or can be made to fluoresce. This is true for the pyridine nucleotides, the flavins, and for other coenzymes and substrates. As a result, it has been possible to apply fluorometry to enzyme measurements in a manner analogous to spectrophotometry. The advantages of such fluorometric procedures are many. They are certainly much more sensitive, by factors of hundreds and thousands, and therefore require much less material. This is particularly important with valuable biochemical preparations. Because of the remarkable sensitivity, fluorometry is particularly suited to studies involving the localization of enzymes and related substrates and coenzymes within organs and even within individual cells. Measurements can be made at such great dilutions that interfering enzymes and metabolites are frequently diluted out. This, and the inherently greater selectivity of fluorescence, endow many fluorometric enzyme methods with far greater specificity than the corresponding spectrophotometric methods. Finally, because of its greater sensitivity to molecular interaction, fluorescence permits measurements of enzyme-coenzyme-substrate interaction which are not possible with other procedures. In addition to changes in light absorption, which are detectable as changes in excitation spectra, molecular interaction can alter the efficiency of the fluorescence process and the degree of polarization of the emitted radiation. These last features are particularly promising and may represent most important advances in unravelling mechanisms of enzyme action.

The applications considered in this section will be of two types: (1) quantitative fluorometric assay of enzymes and substrates of enzymes and (2) enzyme kinetics and mechanisms.

II. QUANTITATIVE ASSAY OF ENZYMES

In Chapter 7 there were discussed methods for measuring a number of coenzymes, and it was pointed out that the fluorescence characteristics of various forms of the coenzymes differed from one another. In the case of the pyridine nucleotides, the reduced forms have a high native fluorescence. On the other hand, the oxidized forms yield fluorescent derivatives upon heating in alkali, alone, or with carbonyl compounds. Flavin fluorescence also disappears on reduction. Wide-spread use has been made of the differences in ultraviolet absorption of oxidized and reduced coenzymes for the quantitative determination of enzymes and intermediary metabolites. Many of the fluorometric procedures for enzyme assay are also based on changes in oxidation or reduction of pyridine nucleotides.

A. Lactic Dehydrogenase in Single Nerve Cells

The extreme sensitivity of fluorometry as applied to enzymology is best exemplified by a method for lactic dehydrogenase devised by Lowry et al.[1] which requires tissue from only a single nerve cell body (less than 0.01 μg. of tissue, dry weight). The method can, of course, be modified to use larger quantities.

Lactic dehydrogenase is measured by the DPN+ formed from added pyruvate and DPNH. Incubations are carried out in a volume of 2 μl. for a period of 30 minutes and the strong alkali method is used for DPN+ assay (Chapter 7). Enzyme activity is expressed as moles per kilogram of dry weight of tissue per hour.

B. 6-Phosphogluconate Dehydrogenase

This enzyme assay was reported along with the procedure for lactic dehydrogenase.[1] The enzyme is measured by the TPNH formed from 6-phosphogluconate and TPN+, incubations being carried out in 1 μl. for 60 minutes. Residual TPN+ is destroyed and TPNH is assayed by the strong alkali method after oxidation. The enzyme activities in individual nerve cell bodies, obtained by these ultramicro fluorometric methods, are shown in Table I.

C. Glutamic Dehydrogenase

Lowry et al.[2, 3] have developed microprocedures and ultramicro procedures for the assay of the enzyme glutamic dehydrogenase. The method is based on the fluorometric assay of the DPN+ formed upon incubation of the tissue with a reagent containing DPNH (0.05–0.1 mM), ketoglutarate

TABLE I

LACTIC DEHYDROGENASE AND 6-PHOSPHOGLUCONIC DEHYDROGENASE IN SINGLE NERVE
CELLS OF THE RABBIT[a]

Cells	Average weight of tissue taken for assay (dry) μg.	Lactic dehydrogenase activity	6-Phosphogluconate dehydrogenase activity
Dorsal root ganglion cell bodies	6.4–13.1	48.5	1.03
Capsules of dorsal root ganglion cell bodies	5.7–20.7	21.1	1.66
Anterior horn cell bodies	5.2–12.1	26.0	0.73
Zona radiata from Ammon's horn	9.7–19.1	54.7	0.81

[a] The values represent the average of 5 to 8 individual assays. Activity is expressed as moles of pyridine nucleotides oxidized or reduced per kilogram of dry weight per hour (after Lowry et al.[1]).

(3 mM), ammonium sulfate (7.5 mM), and nicotinamide (20 mM) in Tris-HCl buffer at pH 7.6. Crystalline bovine serum albumin is added to give a 0.05% protein solution.

To a microtube containing 0.1 μg. of brain tissue (dry weight), add 5 μl. of the above reagent and let soak in the ice-bath for 10 to 15 minutes. Mix by tapping the tube and incubate at 38° for 30 minutes. Return the sample to the ice-bath and add 1 μl. of 1 N HCl to destroy DPNH and to stop the reaction. After mixing, a 5-μl. aliquot is transferred to a fluorometer tube containing 0.1 ml. of 6.6 N NaOH. After 60 minutes at room temperature add 1 ml. of water to the tube, mix, and measure the alkaline fluorescence as described previously. The method is more convenient and more sensitive than spectrophotometric methods. The reaction is also measured in the more favorable direction with the fluorometric assay.

D. Serum Glutamate—Pyruvate Transaminase

The enzyme assay reported by Laursen and Hansen[4] involves coupling the following two reactions:

$$\text{L-Alanine} + \text{ketoglutarate} \underset{}{\overset{\text{transaminase}}{\rightleftharpoons}} \text{pyruvate} + \text{glutamate} \qquad (1)$$

$$\text{pyruvate} + \text{DPNH} \xrightarrow[\text{dehydrogenase}]{\text{lactic}} \text{lactate} + \text{DPN}^+ \qquad (2)$$

The increase in the amount of DPN$^+$ is equivalent to the amount of pyruvate formed and therefore to the transaminase activity.

Add 0.2 ml. of serum, 0.3 ml. of nicotinamide (1%), 0.05 ml. of DPNH (6 mg./ml.), and 0.05 ml. of lactic dehydrogenase* to a 10-ml. centrifuge tube with round bottom and mix. The tube is placed in a water bath at 37° for 10 minutes, then 0.7 ml. of neutralized substrates (20 μmoles α-ketoglutarate and 100 μmoles L-alanine) is added. After exactly 10 minutes, 0.5 ml. of 10% trichloroacetic acid is added to stop the reaction. The tube is centrifuged and 0.5 ml. of supernatant fluid is pipetted into a cuvette containing 0.5 ml. of methyl ethyl ketone. The tube is shaken and 0.2 ml. of 5 N KOH is added and the tube is shaken again. This is followed by 0.8 ml. of 2 N HCl. After the tube has been shaken, it is allowed to stand at room temperature for 5 minutes, then heated in a boiling water bath for 2 minutes to drive off the ketone. Following this, 0.5 ml. of $M/15$ phosphate buffer, pH 7.4, and 6 ml. of water are added and the fluorescence is measured as described in Chapter 7. Comparisons are made with a standard DPN$^+$ curve. Blanks, omitting substrates and enzyme, are run and used to correct the observed values.

The authors report that the values obtained by this method and the colorimetric procedure of Wroblewski and Cabaud[5] do not agree too well, but they believe that the fluorometric procedure is the more reliable. They also point out that the method can be easily modified to take much smaller samples of serum. The same procedure should be applicable to other tissues.

E. Serum Glutamate—Oxaloacetate Transaminase

This procedure by Laursen and Espersen[6] is similar in principle to the previous one. The transamination (3)

$$\text{L-aspartate} + \text{ketoglutarate} \xrightarrow{\text{transaminase}} \text{oxaloacetate} + \text{glutamate} \qquad (3)$$

$$\text{oxaloacetate} + \text{DPNH} \xrightarrow{\substack{\text{malic} \\ \text{dehydrogenase}}} \text{malate} + \text{DPN}^+ \qquad (4)$$

is coupled to malic dehydrogenase (4) and the DPN$^+$ formed is equivalent to the transaminase activity. The procedure is exactly as for the glutamate-pyruvate transaminase, substituting malic dehydrogenase† for lactic dehydrogenase and using the appropriate substrates (20, μmoles ketoglutarate and 50 μmoles aspartate). Ten minutes after addition of the latter,

---

* Lactic acid dehydrogenase, crystalline, can be obtained from commercial sources. A suspension of 5 mg. of enzyme protein per milliliter may be stored in the refrigerator and diluted 1 to 40 with $M/15$ phosphate buffer, pH 7.4, before use.

† A suspension of the crystalline material (commercially available), 1 mg. of enzyme protein per milliliter, is stored in the refrigerator. It is diluted 1 to 10 with $M/15$ phosphate buffer, pH 7.4, before use.

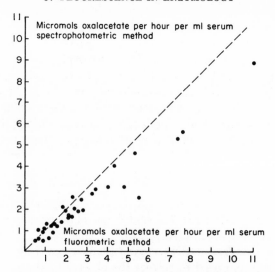

FIG. 1. Comparison between the fluorometric and spectrophotometric methods for the assay of serum glutamate-oxaloacetate transaminase (after Laursen and Espersen[6]).

the reaction is stopped by addition of trichloroacetic acid and DPN+ is assayed fluorometrically as before.

The fluorometric procedure yields slightly higher values for the enzyme than does the spectrophotometric method of Henley and Pollard[7] (Fig. 1). However, Laursen and Espersen[6] point out that it is the spectrophotometric procedure which inherently yields slightly low values. They also state that the fluorometric procedure is sufficiently simple so as to be used routinely in a clinical laboratory running about 20 analyses per day.

F. Serum Lactic Acid Dehydrogenase

This enzyme catalyzes the conversion of pyruvate to lactate, DPNH being oxidized simultaneously.[8a] The appearance of DPN+ is therefore used as a measure of the enzyme activity. The procedure is as described for glutamate-pyruvate transaminase except that no prepared enzyme is added and the reaction is started with the addition of 0.2 μmole of sodium pyruvate. After 10 minutes, the procedure is stopped by the addition of trichloroacetic acid and DPN+ is assayed. The method is simple and reproducible, and, according to the authors, well suited as a routine method for a hospital laboratory. The same procedure should, of course, be applicable to the assay of lactic dehydrogenase in other tissues.

The foregoing procedures for enzyme assay utilizing pyridine nucleotide fluorescence make use of a derived fluorescence and cannot be used in a continuous manner. However, both DPNH and TPNH possess native fluorescence with fairly constant intensity from pH 7 to 12. This fluorescence is about one-tenth as sensitive as the alkaline and carbonyl fluorescences of the oxidized coenzymes. However, it is still much more sensitive than the spectrophotometric method. It should be possible to utilize the native fluorescence as a direct and continuous process without the necessity of protein precipitation and chemical reactions. With purified preparations this is certainly the method of choice where sensitivity is desired or where one wants to conserve precious enzyme or substrate preparations. Although no direct application to quantitative assay of enzyme activity is presented here, native fluorescence of reduced pyridine nucleotides has been widely used in studies on their interaction with various enzymes and substrates. With less pure, or crude, preparations containing much protein and other absorbing and fluorescing materials, the use of native fluorescence in a direct, continuous procedure presents some difficulties. These materials introduce problems of light scattering, inner filter effects, and quenching, which may be difficult to correct even with internal standards. It may be, however, that the sensitivity of the procedure permits the use of such small amounts of a crude preparation as to introduce negligible amounts of interfering material.

G. Succinic Semialdehyde Dehydrogenase

The major route of metabolism for γ-aminobutyric acid in mammalian tissues involves transamination to yield succinic semialdehyde which is subsequently oxidized to succinic acid by a specific DPN-linked dehydrogenase. According to Albers and Koval[9] the enzyme from monkey brain has a K_m of about 3×10^{-6} and exhibits marked substrate inhibition with concentrations of succinic semialdehyde above 2×10^{-5} M. Fluorometric assay based on the production of DPNH is the only procedure which permits quantitative measurement of this enzyme, by making possible precise measurements with microgram quantities of substrate.

There have been many enzyme assays based on fluorophors other than the pyridine nucleotides. Some of these are presented here.

H. Glutamic Acid Decarboxylase

In mammals, glutamic acid decarboxylase is found mainly in brain where it gives rise to γ-aminobutyric acid, a compound which is currently of great interest to the neurochemist. Lowe et al.[8b] devised a fluorometric method for the measurement of γ-aminobutyric acid based on its reaction with ninhydrin (see Chapter 5). They have applied this method to the assay of the enzyme, glutamic acid decarboxylase, by following the appearance of the end product of the reaction. The sensitivity of the fluorometric assay is so great that enzyme measurements can be made on microgram quantities of brain tissue. With the aid of this enzyme assay Lowe et al.[8b] were able to carry out most detailed studies on the localization of the enzyme within the central nervous system.

I. Xanthine Oxidase

It was shown by Lowry et al.[10] that 2-amino-4-hydroxypteridine is readily oxidized by xanthine oxidase to yield isoxanthopterin. Since phosphate buffer quenches the fluorescence of the substrate but has no effect on that of the product, appearance of fluorescence can be used to assay the enzyme.

One milliliter of 0.2 M phosphate buffer pH 7.2, containing 5×10^{-6} M substrate,* is placed in a fluorometer tube and brought to 30°. Sufficient enzyme is added in a volume of 0.01 to 0.1 ml. to oxidize about half the substrate in 30 to 90 minutes (about 1 to 2 mg. of rat liver or kidney). The initial reading is made immediately after mixing, the instrument having been standardized with another solution, preferably quinine. The setting is such that there will be approximately full-scale deflection with the substrate completely oxidized (excitation, 360 mμ; fluorescence, 450–460 mμ; uncorrected). The tube is maintained at 30° and its fluorescence is measured at intervals, mixing each time measurement is made. When 50% of the reaction is complete, the sample is placed in the bath for an interval of time sufficient to complete oxidation. (Purified xanthine oxidase, if available, may be added to speed this up.) The fluorescence, when reaction is complete, represents an internal standard of known concentration. Enzyme activity can then be expressed as micromoles of pteridine oxidized per gram per hour. If the fluorescence change is not linear with time, the initial velocity may be computed by graphical means.

The Michaelis constant for xanthine oxidase with aminohydroxypteridine is extremely small so that the enzyme is always fully saturated. The rate does fall off because of the marked inhibition by the end product. This will be discussed later in this chapter. It should be pointed out that in this assay the fluorophor is being assayed directly and continuously in the incubation mixture. Small amounts of tissue are taken for assay to minimize light scattering and absorption. In addition, each sample acts as its own standard to correct for blank fluorescence or fluorescence losses produced in any way. In such an assay where the reaction proceeds for a long period of time, it is important to maintain constant temperatures for

---

* A stock solution of 2-amino, 4-hydroxypteridine (5×10^{-4} M) in water may be stored indefinitely when frozen. It is diluted 1 to 100 with buffer before use.

all the samples and the quinine standard used for comparison. Obviously, a fluorometer with a temperature regulated sample compartment is necessary for such studies.

J. Fumarase[11, 12]

The malate formed during the reaction is treated with β-naphthol in strong sulfuric acid to yield a highly fluorescent product.

Reagents. Fumaric acid, 0.02 M, in 0.04 M phosphate buffer pH 6.8. This is stored frozen to prevent bacterial growth.

Fluorescence reagent. A stock solution of 56 mg. % β-naphthol in 0.004 N NaOH may be stored in frozen form for long periods of time until it turns very yellow. Just before the assay, 1 volume of the naphthol solution is mixed with 25 volumes of 1:7 sulfuric acid.

Tissue samples, 5–25 μl. volume (equivalent to 50–200 μg. brain or kidney) are transferred to small test tubes in a metal rack in an ice bath. Ice-cold substrate solution, 0.1 ml., is added rapidly with mixing to each, and the rack is transferred to a 38° water bath. After 30 minutes the rack is returned to the ice bath to stop the reaction. Ten microliter aliquots are transferred to fluorometer tubes containing 1 ml. of the β-naphthol-sulfuric acid mixture. After thorough mixing the tubes are stoppered with corks, covered with aluminum, and heated in a boiling water bath for 30 minutes. The tubes are cooled to room temperature, carefully mixed again, and wiped clean. Fluorescence is then assayed using the 365-mμ mercury line for excitation. The secondary filter or monochronomator should be set for maximal transmission at 465 mμ. At these wavelengths quinine is an excellent reference standard. Enzyme activity is reported in terms of micromoles of fumarate formed per hour. Fumarate standards, treated with the β-naphthol reagent, are used for comparison and calculation.

K. Malic Dehydrogenase

Loewus et al.[13] have utilized the β-naphthol assay for malic acid to estimate the enzyme, malic dehydrogenase. Lowry et al.[2, 3] measured enzyme activity by following the formation of DPN$^+$ fluorometrically upon incubation of the tissue with oxaloacetate and DPNH. With their ultramicro method, as little as 1×10^{-5} μg. of brain tissue is sufficient for assay.

L. β-Glucuronidase[14]

Umbelliferone (7-hydroxycoumarin) and 4-methylumbelliferone are highly fluorescent compounds. In the body they are converted to the cor-

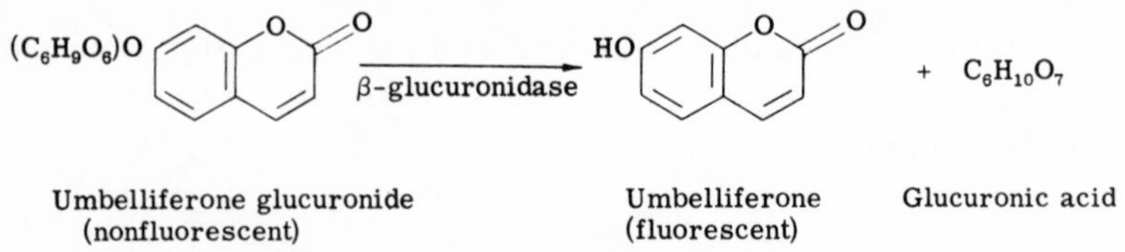

$(C_6H_9O_6)O$ ⟶ (β-glucuronidase) HO ⟶ + $C_6H_{10}O_7$

Umbelliferone glucuronide (nonfluorescent)　　　Umbelliferone (fluorescent)　　　Glucuronic acid

FIG. 2. Conversion of umbelliferone glucuronide to umbelliferone.

responding glucuronides, which are nonfluorescent. In the presence of the enzyme β-glucuronidase, they are split to release the free fluorescent products (Fig. 2). The glucuronide of 4-methylumbelliferone can be prepared in excellent yield by feeding the free compound (commercially available) to rabbits and isolating the derivative from the urine. The procedure described by Mead et al.[14] gives a 20–25% yield (isolated) from an administered dose of 2.5 g.

The enzyme solution (1 ml.) is mixed in a test tube with 3.5 ml. of 0.1 M acetate buffer, pH 4.0–4.5, and 0.5 ml. of 0.001 M 4-methylumbelliferone glucuronide. Immediately after adding substrate and mixing, the tubes are incubated for exactly 30 minutes at 37°. The addition of 20 ml. of 0.1 M glycine buffer, pH 10.3, stops the reaction and provides the proper pH for measurement of fluorescence. Standards of 4-methylumbelliferone (0.02– 5.0 μg.) should be prepared by diluting as for incubation mixtures, omitting

TABLE II

β-GLUCURONIDASE ACTIVITY OF VARIOUS TISSUES DETERMINED FLUOROMETRICALLY[a]

Tissue or fluid	4-Methylumbelliferone liberated μg./mg./hour
Guinea pig liver	8.6
Guinea pig muscle	0.06
Rat liver	22.5'
Rat spleen	14.5
Rat muscle	0.2
Housefly (whole)	2.2
Human urine	0.50–1.8
Human saliva	0.45–0.80
Human plasma	7–13
Rabbit plasma	50–145

[a] All assays were carried out at pH 4.6 except for plasma assays which were made at pH 3.4 for human and pH 4.0 for rabbits. Urine and saliva represent casual samples (after Mead et al.[14]).

the enzyme. Blanks are prepared by incubating with the glucuronide in the absence of enzyme and adding the latter after dilution with the glycine buffer. The excitation maximum is at 340 mμ and the fluorescence maximum at 440 mμ (uncorrected) at pH 10.[15]

The β-glucuronidase content of a variety of tissues, obtained by the fluorescent procedure are shown in Table II. It is apparent that less than milligram quantities of most tissues are required for assay. For this reason there is little interference with the fluorescence emission. Where the activities are lower, as in plasma and urine, care must be taken to detect interference. Small amounts of hemoglobin interfere with the assay so that whole blood cannot be used. Plasma should be free from hemolysis and urine should contain no heme products or other unusual pigments.

M. β-Glucosidase[16]

This is an enzyme which is widely distributed in nature and which splits a variety of β-glucosides. The use of a glucoside of the highly fluorescent umbelliferone makes available an extremely sensitive procedure for assay of this enzyme.

The substrate 4-methylumbelliferone-β-D-glucoside was chemically prepared. A 0.005 M solution was prepared fresh each day since it decomposed appreciably on standing in solution at room temperature.

For assay, 0.5 ml. of substrate solution and 3.5 ml. of 0.2 M acetate buffer, of optimum pH, are added to each of three tubes. One milliliter of enzyme solution (diluted so as to liberate in the order of 1 μg. of umbelliferone in 30 minutes) is added to two of the tubes. To the third tube is added a milliliter of the same enzyme which has been inactivated previously by boiling. Incubation is carried out at 37° for exactly 30 minutes and the reaction is stopped by adding 15 ml. of 0.2 M glycine buffer, pH 10.3. Fluorescence is assayed as described in the glucuronidase procedure and enzyme activity is expressed as micrograms of umbelliferone liberated per gram per hour.

By using this procedure, Robinson[16] showed that the enzyme is widely distributed in plants and animals. The fluorescent procedures were also used in partially successful attempts to separate β-glucosidase from β-glucuronidase by electrophoresis on paper. The sensitivity of this fluorometric procedure is again so high that less than milligram quantities of tissues are sufficient for assay.

N. Phosphatases

Moss[17] has investigated the naphthyl phosphates as substrates of acid and alkaline phosphatase. The fluorescence excitation and emission spectra

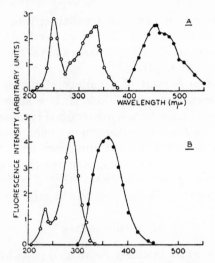

FIG. 3. Corrected excitation (o) and fluorescence (●) spectra of (A) α-naphthol, and (B) β-naphthol, at pH 10 (after Moss[17]).

of α- and β-naphthol and of the corresponding orthophosphate esters are shown in Fig. 3. The excitation maxima for α-naphthol are at 250 and 335 mμ with peak emission at 455 mμ. The corresponding phosphate ester is maximally excited at 235 and 295 mμ, emission being at 365 mμ. Thus, even though the ester is fluorescent, it is possible to select appropriate wavelengths to distinguish the free phenol from the ester. This is achieved at 335 mμ for excitation and 455 mμ for fluorescence. With β-naphthol, excitation at 350 mμ and measurement of fluorescence at 425 mμ permits measurement of the free phenol in the presence of the ester. Both α- and

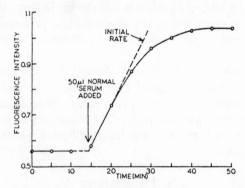

FIG. 4. Hydrolysis of α-naphthyl phosphate by serum alkaline phosphatase; pH 10, 37°, with excitation at 335 mμ and fluorescence at 455 mμ (after Moss[17]).

β-naphthol have maximal and constant fluorescence at pH 10 and above. It is therefore possible to follow alkaline phosphatase continuously (Fig. 4). However, acid phosphatase necessitates alkalinization of the reaction mixture before fluorometric assay.

To assay serum alkaline phosphatase, 10 μl. of serum are diluted with buffer, pH 10. α-Naphthyl phosphate is added and the mixture is incubated for 15 minutes at 37°. The reaction is stopped by adding sufficient N NaOH to bring the pH to 12, thereby inactivating the enzyme. The fluorescence of the liberated phenol is then measured at the appropriate wavelength settings. To measure acid phosphatase in normal serum requires larger volumes of serum (50–100 μl.) for a comparable period of incubation, 15 minutes.

With these fluorometric procedures, Moss[17] investigated the properties of the phosphatases in Paget's disease and in obstructive jaundice.

A histochemical procedure for measurement of phosphatase in tissues, using riboflavin-5′-phosphate as substrate and fluorescence microscopy for detection, has been reported by Takeuchi and Nogami.[18] The procedure was also applied to localization of riboflavin-nucleotidase activity.

O. Sulfatases

In their paper on glucuronidase assay, Mead *et al.*[14] also prepared the ethereal sulfate of 4-methylumbelliferone and investigated its use in the assay of the enzyme sulfatase. They found, however, that unlike the glucuronide the sulfate possessed appreciable fluorescence of its own, thereby giving a large "blank" in an enzyme assay. However, it is highly likely that the ethereal sulfate differs from free 4-methylumbelliferone not only in intensity of fluorescence but also in the location of excitation and fluorescence maxima. With presently available fluorescence spectrometers it should be possible to distinguish the two fluorophors in the same manner as for the naphthols and their phosphates.

P. Aromatic L-Amino Acid Decarboxylase

It is now recognized that tissues contain an enzyme which is capable of converting all the normally occurring aromatic amino acids and histidine to the corresponding amines. 3,4-Dihydroxyphenylalanine (dopa) decarboxylation and 5-hydroxytryptophan decarboxylation are manifestations of the same enzyme, aromatic L-amino acid decarboxylase.[19] The substrates and end products of this reaction are aromatic or cyclic compounds and most of them either possess a high degree of native fluorescence or can be converted to fluorescent derivatives. Since the substrates are amino acids and the end products are amines, the use of simple chromatographic procedures can effect separation before fluorometric assay. The use of a column procedure for following dopa decarboxylation was first described by Dietrich.[20] A modified version[21] of this method, utilizing fluorescence spectrometry, which is applicable to many other amines as well, is presented.

In the enzyme reaction, dopa is converted to dopamine. Following incubation, the mixture is passed over a permutit column. The amine is

$$\text{HO}-\underset{\text{HO}}{\bigcirc}-\text{CH}_2-\text{CHNH}_2-\text{COOH} \xrightarrow[\text{decarboxylase}]{\text{aromatic L-amino acid}} \text{HO}-\underset{\text{HO}}{\bigcirc}-\text{CH}_2-\text{CH}_2\text{NH}_2 \ + \ CO_2$$

Dopa   Dopamine

adsorbed onto the column and the amino acid passes through in the effluent. By washing the column with several volumes of water it is possible to remove all unreacted dopa. The dopamine is then eluted with several milliliters of 20% KCl and assayed fluorometrically, excitation, 285 mμ; fluorescence, 325 mμ. Standards and blanks are prepared in the KCl. The sensitivity of this method is such that with homogenates of guinea pig kidney it is possible to utilize less than milligram quantities of tissue. Although the method is not as rapid as manometry, it is hundreds of times more sensitive and much more specific. It is certainly the procedure of choice in any studies on the localization of the enzyme in tissues. In purification of the enzyme it offers the advantage of requiring much smaller samples for assay at each step of the procedure.

Dopa is not a good substrate with which to assay the activity of the enzyme because this amino acid combines nonenzymatically with the coenzyme, pyridoxal phosphate, leading to substrate inhibition.[22] 5-Hydroxytryptophan is an excellent substrate and it can be used for enzyme assay in the same way as dopa. The resulting amine, serotonin, is even more highly fluorescent than dopamine so that with this indoleamino acid the method becomes even more sensitive. Fluorescence can be measured at neutral pH where excitation is at 295 mμ and fluorescence is at 330 mμ or in 3 N HCl where excitation is also at 295 mμ but fluorescence is at 550 mμ (uncorrected values). Tryptophan, tyrosine, o-tyrosine, and m-tyrosine can all be utilized as substrates by column separation and assay of the native fluorescence of the resulting amine. p-Tyramine can also be assayed by the procedure of Waalkes and Udenfriend.[23] It should be pointed out that since the amino acids and amines have the same fluorescence characteristics the chromatographic separation must be carried out in a highly efficient manner. This is particularly important because the amino acid, being the substrate, is present in much higher concentration than the enzymatically formed amine.

Q. Histidine Decarboxylase

Although L-histidine can serve as a substrate of aromatic L-amino acid decarboxylase, there is also a specific L-histidine decarboxylase present in mast cells.[24] The two enzymes differ markedly in many respects. Their pH optima are so far apart (pH 6 for the specific enzyme and pH 9.0 for the general enzyme) that, by selecting these pH values, each can be measured without interference by the other. As with dopa and 5-hydroxytryptophan, the enzyme activity is assayed by following histamine appearance.

For assay of enzymatically formed histamine a 0.5-ml. aliquot from an incubation mixture is added to a centrifuge tube containing 0.2 ml. of 5% trichloroacetic acid. The tube is placed in a boiling water bath for at least 2 minutes to ensure complete protein precipitation, then 1.5 ml. of water are added and the tube is centrifuged. The entire supernatant solution is poured onto a small column containing about 1 ml. of the anion exchanger, Dowex 1 (carbonate form), to remove the unreacted histidine. Column dimensions of about 5 × 30 mm. have been found practical. The entire effluent is collected in a graduated tube and combined with an additional 1 ml. of water which is passed over the column. The volume is adjusted to 3.2 ml. and a 2-ml. aliquot is assayed for histamine by the o-phthalaldehyde procedure (Chapter 5). The procedure can detect 0.2 μg. of histamine in the presence of 1000 μg. of histidine. The low affinity of enzyme for substrate in the case of the nonspecific enzyme requires large amounts of the amino acid substrate. It should be pointed out that the histidine decarboxylase activity of most tissues is so low that heretofore it was necessary to use radioactive histidine (very highly labeled) and the elaborate isotope dilution technique as developed by Schayer et al.[25] With the fluorescent procedure it is now possible to assay the enzymes in many tissues in a much simpler manner. Tissues such as mast cell tumors of mice contain the specific decarboxylase with an activity of about 100 μg. of histamine formed per gram of tissue per hour. The activity of the nonspecific enzyme in guinea pig kidney is about 50 μg./g./hour.

R. Monoamine Oxidase

This enzyme is widely distributed in animal tissues and is localized largely in the mitochondrial fraction of the cell. It has received a great deal of attention and many procedures are available for its assay—manometric, colorimetric, and spectrophotometric. However, the most sensitive of all procedures is a fluorometric one involving the conversion of tryptamine to indoleacetic acid (Fig. 5) and measurement of the acid as described in Chapter 5. In the presence of excess aldehyde dehydrogenase, the formation of indoleacetic acid from tryptamine becomes proportional to

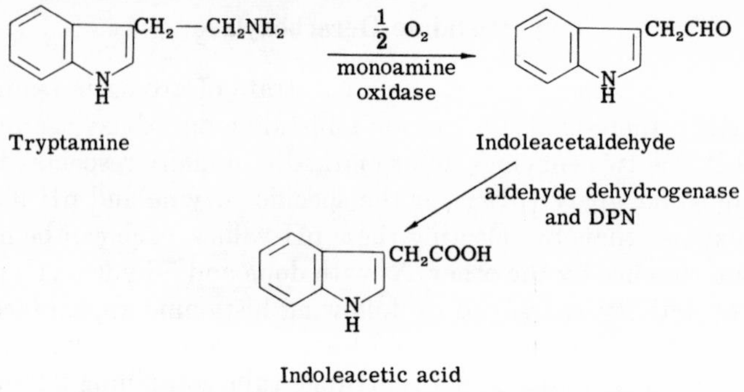

Tryptamine   Indoleacetaldehyde

Indoleacetic acid

FIG. 5. Fluorometric assay of monoamine oxidase.

the monoamine oxidase activity.[26] The sensitivity of the method makes it particularly appropriate for studies on small amounts of tissue as are available from sympathetic nerves and ganglia, blood vessels, and various portions of brain.[27a]

For assay, incubations are carried out in a volume of 3 ml., containing enzyme, 10 μmoles of tryptamine, 10 μmoles of DPN, 240 μmoles of nicotinamide, aldehyde dehydrogenase,* and phosphate buffer, pH 8.0, at a final concentration of 0.08 M. Following incubation, 0.5 ml. of the mixture, containing at least 0.2 μg. of indoleacetic acid, is transferred to a glass-stoppered tube containing 2.0 ml. of 0.1 N HCl and 15 ml. of chloroform. The tube is shaken, centrifuged, and the aqueous layer is removed by aspiration. Another 2-ml. portion of 0.1 N HCl is added and the tube is again shaken and centrifuged. A 10-ml. aliquot of the chloroform layer is transferred to another tube containing 1.5 ml. of 0.1 M phosphate buffer, pH 7.4. After centrifugation, 1 ml. of the buffer is transferred to a quartz cuvette and assayed in a fluorescence spectrometer, excitation, 280 mμ; fluorescence 370 mμ; uncorrected.

With larger amounts of enzyme the monoamine oxidase assay can be run in an even simpler manner using a cation exchange resin to remove the substrate (tryptamine) and measuring the fluorescence of the enzymatically formed indoleacetic acid in the eluate.

S. Diamine Oxidase (Histaminase)

This enzyme is found in many tissues but mainly in liver and kidney. Although it is not specific for histamine, the enzyme is associated mainly

---

* Prepared according to the procedure of E. Racker, *J. Biol. Chem.* **177,** 883 (1947). Each beaker should contain an amount of enzyme which would yield 0.3 mM of DPNH per minute with glycolaldehyde as substrate.

with the metabolism of this active humoral agent. Cohn and Shore[27b] have developed a highly sensitive procedure for assaying this enzyme by measuring the disappearance of histamine or agmatine with the fluorometric o-phthalaldehyde procedure for the amine (see Chapter 5).

With kidney preparations, an amount equivalent to about 50 to 150 mg. of fresh tissue is placed in an incubation beaker and the volume is made up to 3.0 ml. with 0.2 M sodium phosphate buffer, pH 7.2. The beaker is incubated in a Dubnoff metabolic shaker at 37° in air. Following a 15-minute preincubation period, 0.27 μmole of histamine is added. The enzymatic reaction is terminated at the end of an hour by adding 1 ml. of 2 N perchloric acid to the beaker or by transferring a 1-ml. aliquot of the incubation mixture to a tube containing 4 ml. of 0.4 N perchloric acid. The histamine in the protein-free perchloric acid solutions is then assayed fluorometrically as described in Chapter 5. Diamine oxidase activity is expressed as micromoles of histamine metabolized per gram of tissue per hour. Tissue blanks without added histamine are prepared and incubated with the other vessels.

The sensitivity of the fluorometric assay for histamine permits the determination of diamine oxidase in small amounts of tissue or in tissues with relatively low enzyme activity. The determination of the K_m for the guinea pig kidney enzyme required such a sensitive method since it was found to be approximately 5 $\times$ 10^{-6} M.[27b]

T. Catechol-O-Methylpherase

This enzyme is found in many tissues and is present in highest concentration in liver and kidney. It is certainly of greatest importance in the metabolism of circulating epinephrine and norepinephrine. A fluorometric procedure for measuring the activity of this enzyme by assaying the amount of metanephrine formed from epinephrine (Fig. 6) was reported by Axelrod and Tomchick.[28]

A mixture containing the enzyme preparation, 0.3 μmole of l-epinephrine, 10 μmoles of magnesium chloride, 0.1 μmole of S-adenosylmethionine, 50 μmoles of phosphate buffer at pH 7.8, and water to make a final volume of 1 ml. is incubated in a 40-ml. glass-stoppered centrifuge tube at 37°.

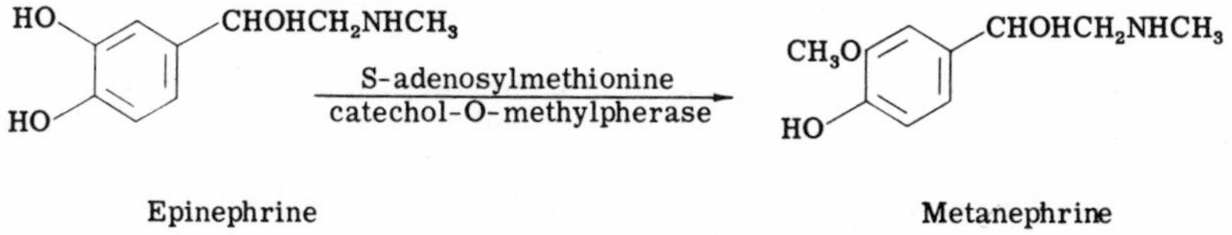

FIG. 6. Fluorometric assay of catechol-O-methylpherase.

TABLE III

CATECHOL-O-METHYL TRANSFERASE IN SOME ANIMAL TISSUES[a]

Tissue	Metanephrine formed $\mu g./g./hour$
Rat liver	1182
Rat kidney	414
Rat lung	60
Rat brain	39
Rat muscle	39
Cat liver	60
Rabbit liver	20
Human liver[b]	78

[a] After Axelrod and Tomchick.[28]
[b] The sample of human liver had been stored at $-10°$ for 3 months before assay.

After 30 minutes, 0.5 ml. of 0.5 M borate buffer, pH 10.0, is added and the incubation mixture is assayed for metanephrine by the procedure involving solvent extraction and fluorometric assay as described in Chapter 5. Excitation is at 285 mμ and fluorescence at 335 mμ.

With this procedure Axelrod and Tomchick[28] determined many of the properties of mammalian catechol-O-methylpherase and measured its distribution in various tissues (Table III). Since the procedure for metanephrine is sensitive to the microgram level, it is apparent that many tissues require only milligram quantities of material for enzyme assay.

U. Kynureninase

Jakoby and Bonner[29] used a fluorometric procedure to assay the enzyme kynureninase (Fig. 7). The assay takes advantage of the differences in

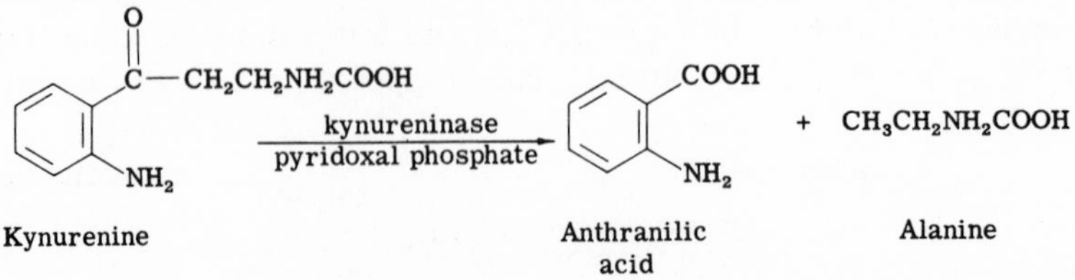

FIG. 7. Fluorometric assay of kynureninase. Following incubation with enzyme and deproteinization, pH is adjusted to 5.5 and fluorescence emitted at 405 mμ is measured when excitation is at 300 mμ (after Jakoby and Bonner[29]).

fluorescence characteristics between the substrate and end product. Anthranilic acid fluoresces maximally in neutral or slightly acid solution whereas kynurenine fluorescence is maximal at pH 11. The substrate also fluoresces less intensely and at different wavelength than the product. Excitation and fluorescence maxima for anthranilic acid are 300 and 405 mμ, respectively; for kynurenine they are 370 and 490 mμ.[30a] Following incubation with enzyme, the samples are deproteinized and aliquots are buffered to pH 5.5 where kynurenine fluorescence is minimal. With the monochromators set to the appropriate wavelengths for anthranilic acid, or with suitable filters, the fluorescence is a linear function of anthranilic acid concentration. This is, in turn, directly proportional to kynureninase activity. Comparable assays have been carried out with hydroxykynurenine and formylkynurenine as substrates of kynureninase.[29]

V. 3-Hydroxyanthranilic Acid Oxidase

Bokman and Schweigert[30b] have studied the enzyme systems responsible for converting 3-hydroxyanthranilic acid to quinolinic acid by measuring the disappearance of 3-hydroxyanthranilic acid fluorometrically. With mammalian enzymes, only quinolinic acid is formed. With less pure systems, N'-methylnicotinamide and nicotinic acid also appear. However, none of these compounds possess native fluorescence comparable to that of 3-hydroxyanthranilic acid. It was also shown that the substrate can be recovered quantitatively from heat inactivated enzyme preparations so that its disappearance can be used as a direct measure of enzyme activity. 3-Hydroxyanthranilic acid fluoresces maximally at pH 7. The excitation maximum is at 320 mμ and the fluorescence maximum at 415 mμ (uncorrected).[30a]

W. Fibrinolysin

Lüscher and Käser-Glanzmann[30c] have developed a method for the measurement of fibrinolysin with the fluorescent dye Lissamine Rhodamine B 200 which is chemically bound to fibrinogen and which is liberated on lysis of the fibrin clot. The fluorescent fibrinogen is prepared by treatment of the protein with the sulfonyl chloride of the dye and purified by procedures such as those described in Chapter 6, Section VII. To measure fibrinolytic activity of a tissue sample the dye-labeled protein is incubated with it and other components required for clot formation. The fluorescence liberated into the fluid on clot retraction is proportional to the fibrinolytic activity. The number of fluorescent groups combined with a mole of protein can be ascertained as described in Chapter 6.

III. THE USE OF ENZYMES AND FLUOROMETRY TO DETERMINE SUBSTRATES AND COFACTORS

Just as an enzyme can be assayed by making it the rate limiting component of a reaction, so can a substrate or a coenzyme be assayed. Again, pyridine nucleotide enzymes can be used most effectively by utilizing an excess of enzyme and coenzyme and measuring the formation of the oxidized or reduced coenzyme. Greengard[31-33] has published a large number of such fluorometric assays for substrates including methods for pyruvate, adenosine diphosphate, adenosine triphosphate, phosphoenol pyruvate, and glucose-6-phosphate. The methods were applied in studies on metabolism in nerve fibers.

A. Pyruvate[31]

This compound can be assayed as in the spectrophotometric method of Ochoa *et al.*[34] by following DPN$^+$ formation in the following reaction:

$$\text{Pyruvate} + \text{DPNH} \xrightarrow[\text{dehydrogenase}]{\text{lactic}} \text{lactate} + \text{DPN}^+$$

The following are added to a fluorometer tube: the sample, 10 μmoles of triethanolamine, pH 7.7, 0.5 μmole of ethylenediaminetetraacetate, pH 7.4, and 1.6 mμmoles of DPNH. The volume is adjusted to 1 ml. and the

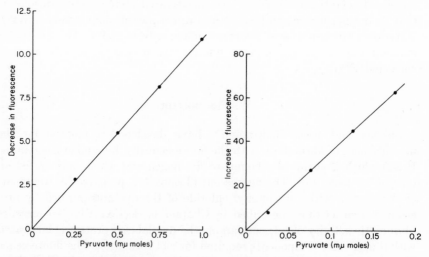

FIG. 8. Assay of pyruvate according to Greengard[31]; at left by measurement of the decrease in native fluorescence of DPNH, at right by measurement of enzymatically formed DPN$^+$ with the ketone condensation method.

samples are placed in a constant temperature bath. The reaction is started by adding 0.01 ml. of lactic dehydrogenase (about 40 Bücher units) and readings of the native fluorescence of DPNH are made at 1-minute intervals until a stable value is obtained. Blanks and standards can be prepared by substituting water and known amounts of pyruvate for the sample. The use of the native fluorescence permits continuous measurement of the activity and requires no additional chemical manipulations. This permits greater precision and makes it possible to use an aliquot of enzyme for more than one assay. Yet, it is sufficiently sensitive so that pyruvate can be estimated in amounts of 2.5×10^{-10} mole in a volume of 1 ml. If greater sensitivity is desired, the reaction can be stopped and the enzymatically formed DPN^+ may be measured as the ketone condensation product (Chapter V). With the latter procedure the assay can be used to estimate 2.5×10^{-11} mole of pyruvate per milliliter of incubation mixture. Standard curves obtained for pyruvate by both methods are shown in Fig. 8.

B. Glucose-6-Phosphate[31]

This compound is assayed as in the spectrophotometric method of Ochoa et al.[35]:

$$\text{Glucose-6-phosphate} + TPN^+ \xrightarrow[\text{dehydrogenase}]{\text{glucose-6-phosphate}} \text{6-phosphogluconate} + TPNH$$

With excess TPN^+ one molecule of the pyridine nucleotide is reduced for each molecule of glucose-6-phosphate present. Again, enzymatically formed TPNH can be assayed by its native fluorescence, in a continuous fashion, or by employing the ketone condensation procedure.

C. Adenosine Diphosphate[31]

The sample, 0.25–2.5 mμmoles of ADP in several microliters, is added to a fluorometer tube containing 10 μmoles triethanolamine, pH 7.7, 0.5 μmole ethylenediaminetetraacetate, pH 7.4, 1.6 mμmoles of DPNH, 4 μmoles magnesium sulfate, 80 μmoles KCl, 1.6 mμmoles of phosphoenolpyruvate, and the volume is adjusted to 1 ml. by the addition of water. The reaction may be run at room temperature and is started by the addition of 0.04 ml. of enzyme mixture containing 40 Bücher units each of lactic dehydrogenase and pyruvate phosphokinase. Fluorometric measurements of DPNH are made at regular intervals until stable values are obtained. Blanks and standards are run by substituting water and known amounts of adenosine diphosphate for the sample. Less than 2×10^{-10} mole of adenosine diphosphate can be assayed in 1 ml. of solution.

D. Phosphoenolpyruvate[31]

This is assayed with the same system as above, but with the addition of excess adenosine diphosphate. Under these conditions phosphoenolpyruvate becomes limiting and is equivalent to DPNH disappearance.

E. Adenosine Triphosphate

Greengard[31, 33] has reported two procedures for the enzyme-fluorometric assay of this nucleotide. The first utilizes the enzyme systems corresponding to those used in the spectrophotometric procedure of Kornberg[36]:

$$\text{ATP} + \text{glucose} \xrightarrow{\text{hexokinase}} \text{ADP} + \text{glucose-6-phosphate}$$

$$\text{Glucose-6-phosphate} + \text{TPN}^+ \xrightarrow[\text{dehydrogenase}]{\text{glucose-6-phosphate}} \text{6-phosphogluconate} + \text{TPNH} + \text{H}^+$$

The following procedure was designed to determine adenosine triphosphate in peripheral nerve fibers.[33] A nerve bundle (2–20 mg.) is plunged into a centrifuge tube containing 1.5 ml. of 0.1 M triethanolamine buffer, pH 8.1, which has been preheated to 100° in a water bath. After 40 seconds, the tube is cooled rapidly to 0° and the contents are homogenized. Following this, 1.5 ml. of alcohol-free* chloroform are added, the tube is stoppered with a ground-glass stopper, shaken vigorously for 3 minutes, and then centrifuged at 3000 × g for 10 minutes. An aliquot of the clear supernatant fluid is transferred to a fluorometer tube containing 10 μmoles of glucose, 5 × 10^{-3} μmoles of TPN$^+$, 1.5 μmoles of MgCl$_2$, 0.6 μmoles of ethylenediamine tetraacetate, 12 μmoles of triethanolamine, pH 8. The volume is adjusted to 0.95 ml. and, after mixing, an initial reading is taken. Following this, 0.05 ml. of glucose 6-phosphate dehydrogenase (0.08 units)[37] is added and the sample is mixed again. Readings of native TPNH fluorescence are made at suitable intervals until the reduction is maximal. The peak reading is proportional to the adenosine triphosphate content. There is sufficient hexokinase activity in most of the nerve extracts to permit completion of the reaction in 20 to 30 minutes. A reagent blank and a calibration curve are used for calculation.

A second fluorometric procedure for adenosine triphosphate, based on

---

* Chloroform frequently contains 1% ethanol as a preservative. The presence of this alcohol will introduce an error from the alcohol dehydrogenase present in the tissues. The chloroform can be freed of alcohol by washing it several times with equivalent volumes of water.

the spectrophotometric procedure of Thorn *et al.*,[38] is also available[33]:

$$\text{ATP} + \text{3-phosphoglyceric acid} \xrightleftharpoons{\text{kinase}} \text{ADP} + \text{1,3-diphosphoglyceric acid}$$

$$\text{1,3-diphosphoglyceric acid} + \text{DPNH} + \text{H}^+ \xrightleftharpoons[\text{phosphate dehydrogenase}]{\text{glyceraldehyde-3-}}$$

$$\text{glyceraldehyde-3-phosphate} + \text{inorganic phosphate} + \text{DPN}^+$$

Modifications of both methods to permit assay of creatine phosphate, and adenosine monophosphate are also described.

Greengard and Straub[39] applied these microfluorometric assay procedures to the measurement of the various organic phosphates and other metabolites in normal and stimulated nerves. The values for some normal metabolites under resting conditions are shown in Table IV. They were also able to demonstrate an exact balance between adenosine triphosphate disappearance and appearance of adenosine diphosphate and monophosphate during electrical stimulation of the nerves.

It should be pointed out that the same instruments which are used to measure fluorescence can be used to measure luminescence, light emitted through chemical reaction. Thus, adenosine triphosphate can be measured in the standard fluorometer or fluorescence spectrometer by making use of the reaction involving luciferin and luciferase[40]:

$$\text{ATP} + \text{Mg}^{++} + \text{O}_2 + \text{reduced luciferin} + \text{luciferase} \rightarrow \text{light}$$

TABLE IV

CONCENTRATIONS OF METABOLITES IN RESTING CERVICAL NERVES OF RABBITS DETERMINED BY FLUOROMETRIC ASSAY[a]

Metabolite	Concentration in resting nerve mμmole/mg. fresh wt.
Creatine phosphate	1.96
Adenosine triphosphate	2.11
Adenosine diphosphate	0.28
Adenosine monophosphate	0.08
Pyruvate	0.22
Glucose-6-phosphate	<0.03
1,3-Diphosphoglycerate	<0.03
Phosphoenolpyruvate	<0.03
α-Ketoglutarate	<0.03

[a] After Greengard and Straub.[39]

The reaction, when carried out in a commercial fluorometer with a photomultiplier detector (without the exciting lamp), is capable of measuring microgram quantities of adenosine triphosphate with an accuracy of 5%. It is, of course, possible to couple many other reactions to this one and, with excess adenosine triphosphate, assay other metabolites which are made limiting.

F. Hexose and Triosephosphates

Seraydarian et al.[41] have adapted the Slater procedures[42] for the enzymatic determination of hexose phosphate and triose phosphate intermediates to fluorometry, utilizing the disappearance of native DPNH fluorescence during oxidation. The methods are based on the following reactions:

Enzyme preparation A

Fructose-diphosphate $\rightleftarrows$ glyceraldehyde-phosphate + dihydroxyacetone phosphate
$\Updownarrow$
dihydroxyacetone phosphate

Dihydroxyacetone-phosphate + DPNH $\rightarrow$ glycerolphosphate + DPN$^+$

Enzyme preparation B

Glucose-6-phosphate$\rightarrow$fructose-6-phosphate

Fructose-6-phosphate + ATP $\rightarrow$ fructose-diphosphate

Enzyme preparation A determines the sum of fructose-disphosphate and triose phosphates. Subsequent addition of enzyme preparation B yields the additional determination of total hexose monophosphates (including glucose-1-phosphate).

In a typcial assay of hexose phosphates in muscle extracts, the following additions should be made to the cuvette: 0.5 ml. of reaction mixture (prepared by combining 8 ml. of 0.05 M triethanolamine buffer, pH. 7.6, containing 0.005 M ethylenediaminetetraacetate, 0.34 ml. of 0.1 M MgCl$_2$, 0.34 ml. of 0.005 M ATP, 4 ml. of 0.007 mM DPNH and 1.32 ml. of H$_2$O) 0.50 ml. of unknown extract or standard solution and 0.05 ml. of enzyme preparation A to start the reaction. Following completion of reaction A, enzyme preparation B is added. A blank, substituting water for extract or standard, is run together with the analyses. A typical analysis is shown in Fig. 9.

Solutions of DPNH may be standardized by measurement of absorbancy at 340 mμ in a spectrophotometer. Appropriate dilutions of the standardized DPNH solutions are then measured fluorometrically to permit

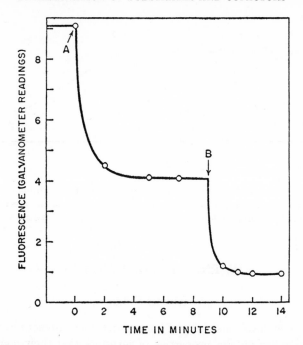

FIG. 9. Fluorometric assay of hexose and triose phosphates in muscle extracts according to Seraydarian et al.[41] At A, enzyme preparation A is added to determine hexose diphosphate and triose phosphates. When this reaction is completed enzyme preparation B is added (at B) to determine hexose monophosphates. See text for composition of enzyme preparations.

calculations. Daily calibrations should also be performed, using standard solutions of fructose-6-phosphate and fructose-diphosphate.

Comparison of this fluorometric procedure with spectrophotometric methods was found to be excellent and applications to studies on muscle contraction were found to yield data comparable to those obtained by Lundsgaard[43] and consistent with the results of other investigators; large increases in hexose monophosphates and diphosphates accompanying the disappearance of creatine phosphate and adenosinetriphosphate.

It should be pointed out that prior to assay muscle extracts are freed of glucose by a simple chromatographic procedure. This is necessary because enzyme preparation B usually contains some hexokinase. This can be shown by adding glucose to an incubation mixture (Fig. 10). It should be possible to assay glucose and hexokinase by some modifications of the fluorometric procedure.

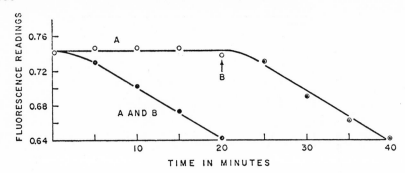

FIG. 10. Disappearance of DPNH fluorescence upon the addition of glucose to enzyme preparations, according to Seraydarian *et al.*,[41] indicating the presence of hexokinase activity. Only preparation B (see text) contains appreciable amounts of hexokinase.

IV. ENZYME MECHANISMS

A. Kinetic Studies

The extremely high sensitivity of fluorescence measurement has also been applied to studies on the properties of enzymes and their combinations with substrate and coenzyme. Spectrophotometry and other procedures are frequently not sufficiently sensitive to measure the dissociation constants and Michaelis-Menten quotients (K_m) of enzyme substrate or enzyme coenzyme complexes. Thus, Lowry *et al.*,[10] in studying the enzyme xanthine oxidase, were able to make precise measurements of the K_m of xanthopterin, which is fluorescent. However, they could only approximate the value for xanthine, which does not emit detectable fluorescence. The high sensitivity was needed because the dissociation constants of these substrates with xanthine oxidase are of the order of 10^{-6} M. One of the pteridines had a K_m of 5×10^{-7} M. When the K_m of an enzyme for a metabolite or coenzyme is about 10^{-6} or lower, procedures more sensitive than spectrophotometry are required. When the substrate is a fluorophor or can be measured fluorometrically, it· is even possible to extend K_m measurements several orders of magnitude lower. Lowry *et al.*[1] also applied fluorescence assay to determine the Michaelis constant of brain glucose-6-phosphate dehydrogenase for TPN+. As shown in Fig. 11, this turned out to be 2.79×10^{-6}. Lovenberg *et al.*[21] utilized the fluorescence of a number of amines to determine their K_m values with aromatic L-amino acid decarboxylase.

However, fluorescence offers to enzyme studies much more than new microtechniques. The marked sensitivity of fluorescence to intramolecular

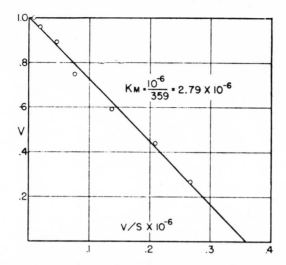

FIG. 11. Plot for estimating the Michaelis constant of TPN$^+$ with glucose-6-phosphate dehydrogenase from rabbit brain. Initial velocities were determined from direct readings of the native fluorescence of TPNH as formed in the incubation vessel. The initial TPN$^+$ concentrations ranged from 10^{-4} to 10^{-6} M (after Lowry *et al.*[1]).

and intermolecular changes permit the detection of molecular interactions which are unique to fluorescence. Ionizations or molecular interactions which are not readily detectable by spectrophotometric methods can alter the efficiency of fluorescence. Fluorophors, such as DPNH or flavin adenine dinucleotide, upon interaction with a specific enzyme may yield more or less fluorescence, the degree of enhancement or quenching being indicative of the binding mechanism. Molecular interactions can also lead to changes in excitation and fluorescence spectra. Such changes are indications of alterations in the structure of the fluorophor accompanying the molecular interaction.

When two different molecules are sufficiently close to one another, they can influence each other's fluorescence. Thus, if one happens to absorb radiant energy which the second emits as fluorescence, it is possible to demonstrate a fairly efficient transfer of this energy from one molecule to the other when the molecular complex is irradiated. Such interaction can involve the aromatic amino acids in a protein enzyme and fluorescent coenzymes. The distances between such molecules can, therefore, be estimated. The fluorescence emitted from individual molecules of a species also occur at a definite angle and direction in relation to the molecular parameters. Fluorescence from rigid molecules, as in a solid, is therefore highly polarized. Small molecules in solution in nonviscous solvents show little fluorescence polarization because their positions are randomized by the rapid Brownian motion. However, large molecules, such as proteins, have a much slower Brownian motion thereby maintaining some degree of polarization during the lifetime of the fluorescence emission. Fluorophors within a protein molecule or combined with a protein as coenzyme or substrate complexes will show fluorescence polarization. The degree of polarization of such complexes and the effects of various factors on this polarization can provide valuable information concerning enzyme mechanisms. All of these aspects of fluorescence are not unique to the field of enzymes but pertain to proteins in general. They have, therefore, been combined with the discussions on fluorescence and protein structure (Chapter 6).

B. Studies in Intact Cells and Subcellular Particles

Methods and instrumentation for measurement of fluorescence changes in intact cells and in subcellular particles were described by Duysens and his collaborators.[44, 45] This group was probably the first to study pyridine nucleotide changes in whole cells by means of fluorometry. As was pointed out before, reduced pyridine nucleotide, DPNH, is fluorescent whereas the oxidized form, DPN^+, is not. The fluorescence spectrum of free DPNH is characterized by a maximum at 460 mμ. This maximum is shifted to

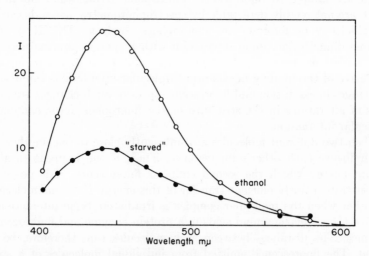

FIG. 12. Fluorescence spectra of "starved" yeast cells and of the same cells in 10% ethanol; excitation is at 366 mμ. Both spectra have maxima at 443 mμ which is characteristic of enzyme bound reduced pyridine nucleotide (after Duysens and Amesz[44]). I represents fluorescence intensity in arbitrary units.

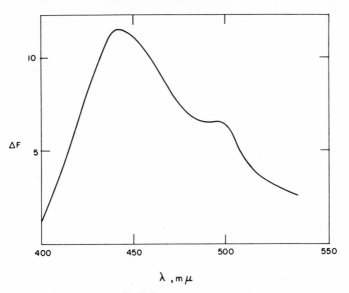

λ , m μ

Fig. 13. Difference spectrum of fluorescence (ΔF) for Rhodospirillum rubrum, representing the enzymatic reduction of bound pyridine nucleotide on illumination (after Duysens and Sweep[45]).

440 mμ when the DPNH is bound to certain enzymes, the shift being accompanied by a change in fluorescence yield.* Duysens and Amesz[44] found that the fluorescence spectra of yeast, luminescent bacteria, and *Chlorella* were similar to the fluorescence spectrum of bound DPNH (Fig. 12). Duysens and Sweep[45] also showed that when photosynthetic organisms (*Rhodospirillum rubrum* and *Anacystis nidulans*) are illuminated, fluorescence increases and the spectrum of this increased fluorescence was again found to be the same as that of "enzyme-bound" DPNH (440-mμ maximum). The data shown in Fig. 13 represent such a difference spectrum and were taken by Duysens and Sweep[45] to represent the reduction of bound pyridine nucleotide upon illumination. Such data indicate that pyridine nucleotide reduction accompanies the reduction of CO_2 during photosynthesis. As shown in Fig. 14, adenine nucleotide is also required.

Olson[46] was able to further identify the fluorescence changes of the photosynthetic organisms by determining their excitation spectra. As can be seen from Fig. 15, the excitation spectra for the light-induced changes in fluorescence in the organisms are practically the same as the spectrum of enzyme-bound DPNH.

---

* A detailed discussion concerning the effects of enzyme binding on DPNH fluorescence is presented in Chapter 6.

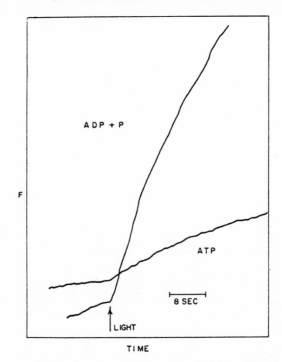

TIME

FIG. 14. The fluorescence (F) of spinach chloroplasts in the presence of 0.002 M TPN+. When red actinic light is turned on (arrow), the slope of the fluorescence curve becomes steeper indicating an increased rate of TPNH formation. The rate is much higher with ADP and phosphate than with ATP. The low rate in the "dark" may result from the weak fluorescence-exciting radiation (after Duysens[47]).

Having established the identity of this fluorescence from photosynthetic microorganisms as arising from reduced pyridine nucleotide, Duysens,[47] Olson,[46] and others were able to study the role of pyridine nucleotides during the photosynthetic process. Some of their results will be discussed in a later section on photosynthesis.

Chance and Baltscheffsky[48] have applied microfluorometric techniques to the measurement of pyridine nucleotide changes in mitochondria. They were able to show that the excitation spectrum of a suspension of rat liver mitochondria, to which substrates had been added to yield maximal reduction of pyridine nucleotides, had a maximum at 345 mμ (corrected) (Fig. 16). This values agrees quite well with the known absorption maximum of reduced pyridine nucleotide, 340 mμ. Under these conditions, the spectrum of the mitochondrial fluorescence yields a maximum at 440 mμ (Fig. 17). This fluorescence maximum agrees well with the values for

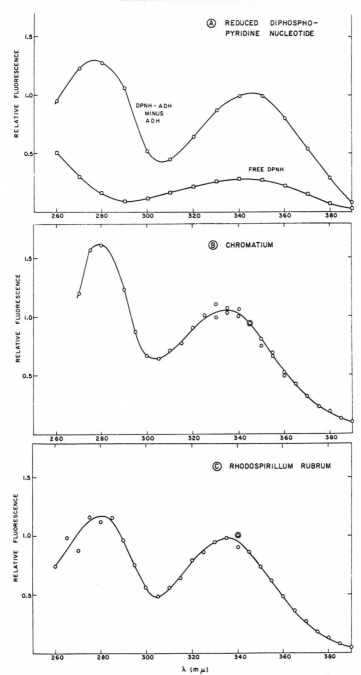

FIG. 15. Excitation spectra of light-induced changes in some photosynthesizing organisms. *A*, DPNH, free and bound to yeast alcohol dehydrogenase (ADH); *B*, Chromatium; *C*, Rhodospirillum rubrum (after Olson[46]).

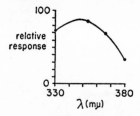

FIG. 16. Excitation spectrum of mitochondria suspended in a phosphorylating medium and supplied with succinate. Measurements were made in the microfluorometer of Chance and Baltscheffsky[48] with fluorescence being measured at 452 mμ.

"enzyme-bound" DPNH reported by Boyer and Theorell[49] and by Duysens and Amesz.[44] When the mitochondria were treated with adenosine diphosphate to oxidize pyridine nucleotides, the 440-mμ fluorescence did indeed diminish. Under these conditions, they were able to demonstrate a slight contribution of flavin to the mitochondrial fluorescence. However, it appears that the flavoprotein of the respiratory chain does not contribute appreciably to the fluorescence spectra measured by excitation in the region 330–380 mμ. Fluorescence spectra of mitochondria in various states

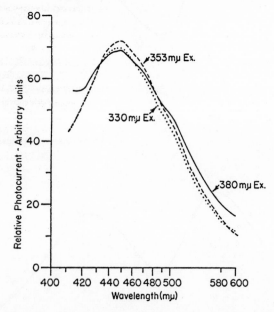

FIG. 17. Fluorescence spectra (corrected) of reduced pyridine nucleotide in mitochondria suspended in a phosphorylating medium containing succinate (after Chance and Baltscheffsky[48]). Measurements were made with excitation at each of the three wavelengths indicated.

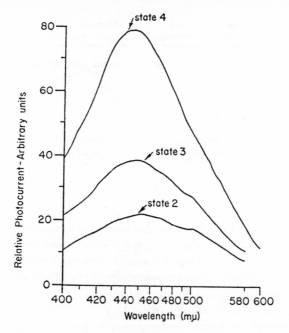

FIG. 18. Fluorescence spectra of mitochondria in three metabolic states; state 4 treated with succinate, state 3 treated with succinate and ADP, state 2 treated with ADP only. Measurements were made with the microfluorometer with excitation at 330 mμ (after Chance and Baltscheffsky[48]).

of reduction are shown in Fig. 18. Under the most oxidized conditions (state 2) there is not only much less fluorescence but a shift towards higher wavelengths indicating other contributions to the fluorescence. However, some of the fluorescence in state 2 is still due to reduced pyridine nucleotide because further reduction is possible. The use of the fluorescent technique to study enzyme kinetics in intact mitochondria is shown in

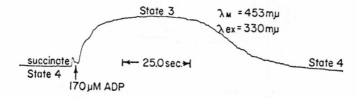

FIG. 19. Kinetics of fluorescence change in mitochondria due to the addition of ADP. Pyridine nucleotides are brought to a reduced state by the addition of succinate. ADP is added at the time indicated (arrow) and fluorescence intensity falls (inverse of the recording) and returns to control values when the ADP is exhausted (after Chance and Baltscheffsky[48]).

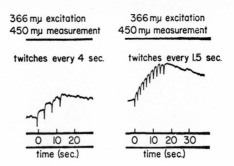

FIG. 20. Fluorescence response of frog sartorius muscle to electrically induced contraction (after Chance and Jöbsis[50a]).

Fig. 19. Chance and Jöbsis[50a] have also been able to apply the fluorometric procedure to follow changes in reduced pyridine nucleotide fluorescence in frog sartorius muscle following electrically induced contraction (Fig. 20). The procedure reflects changes in adenosine diphosphate and is applicable to thick layers of muscle. According to Chance and Jöbsis, the procedure is relatively insensitive to hemoglobin and should therefore be more readily applicable to intact tissues than the spectrophotometric method. Packard[50b] has used fluorometry to follow pyridine nucleotide changes during microsomal swelling.

Chance and Legallais[51] described a differential microfluorometer (described in Chapter 3) which can be used for following metabolic changes in cytologically defined parts of living cells. Figure 21 represents the fluorescence changes of mitochondria in the cytoplasm of a single ascites tumor cell in a drop of liquid under a paraffin-sealed cover slip and indi-

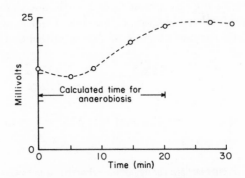

FIG. 21. The kinetics of fluorescence changes (millivolts) in the mitochondria of a single ascites tumor cell. Measurements were made with the differential microfluorometer and represent the reduction of pyridine nucleotide resulting from the exhaustion of oxygen under the coverslip (after Chance and Legallais[51]).

cates the onset of anaerobiosis. Since filters were used to limit excitation to 366 mμ and fluorescence to the region between 410 and 500 mμ, the increased fluorescence is interpreted as a conversion of oxidized pyridine nucleotide, which is nonfluorescent, to the reduced form which is highly fluorescent. It is apparent that the initially aerobic material maintains a constant fluorescence until the oxygen has been used up by cell respiration (about 10 minutes). When this happens the fluorescence intensity rises. This response, observed in a single cell, is characteristic of the increased reduction of pyridine nucleotide observed, by macroprocedures, in suspensions of large numbers of cells.

In a subsequent paper, Chance and Thorell[52] applied the differential microfluorometer to studies on an aggregate of mitochondria found in grasshopper sperm, the nebenkern. The time course of the fluorescence changes in the nebenkern and the cytoplasm of a grasshopper spermatid during the transition from aerobic to anaerobic conditions is shown in Fig. 22. The increase in fluorescence of the nebenkern is characteristic of pyridine nucleotide reduction and consistent with such observations in many whole cells. It would appear, however, that there is no rapid change in the oxidation-reduction state of the cytoplasm. Since pyridine nucleotide is present in the cytoplasm, the failure to observe increased reduction of the nucleotide in this portion of the cell when that in the mitochondria (nebenkern) is markedly reduced, is a demonstration of the impermeability of mitochondria to reduced pyridine nucleotide. These findings *in vivo* are consistent with *in vitro* observations made by other procedures.[53] The differential microfluorometer promises to be a most useful tool with which to follow the kinetics of changes in the amounts of reduced pyridine

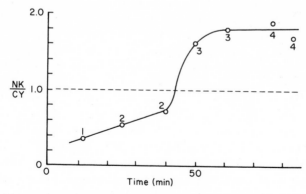

FIG. 22. Kinetics of an intracellular enzyme reaction recorded with the differential microfluorometer. The ordinate represents the ratio of fluorescence in the nebenkern to that in the cytoplasm, NK/CY (after Chance and Thorell[52]).

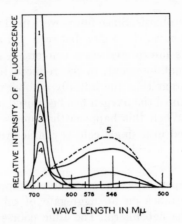

FIG. 23. Sequence of typical changes in the fluorescence spectrum of *Chlorella* chloroplasts during photochemical bleaching (after Olson[55]).

nucleotide and other fluorescent cofactors and metabolites, in various regions of the cell *in vivo*. This represents another application of fluorometry which would be difficult to achieve by any other methods presently available.

The differential microfluorometer developed by Chance and Legallais[51] is a filter instrument which does not permit spectral identification of the fluorescent material. More recently, Olson[54] has reported the development of a microspectrophotofluorometer (Chapter 3) which can be used to determine the spectral characteristics of emitted fluorescence in a microscopic field. The data obtained can also be quantitated by comparison with appropriate standards. The instrument was applied to measurement of chlorophyll and products of its photochemical bleaching in chloroplasts from *Chlorella*. The data shown in Fig. 23 represent the time course of photooxidation of chlorophyll as viewed by the microspectrophotofluorometer. It can be seen that chlorophyll gradually disappears, as shown by the fall in 680-mμ fluorescence, and is replaced by a substance which is also fluorescent but which emits maximally at 545 mμ. The process is, to a large extent, reversible so that in the dark the original 680-mμ fluorescence reappears. The present sensitivity of the instrument is such that with chlorophyll less than 25 chloroplasts, in a field approximately 294 μ^2, are required for measuring the time course of spectral changes. The amounts of chlorophyll which must be present in the optical field were estimated to be about 10^{-5} μg. With the microspectrophotofluorometer it shou'd be possible to confirm and extend the studies with pyridine nucleotide enzymes in intact cells. One interesting possibility is that, with the addition of a second monochronomator to the apparatus, it should be possible to obtain excitation spectra from a microscope field. The combination of all these features should make it possible to localize and quantitate with a fair degree of certainty many enzyme reactions within the living cell.

REFERENCES

1. Lowry, O. H., Roberts, N. R., and Kapphahn, J. I., *J. Biol. Chem.* **224**, 1047 (1957).
2. Lowry, O. H., Roberts, N. R., and Lewis, C., *J. Biol. Chem.* **220**, 879 (1956).
3. Lowry, O. H., Roberts, N. R., and Chang, M.-L. W., *J. Biol. Chem.* **222**, 97 (1956).
4. Laursen, T., and Hansen, P. F., *Scand. J. Clin. & Lab. Invest.* **10**, 53 (1958).
5. Wroblewski, F., and Cabaud, P., *Am. J. Clin. Pathol.* **27**, 235 (1957).
6. Laursen, T., and Espersen, G., *Scand. J. Clin. & Lab. Invest.* **11**, 61 (1959).
7. Henley, K. S., and Pollard, H. M., *J. Lab. Clin. Med.* **46**, 185 (1955).
8a. Laursen, T., *Scand. J. Clin. & Lab. Invest.* **11**, 134 (1959).
8b. Lowe, I. P., Robins, E., and Eyerman, G. S., *J. Neurochem.* **3**, 8 (1958).
9. Albers, R. W., and Koval, G. K., *Biochim. Biophys. Acta* **52**, 29 (1961).
10. Lowry, O. H., Bessey, O. A., and Crawford, E. J., *J. Biol. Chem.* **180**, 399 (1949).
11. Lowry, O. H., Roberts, N. R., Wu, M.-L., Hixon, W. S., and Crawford, E. J., *J. Biol. Chem.* **207**, 19 (1954).
12. Lowry, O. H., *in* "Methods in Enzymology" (S. P. Colowick and N. O. Kaplan, eds.), p. 374. Academic Press, New York, 1957.
13. Loewus, F. A., Tchen, T. T., and Vennesland, B., *J. Biol. Chem.* **212**, 787 (1955).
14. Mead, J. A. R., Smith, J. N., and Williams, R. T., *Biochem. J.* **61**, 569 (1955).
15. Williams, R. T., *J. Roy. Inst. Chem.* **83**, 611 (1959).
16. Robinson, D., *Biochem. J.* **63**, 39 (1956).
17. Moss, D. W., *Clin. Chim. Acta.* **5**, 283 (1960).
18. Takeuchi, T., and Nogami, S., *Acta. Pathol. Japon.* **4**, 277 (1954).
19. Udenfriend, S., Lovenberg, W., and Weissbach, H., *Federation Proc.* **19**, 7 (1960).
20. Dietrich, L. S., *J. Biol. Chem.* **204**, 587 (1953).
21. Lovenberg, W., Weissbach, H., and Udenfriend, S., *J. Biol. Chem.* **237**, 89 (1962).
22. Schott, H. F., and Clark, W. C., *J. Biol. Chem.* **196**, 449 (1952).
23. Waalkes, T. P., and Udenfriend, S., *J. Lab. Clin. Med.* **50**, 733 (1957).
24. Weissbach, H., Lovenberg, W., and Udenfriend, S., *Biochim. et Biophys. Acta* **50**, 177 (1961).
25. Schayer, R. W., Rothschild, Z., and Bizony, P., *Am. J. Physiol.* **196**, 295 (1959).
26. Weissbach, H., Smith, T. E., and Udenfriend, S., *Biochemistry* **1**, 137 (1962).
27a. Lovenberg, W., Levine, R., and Sjoerdsma, A., *Federation Proc.* **20**, 318 (1961).
27b. Cohn, V. H., Jr., and Shore, P. A., *Federation Proc.* **20**, 258 (1961).
28. Axelrod, J., and Tomchick, R., *J. Biol. Chem.* **233**, 702 (1958).
29. Jakoby, W. B., and Bonner, D. M., *J. Biol. Chem.* **205**, 699 (1953).
30a. Duggan, D. E., Bowman, R. L., Brodie, B. B., and Udenfriend, S., *Arch. Biochem. Biophys.* **68**, 1 (1957).
30b. Bokman, A. H., and Schweigert, B. S., *J. Biol. Chem.* **186**, 153 (1950).
30c. Lüscher, E. F., and Käser-Glanzmann, R., *Vox Sang* **6**, 116 (1961).
31. Greengard, P., *Nature* **178**, 632 (1956).
32. Greengard, P., *Photoelec. Spectrometry Group Bull. No.* **11**, 292 (1958).
33. Greengard, P., *in* "Methoden der Enzymatische Analyse" (H.-U. Bergmeyer, ed.), p. 551. Verlag Chemie, Weinheim/Bergstr. Germany, 1962.
34. Ochoa, S., Mehler, A. H., and Kornberg, A., *J. Biol. Chem.* **174**, 979 (1948).

35. Ochoa, S., Salles, J. B. S., and Ortiz, P. J., *J. Biol. Chem.* **187**, 863 (1950).
36. Kornberg, A., *J. Biol. Chem.* **182**, 779 (1950).
37. Kornberg, A., *J. Biol. Chem.* **182**, 805 (1950).
38. Thorn, W., Pfleiderer, G., Frowein, R. A., and Pors, I., *Arch. ges. Physiol. Pflüger's* **261**, 334 (1955).
39. Greengard, P., and Straub, R. W., *J. Physiol. (London)* **148**, 353 (1959).
40. Strehler, B. L., and McElroy, W. D., *in* "Methods in Enzymology" (S. P. Colowick and N. O. Kaplan, eds.), Vol. III, p. 871. Academic Press, New York, 1957.
41. Seraydarian, K., Mommaerts, W. F. H. M., and Wallner, A., *J. Biol. Chem.* **235**, 2191 (1960).
42. Slater, E. C., *Biochem. J.* **53**, 157 (1953).
43. Lundsgaard, E., *Biochem. Z.* **227**, 51 (1930).
44. Duysens, L. N. M., and Amesz, J., *Biochim. et Biophys. Acta.* **24**, 19 (1957).
45. Duysens, L. N. M., and Sweep, G., *Biochim. et Biophys. Acta.* **25**, 13 (1957).
46. Olson, J. M., *in* The Photochemical Apparatus, Its Structure and Function. *Brookhaven Symposia in Biol. No.* **11**, 316 (1959).
47. Duysens, L. N. M., *in* The Photochemical Apparatus, Its Structure and Function. *Brookhaven Symposia in Biol. No.* **11**, 10 (1959).
48. Chance, B., and Baltscheffsky, H., *J. Biol. Chem.* **233**, 736 (1958).
49. Boyer, P. D., and Theorell, H., *Acta. Chem. Scand.* **10**, 477 (1956).
50a. Chance, B., and Jöbsis, F., *Nature* **184**, 195 (1959).
50b. Packer, L., *J. Biol. Chem.* **236**, 214 (1961).
51. Chance, B., and Legallais, V., *Rev. Sci. Instr.* **30**, 732 (1959).
52. Chance, B., and Thorell, B., *Nature* **184**, 931 (1959).
53. Lehninger, A., *Harvey Lectures Ser.* **49**, 176 (1955).
54. Olson, R. A., *Rev. Sci. Instr.* **31**, 844 (1960).
55. Olson, R. A., "Progress in Photobiology " (B. C. Christensen and B. Buchman eds.), p. 473. Elsevier, Amsterdam, 1961.

10

STEROIDS

I. INTRODUCTORY REMARKS

THERE are many different steroids divided into several important classes. Although many are intimately involved metabolically, they can be sorted into groups on structural and physiological grounds.

Cholesterol
Estrogens-female sex hormones
Adrenal steroids and androgens
Bile acids

Only the estrogens possess an unsaturated ring and phenolic substitution which results in ultraviolet absorption and fluorescence of the native molecules. This fluorescence is also in the ultraviolet region. However, it has long been known that in the presence of concentrated sulfuric acid (Salkowski reaction) or in sulfuric acid-chloroform-acetic anhydride mixtures (Lieberman-Burchard reaction) steroids give rise to products which absorb in the visible (colored) and ultraviolet. At lower concentrations these light-absorbing products are also fluorescent, and there have been a great many reports concerning the properties of these steroid fluorophors and conditions for their formation and assay. At present, all that can be concluded is that fluorescence in acid media occurs with almost all steroids, the relative intensities varying with the specific type of acid, concentration of acid, temperature, and many other factors. According to Kalant,[1] each steroid gives rise to several absorbing and fluorescing species, the relative amounts of each component varying with the conditions used. The measured excitation and fluorescent spectra represent composites due to several species which can be reproduced only under very specific conditions. Braunsberg and James[2] have reviewed much of the literature concerning the production of steroid fluorescence and agree with Kalant[1] that at present the best conditions for developing fluorescence of a given steroid must be arrived at empirically. Because the appearance of fluorescence in acid solution is a property of almost all steroids, there have appeared countless papers describing the nature of the fluorescent products, instrumental requirements for their assay and many other findings of a general nature. However, there have been relatively few reports of methods devised for specific compounds and applied to blood, tissues, or urine. In the remainder of this section we shall discuss the available methodology for each class of steroid attempting to emphasize mainly methods which have actually been applied to tissues.

II. CHOLESTEROL

Cholesterol

Cholesterol is the most abundant sterol found in animal tissues and its disposition and metabolism influences many body functions. A large part of cholesterol (free and esterified) is found in plasma and lymph, dispersed in chylomicra. Measurement of the circulating cholesterol has been a routine practice of clinical laboratories for many years. Although the quantities of cholesterol present in blood are sufficiently high for the available macroprocedures, investigations at the histochemical level, on finger-tip blood, or on the small blood samples obtainable in experiments on rats and mice have led to the development of microprocedures for cholesterol. Albers and Lowry[3] have modified the colorimetric Liebermann-Burchard reaction to make it applicable to histochemical studies.* The procedure is applicable to 1 to 25 μg. of frozen-dried sections of tissue containing 0.1–2 μg. of cholesterol. The sterol is extracted from frozen-dried or wet tissue samples with absolute ethanol, the ethanol extracts are evaporated to dryness and fluorescence is developed in the following manner. The dried residue is dissolved in 0.15 ml. of a freshly prepared mixture of trichloroethane† and acetic anhydride (5:1). After 20 minutes, 0.006 ml. of concentrated H_2SO_4‡ is added to each sample with prompt and thorough mixing. The tubes are capped and, after 1 to 2 hours, are

---

* Microtechniques and equipment devised by Lowry and his colleagues were used; O. H. Lowry, in "Methods in Enzymology" (S. P. Colowick and N. O. Kaplan, eds.), p. 366. Academic Press, New York, 1957.

† Purified by distillation (111°–113°), washing several times with concentrated H_2SO_4 (one-tenth volume), then with water, and dried over anhydrous Na_2SO_4.

‡ Analytical grade sulfuric acid may contribute to the reagent blank. If so, it can be purified by heating with 5% by volume of 70% perchloric acid until the solution becomes colorless and fuming subsides.

TABLE I

FLUORESCENCE OF OTHER STEROIDS RELATIVE TO THAT OF CHOLESTEROL[a,b]

Compound	Relative fluorescence
Cholesterol	100
Dihydrocholesterol	30
Δ^7-Cholestenol	135
Cortisone	0
Pregnandiol	1
Dihydroepiandrostane	10
Alloprenanolone	12
Androsterone	22
Estrone[c]	26
Progesterone	35
Desoxycorticosterone	66
Δ^5-Pregnenolone	115

[a] The exciting wavelength was 546 mμ.

[b] After Albers and Lowry.[3]

[c] Estrone fluorescence fades rapidly on irradiation.

read in a fluorometer provided with an adapter for microcuvettes.* Standards and reagent blanks are carried through the entire procedure. Filters are used to isolate the 546-mμ mercury line for excitation. The secondary filter is a cutoff filter transmitting appreciably at 590 mμ in order to detect the red-orange fluorescence. It is apparent that the reaction yields a mixture of products since excitation at 365 mμ yields a blue fluorescence. Although the blue fluorescence is more sensitive, the red-orange fluorescence shows higher specificity and is therefore more desirable. As shown in Table I, other steroids fluoresce under these conditions. In each case the fluorescence differs from that of cholesterol in intensity and in spectral characteristics so that, if necessary, spectra can be identified with fluorescence spectrometers. However, in most extracts the amount of cholesterol present is several orders of magnitude greater than other steroids so that interference is not a serious problem.

Fluorescence is proportional to concentration up to 25 μg./ml. and the same molar fluorescence intensities are given by cholesterol and cholesterol stearate. Recoveries of added amounts of cholesterol are quantitative and values obtained by the fluorometric procedure on rabbit brain compared well with standard colorimetric assay.

McDougal and Farmer[4] have modified the Albers and Lowry[3] method to permit rapid assay of cholesterol in the small amounts of blood available from the tip of a rat's tail.

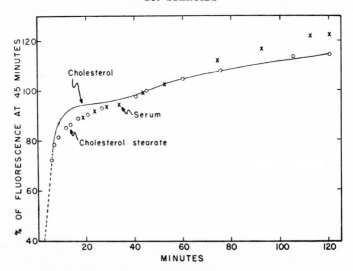

FIG. 1. Development of fluorescence with time for cholesterol, cholesterol stearate, and material extracted from serum (after McDougal and Farmer[4]).

Four microliters of serum are added to a tube (10 × 75 mm.) containing 40 μl. of a mixture of glacial acetic acid and trichloroethane (3:2).* A precipitate forms which is redissolved by mixing. This is essential otherwise results may be erratic and low. To the clear solution is added 1 ml. of a freshly prepared mixture of trichloroethane (5:1) and the solution is mixed. Following this, 40 μl. of concentrated sulfuric acid are added with prompt and thorough mixing. The mixture is centrifuged and the clear supernatant solution is transferred to a cuvette. Fluorescence intensity increases over a period of time (Fig. 1) and is measured after 40 minutes when values for the free and esterfied form are comparable. Fluorescence is assayed with comparable equipment and filters as were used by Albers and Lowry.[3] However, since the final volume is over 1 ml., standard fluorometers can be used. Values for serum cholesterol in rats and humans are comparable to those obtained by standard colorimetric procedures. For those interested in speed and convenience, as many as 60 samples can be completed in 3 hours.

Carpenter et al.[5] have also reported a modification of the Albers and Lowry procedure for serum cholesterol which utilizes 0.02 ml. of serum and can be run with standard laboratory apparatus.

---

* If larger amounts of serum are available, they may be diluted with 10 volumes of acetic acid-trichloroethane mixture and 40 ml. of this mixture taken for subsequent assay.

III. ESTROGENS AND TESTOSTERONE

The estrogens differ from other steroids in having an aromatic ring. The hydroxyl group on that ring is therefore phenolic, endowing the estrogens with unique properties among the steroids.

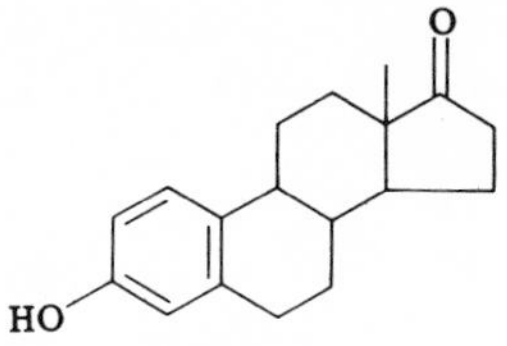

Estrone

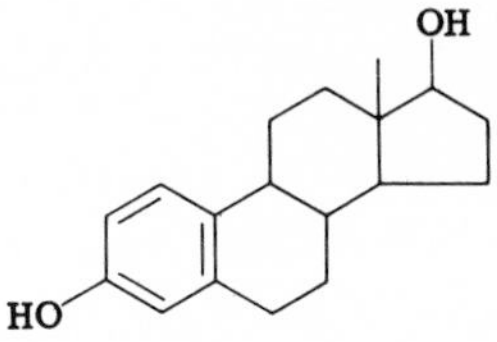

β-Estradiol

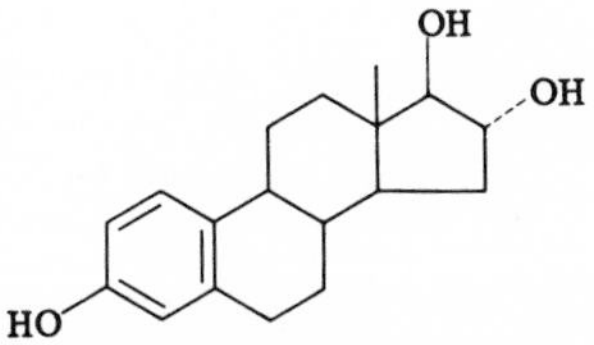

Estriol

Two additional estrogenic substances are found in mare urine only, equilin and equilenin. These are also phenolic and even more aromatized.

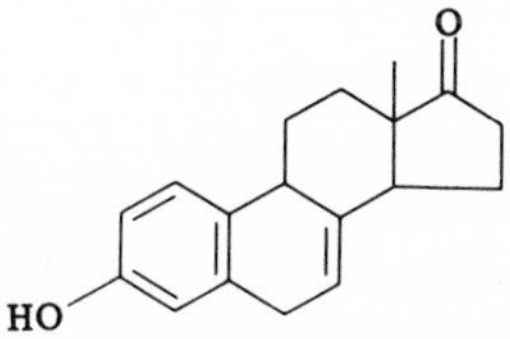

Equilin

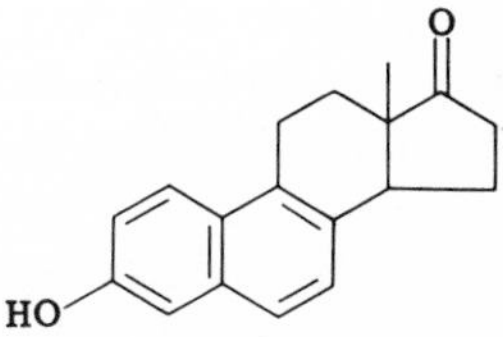

Equilenin

The phenolic nature of these compounds makes it possible to separate them from other steroids by extraction from organic solvents into aqueous alkali. As we shall see, this is an important part of all methods for estrogen assay. An additional attribute of phenols is that they are fluorescent so that unlike other steroids the estrogens possess native fluorescence.[6] This fluorescence (Table II) occurs in the ultraviolet and is sufficiently intense to permit detection of quantities well below the microgram level. As yet, the native estrogen fluorescence has not been utilized for analytical purposes. However, the availability of suitable instrumentation should now make this possible. An important consideration is that the compounds

10. STEROIDS

TABLE II

NATIVE FLUORESCENCE OF ESTROGENS[a]

Compound	Excitation maxima mμ	Fluorescence maxima mμ	Relative intensity
β-Estradiol	285	330	100
Estrone	285	325	13
Estriol[b]	285	325	—
Equilin	290	345 (420)[c]	10
Equilenin	250, 290, 340	370	1000

[a] Measurements were made in 99% ethanol with the Aminco-Bowman spectrophotofluorometer and values have not been corrected for instrumental response. The intensity of β-estradiol is arbitrarily set at 100. As little as 0.1 μg./ml. of β-estradiol gives fluorescence several times that of the reagents (after Duggan et al.[6]).

[b] Excitation and fluorescent maxima of estriol were not actually measured but are assumed on structural grounds.

[c] The 420-mμ fluorescence may have been due to an impurity in the sample of equilin.

differ markedly in biological activity so that there is a need for measuring them individually.

Many fluorescent procedures have been described for estrogen assay based on solvent extraction followed by treatment with sulfuric or phosphoric acid to yield visible fluorescence. Starting with the independent reports of Bates and Cohen,[7] of Jailer,[8a] and of Finkelstein et al.,[8b] there have been scores of reports concerning fluorometric assay of estrogens. Most of these have been concerned with optimal conditions for forming the fluorophor and with appropriate instrumentation. A summary and review of the procedures reported through 1954 was presented by Bates.[9] He pointed out that the absorption spectra of all the acid products of the estrogens exhibit maxima with broad peaks ranging from 410 to 450 mμ (somewhat higher for equilenin). With filter instruments it is therefore desirable to utilize the 436-mμ mercury line. The resulting fluorescence is a broad band with a peak near 500 mμ.

A detailed investigation of the absorption and fluorescence characteristics of the estrogen acid products has recently appeared.[10] Figure 2 shows the absorption and fluorescence spectra of estrone and β-estradiol formed in 75% sulfuric acid, with and without the addition of toluene-ethanol. All three estrogens have similar spectra with an absorption maximum at 453 to 456 mμ and a fluorescence maximum at 479 to 484 mμ. The marked overlapping of the absorption and fluorescent spectra makes it necessary to select filters carefully when a filter instrument is employed.

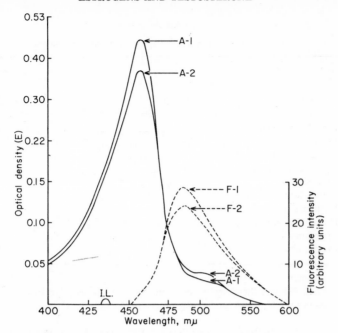

Fig. 2. Absorption (A) and fluorescence (F) spectra of estrone in 75% (v/v) sulfuric acid; with (A–1, F–1) and without (A–2, F–2) the addition of toluene-ethanol. I.L. = incident light (after Bauld et al.[10]).

It is apparent that excitation cannot be carried out at the peak, but must be done at some point further removed from the fluorescence. That is the reason for choosing the 436-mμ mercury line. The relatively high absorption overlapping the fluorescence band indicates that an inner filter effect should appear at relatively low concentrations so that the relationship between concentration and fluorescence will not remain linear.

The method for measuring combined estrogens in urine, as described by Jailer,[8] involves hydrolysis, extraction from the alkalinized hydrolysate into benzene, several transfers back and forth between solvent and alkali, and washing of the solvent with water and dilute Na_2CO_3. Following this tedious extraction, the solvent is dried, evaporated to dryness, and the residue taken up in alcohol. Aliquots of the alcohol are treated with sulfuric acid and measured fluorometrically. Although the extraction procedure appears more complicated than necessary and recoveries were only 50–60%, Jailer[8] reported that the method was reproducible and did not respond to corticosteroids, cholesterol, and androgens. Furthermore, estriol was removed in the extraction. Although the values obtained with the method did not agree with those obtained by bioassay and other evi-

TABLE III

URINARY ESTROGEN EXCRETION IN VARIOUS STATES[a]

Subject	Estrogen excretion μg./24 hours
Females, normal[b]	15–25
Females, normal[b]	35–50
Females, menopause	10–25
Females, post menopause	0–6
Children	3–6
Males (adult)	20–26
Virilizing tumors	114–296
Pregnancy (8th month)	365
Pregnancy (9th month)	592
Pregnancy (2 days before delivery)	1500

[a] After Jailer.[8]

[b] In normal females there are cyclic increases and decreases in estrogen excretion. The two values are the lower and upper ranges of the cycle.

dence of nonspecificity was apparent, the method is satisfactory for distinguishing normal urine from that obtained in pregnancy and in certain pathologic conditions (Table III). More recently, Smith et al.[11] applied a modification of this procedure to the assay of estrogens in bovine urine. They also used partition chromatography in an attempt to separate the estrogens.

Since the various estrogens differ in biological activity and in fluorescence intensity, procedures for their combined assay yield rough approximations, at best. Preliminary separations of the estrogens before fluorometric assay have been developed in many laboratories. Braunsberg et al.[12] reported the use of partition chromatography with a celite column holding 3.1 N NaOH as the stationary phase; benzene-petroleum ether mixtures and butanol-chloroform were used to elute the three estrogens. Although separations of estrone and estradiol were satisfactory, recoveries of estriol were erratic and the method was found to be time-consuming and subject to various interferences.

The method which appears to be the best and simplest of those currently available is the one reported by Nakao and Aizawa.[13] This procedure is applicable to both blood and urine and utilizes chromatography to determine the individual estrogens.

Blood: Ten milliliters of oxalated blood are added to 20 ml. of water and mixed to effect hemolysis. The solution is extracted 3 times with 30-ml.

portions of ether and the combined ether extracts are washed twice with 10 ml. of saturated sodium bicarbonate solution. The ether is then extracted with two 20-ml. portions of 10% NaOH. The combined alkaline extract is acidified with concentrated HCl and again extracted with 20 ml. of ether. The ether is washed with 10 ml. of water and evaporated to dryness. The residue is taken up in one ml. of benzene.

Urine: To 50 ml. of urine are added 7 ml. of concentrated HCl and the solution is heated at 100° for 30 minutes to effect hydrolysis. The hydrolysate is then extracted 3 times with 10 ml. of ether and the combined ether extracts are treated as described for blood.

Chromatography: A glass column, 2 mm. i.d. and 8 cm. high, is filled with 0.3 g. of activated alumina. The column is washed with several milliliters of benzene and then the benzene solution obtained from the tissue or urine is placed on the column and permitted to flow through at a rate of about 13 to 15 drops per minute. When the entire sample has penetrated into the column, elution is carried out with 5-ml. portions of 1, 5, and 30% methanol in benzene. As shown in Fig. 3, this system effects a complete separation of the three compounds. Estrone is eluted first, then estradiol and estriol. Temperature regulation of the column is suggested.

The eluates are evaporated to dryness and taken up in a measured amount of ethanol. To 0.1 ml. of ethanol extract is added 1 ml. of 90% sulfuric acid, and the solution is heated at 80° for 10 minutes. The sample is then diluted with 10 ml. of 65% sulfuric acid and fluorescence is measured using appropriate instrumentation as described earlier in this chapter.

The sensitivity is such that the final solution requires as little as 0.006 µg. of estrone or β-estradiol or 0.02 µg. of estriol. Recoveries range from

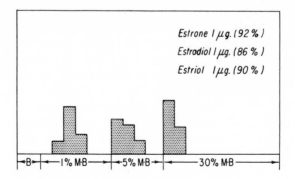

FIG. 3. Chromatographic separation of a mixture containing 1 µg. each of estrone, estradiol, and estriol. B = benzene, M–B = methanol in benzene, figures in parenthesis represent per cent recovery (after Nakao and Aizawa[13]).

TABLE IV

ESTROGEN CONTENT OF HUMAN PLASMA

Subject	Estrogen, μg./10 ml.			Reference
	Estrone	Estradiol	Estriol	
Normal female[a]	0.18–0.53	0.03–0.09	0.05–0.19	(13)
Latter period of pregnancy[b]	0.24–0.59	0.24–0.30	0.35–0.48	(13)
Latter period of pregnancy[c]	0.51	0.32	0.30	(14)
Toxemia of pregnancy	1.3	0.44	2.3	(13)

[a] Six subjects.
[b] Three subjects.
[c] A pooled sample collected from 19 women.

about 70 to 90%. The somewhat low values are apparently due to significant color in the samples leading to absorption at the excitation and fluorescence wavelengths. If this is so, internal standards should allow simple correction. Treatment with H_2O_2 reveals a small amount of fluorescence which persists when steroid fluorescence is destroyed. This is a small correction and normally is not required.

Values for the three estrogens in blood under varying conditions are shown in Table IV. The plasma levels obtained in late pregnancy have been compared with comparable data obtained by Oertel et al.[14] The latter group identified the individual estrogens in plasma by countercurrent distribution, paper chromatography, infrared spectrophotometry, color reactions, and isotope dilution.

Strickler et al.[15] have utilized a combination of extraction, counter current distribution, and adsorption for assay of urinary estrogens and have also corrected for differential decay of steroid fluorescence and blank fluorescence. They claim to find lower values than do other authors. However, it is difficult to compare their data with that reported by the others. Eberlein et al.[16a] have reported a method for urinary estriol similar to the one used in the plasma procedure of Nakao and Aizawa.[13] More recently, Finkelstein et al.[16b] have devised a method for the determination of urinary estrogens consisting of hydrolysis, extraction, partition between solvents, paper chromatography, and fluorometry in phosphoric acid. As little as 1 μg. of estrogen, added to a 24-hour urine sample, can be recovered quantitatively. The values reported by them for excretion in normal nonpregnant women are: estrone, 3–5 μg., estradiol, 2–4 μg., and estriol, 5–10 μg. per 24 hours.

Testosterone can be estimated in human plasma by extraction, enzymatic conversion to estradiol and estrone, followed by fluorometric assay of the estrogens[16c] as described by Finkelstein et al.[16b]

IV. ADRENAL CORTICAL STEROIDS

The adrenal cortex gives rise to a large number of steroids. Cholesterol is present probably as a precursor, and androgenic and luteal hormones are found in the gland. However, the most important adrenal steroids are those which can be used to replace the adrenal gland in adrenalectomized animals (having adrenal cortical activity). The most active of these agents are:

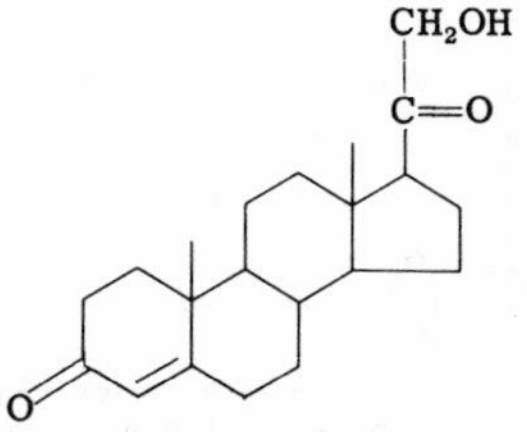

Deoxycorticosterone

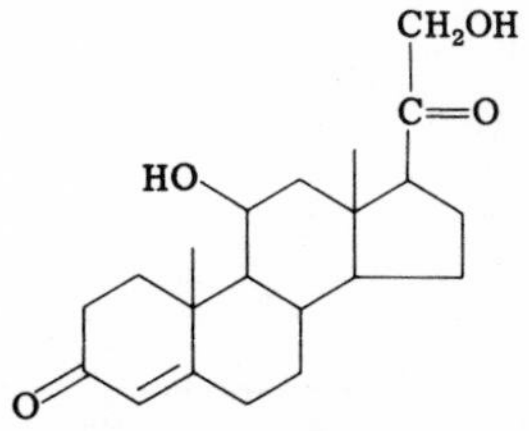

Corticosterone

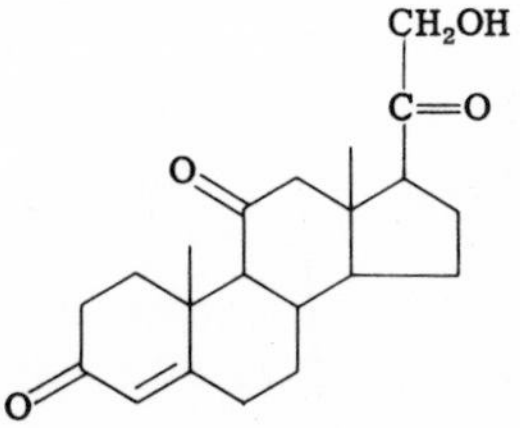

11-Dehydrocorticosterone

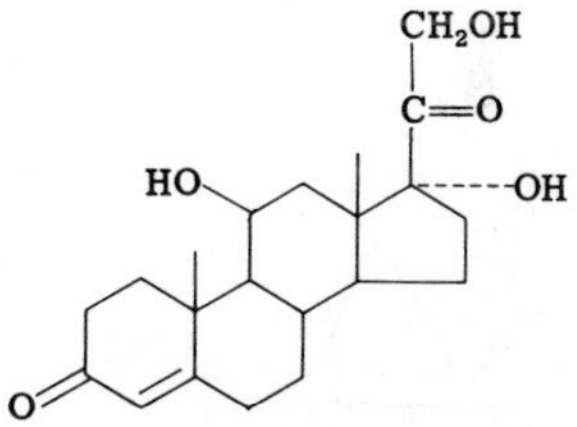

17α-Hydroxycorticosterone
(cortisol; hydrocortisone)

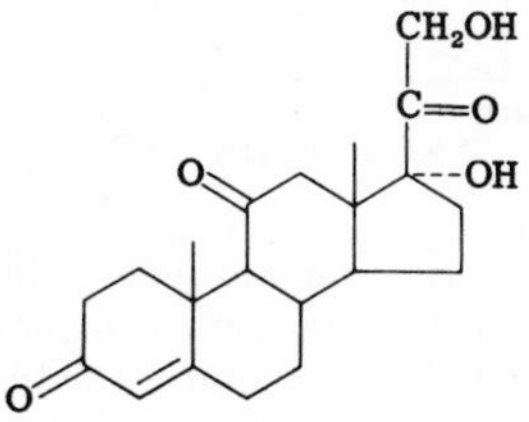

11-Dehydro-17α-hydroxycorticosterone
(cortisone)

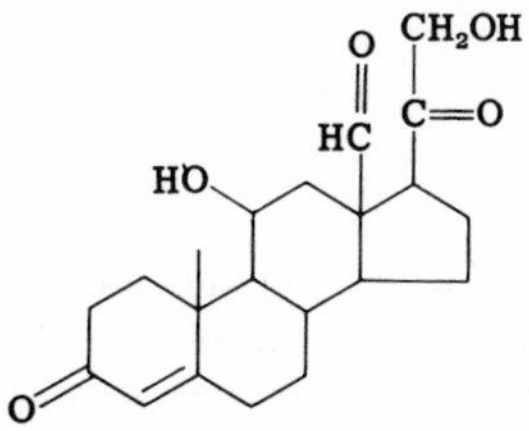

Aldosterone

None of these compounds possesses appreciable native fluorescence. However, they can be converted to fluorescent products in two ways.

A. Fluorescence Produced in Strong Acid

By treatment with strong sulfuric or phosphoric acid, those adrenal steroids possessing an 11 hydroxy group yield characteristic fluorescent derivatives, except for aldosterone. The characteristics of the resulting fluorophors are 470- to 475-mμ excitation and 520- to 530-mμ fluorescence.[1, 17, 18] As shown in Fig. 4, both corticosterone and hydrocortisone

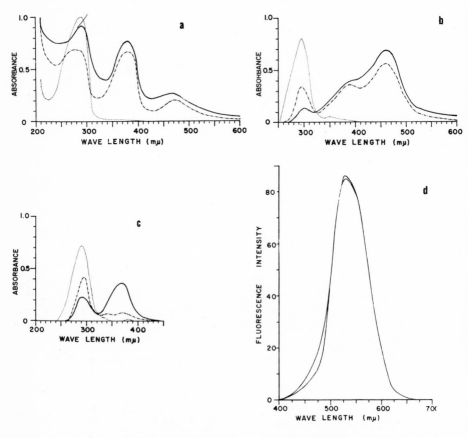

FIG. 4. Curves a–c represent absorption spectra of corticosterone, hydrocortisone, and cortisone, respectively, in polyphosphoric acid. The dotted lines were at room temperature for 2 hours; broken lines were incubated at 100° for 20 minutes; solid lines were incubated at 100° for 60 minutes. Curve d is the fluorescence spectrum of corticosterone in sulfuric acid (after Goldzieher and Besch[18]).

yield compounds with absorption bands at 470 mμ whereas cortisone under these conditions shows no absorption above 410 mμ. The fluorescence spectrum shown for corticosterone is qualitatively similar to that for hydrocortisone. Of large numbers of other compounds investigated,[19-21] relatively few yield appreciable fluorescence. The latter include mainly estrone and estradiol, although pregnanediol and tetrahydrocortisone yield slight fluorescence. Steroid drugs such as prednisone and prednisolone do not yield fluorescence. Cholesterol yields only slight fluorescence of this type. However, since it is present in such huge amounts compared to the hormones, it must be considered as an interfering substance.

The blood of most animal species contains mainly corticosterone and hydrocortisone. In the rat, there is little, if any, hydrocortisone in the blood. On the other hand, man, monkey, and guinea pig contain mainly hydrocortisone. Quite similar procedures for the assay of combined circulating 11 hydroxyadrenal steroids have been reported by Silber et al.[19] and by Zenker and Bernstein.[20] These procedures are relatively simple and are particularly useful for laboratory studies which utilize the rat. The method of Zenker and Bernstein[20] is presented in detail.

An aliquot of 0.3 to 3.0 ml. of peripheral, or 0.02 to 0.1 ml. of adrenal venous, rat plasma is transferred to a 50-ml. glass-stoppered conical centrifuge tube and the volume is made up to 3 ml. with water. Following the addition of 45 ml. of chloroform, the tube is shaken and then centrifuged to separate the phases. The aqueous phase is removed by aspiration and 4.5 ml. of 0.1 N NaOH are added and the tubes are shaken again for 15 seconds and centrifuged. This step removes the phenolic estrogens, but too long contact with the alkali can lead to destruction of the adrenal steroids. The alkali is removed and discarded, and two 15-ml. aliquots of chloroform are transferred to clean 50-ml. glass-stoppered centrifuge tubes containing 3.0 ml. of sulfuric acid reagent (2.4 volumes of concentrated sulfuric acid added to 1.0 volume of 50% aqueous ethanol). The tubes are shaken vigorously for 15 seconds and centrifuged. The chloroform is removed and each supernatant acid layer is transferred to a cuvette and permitted to stand at room temperature for 2 hours. The fluorescence is then measured using appropriate instrumentation for the characteristics of the fluorophor. Standards and blanks are carried through the entire procedure, substituting water and known amounts of corticosterone for the plasma.

The material isolated in this manner was shown to have the same fluorescence characteristics as authentic corticosterone and to increase in fluorescence at the same rate as the standards upon treatment with the sulfuric acid reagent. Recoveries of corticosterone added to plasma averaged about 90%. Values obtained for rat plasma by these simple extrac-

TABLE V

CORTICOSTERONE IN RAT PLASMA UNDER VARIOUS EXPERIMENTAL CONDITIONS

	Plasma corticosterone (μg./100 ml.)	
Condition	Method A[a]	Method B[b]
Control	30	17
Adrenalectomized	1–2	0
After ACTH	—	56

[a] Zenker and Bernstein.[20]
[b] Silber et al.[19]

tion procedures are shown in Table V. The values by Silber et al.[19] are lower because of an applied correction utilizing 36 N H_2SO_4 on duplicate aliquots to reveal interfering substances in the extracts (the 11 hydroxy steroids show no fluorescence in 36 N H_2SO_4).

More precise estimations of corticosterone and hydrocortisone require separation of the two steroids from each other and from cholesterol and other products. Sweat[22] reported the use of a silica gel microcolumn for the chromatographic separation of the steroids following extraction from the blood with chloroform. More recently, McLaughlin et al.[21] modified the procedure and applied it to human plasma.

Heparinized plasma (2–8 ml.) is extracted with a mixture of chloroform-ethanol (50:5 v/v). The organic solvent is transferred to another vessel, evaporated to complete dryness and the residue is taken up in 10 ml. of absolute chloroform.* Chromatographic columns are prepared by pouring 1 g. of purified silica gel, suspended in 10 ml. of absolute chloroform, onto the column. Air bubbles are removed with a stirring rod and after the chloroform has passed through the plasma extract is placed on the column (in 10 ml. absolute chloroform). When this has passed through, the column is washed with 20 ml. of 1.8% ethanol in chloroform. Cholesterol and other contaminants are removed in these washings and are discarded. Corticosterone is eluted with 40 ml. of 3% ethanol in chloroform and the hydrocortisone with 30 ml. of 7% ethanol in chloroform. To the corticosterone fraction is added 0.9 ml. of ethanol to bring the total amount of alcohol to 2.1 ml. Both eluates are then shaken with 8 ml. of concentrated sulfuric

---

* Commercial chloroform contains 0.5–1.0% of ethanol. The alcohol can be removed by passing the chloroform over a silica gel column which can remove 8% of its weight of ethanol. The purified material contains about 0.01% ethanol as determined by infrared absorption at 9.58μ. It can be stored indefinitely in brown bottles at 2°.

TABLE VI

Corticosterone and Hydrocortisone in Human Plasma[a]

Subjects	Hydrocortisone	Corticosterone	
	McLaughin et al. (21)	McLaughin et al. (21)	Peterson (23)
Normal (Adults)_____	12–14	2.0	0.5–2.0
Children_____	7.4	10.7	—
Hypopituitarism_____	—	0.4	0–0.2
Normal after ACTH[b]_____	47.3	7.1	7.5–10

[a] μg./100 ml.
[b] ACTH = adrenocorticotropic hormone.

acid.* After separation of the phases, 2 ml. of the acid layer are transferred to a cuvette for comparison with appropriate blanks and standards. Although hydrocortisone fluorescence is relatively stable, corticosterone fluorescence deteriorates rapidly.* Measurements are therefore made as rapidly as possible. Recoveries of both steroids, at the level of 4 μg./5 ml., were 80–95%. Unfortunately, the recovered amounts were, in the case of corticosterone, 40–50 times higher than normal values so that the reported recoveries may not be indicative of the actual performance of the method. Specificity seems to be better than in the methods which do not utilize chromatography. Values for human plasma are shown in Table VI.

Corticosterone is present in relatively small amounts in human blood, so that adequate recovery and specificity are difficult to ascertain. To overcome these problems, Peterson[23] introduced the use of radioactive corticosterone for identification and as an indicator of completeness of recovery. Introduction of an isotopic indicator permits quantitative analysis without the need for complete recovery. The procedure for corticosterone in plasma is presented in detail because it represents an approach which is now widely used in steroid assay.

To 30 ml. of heparinized plasma in a 125-ml. glass-stoppered reagent bottle is added a measured amount of corticosterone-C[14]† dissolved in about 0.1 ml. of 50 ethanol.‡ After careful mixing, 75 ml. of extracting

---

* It would appear that evaporation of the eluates at this stage would be more desirable. This would permit measurement in smaller volumes and make it possible to control the proportions of ethanol and sulfuric acid. The instability of corticosterone fluorescence would also be avoided.

† The total radioactivity should be at least 4000 c.p.m. and the specific activity sufficiently high so that the amount of corticosterone-C[14] added is much less than the amount present in the blood sample.

‡ All organic solvents must be purified by adsorption and distillation procedures and filter paper for chromatography should be washed with reagent grade methanol.

solvent (a mixture of equal volumes of carbon tetrachloride and dichloromethane*) are added and the mixture is shaken. The bottle is then centrifuged the plasma layer is removed by aspiration and 5.0 ml. of 0.01 N NaOH are added and the mixture is shaken again. The alkali wash is followed by washes with 0.02 N acetic acid and with water. Following removal of the final aqueous phase, the solvent is transferred to another vessel and evaporated to dryness at 40° to 45° under a stream of nitrogen. The residue is dissolved in dichloromethane and the solution is transferred to a test tube and evaporated to a small volume. This is transferred to a small spot on Whatman No. 1 filter paper* and the chromatogram is developed with cyclohexane-benzene-methanol-water (4:4:2:1, organic phase) for 8 to 12 hours. Reference standards containing 5–10 µg. of corticosterone are run adjacent to the plasma extract. After development, the corticosterone area is located by the reference standards (with an ultraviolet lamp) and a small transverse band of paper containing the plasma corticosterone is cut out. This need not be quantitative, specificity being the more desirable goal. The steroid is eluted from the paper with 3 ml. of 95% ethanol and the eluate is evaporated to dryness with a stream of nitrogen. The residue is dissolved in 6.5 ml. of dichloromethane. Two milliliters of this are taken for radioactive assay. Of the remainder, 4 ml. are added to a glass-stoppered tube containing 1 ml. of sulfuric acid reagent (65 volumes of concentrated sulfuric acid plus 35 volumes of ethanol) and shaken immediately. The solvent (upper) layer is removed by aspiration and after 2 to 4 hours fluorescence is assayed with the appropriate filters. Blanks and recoveries are carried through the entire procedure.

Measurement of both radioactivity and fluorescence yield a specific activity of the isolated corticosterone (I). The quantity of corticosterone in the plasma sample, A, can be calculated from the simple proportion:

$$I_0 a = I(A + a) \tag{1}$$

or

$$A = a\left(\frac{I_0}{I} - 1\right) \tag{2}$$

where I_0 is the specific activity of the added corticosterone, I is the specific activity of the isolated steroid, and a is the amount of added corticosterone (µg.).

The specificity of the assay was well established by absorption and fluorescence spectra (Fig. 5), by comparing sulfuric acid fluorescence with authentic corticosterone (Fig. 6) and by demonstrating constant specific

---

* All organic solvents must be purified by adsorption and distillation procedures and filter paper for chromatography should be washed with reagent grade methanol.

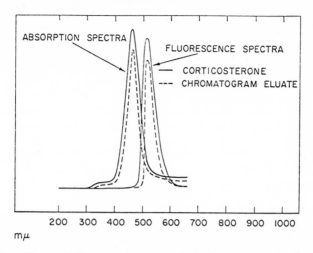

FIG. 5. Absorption and fluorescent spectra of corticosterone and plasma steroid eluted from a paper chromatogram; in 65 parts of sulfuric acid and 35 parts of ethanol (after Peterson[23]).

activity of the isolated C^{14} material from plasma using a variety of derivatives and isolation procedures.

Recoveries of added corticosterone are essentially quantitative and as little as 0.2 μg. % can be detected. The precision is about $\pm 20\%$ at levels of 0.5 μg./100 ml. but is $\pm 5\%$ at levels of 2 μg./100 ml. and over. With

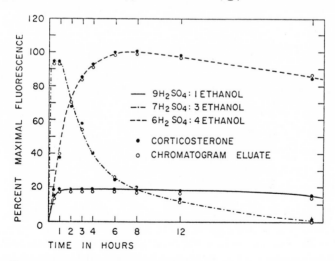

FIG. 6. Comparison of the fluorescence characteristics of authentic corticosterone and the plasma steroid eluted from paper chromatograms (after Peterson[23]).

more highly radioactive corticosterone, it should be possible to add so little to the plasma that a, in equation (1), is negligible compared to A and the equation reduces to

$$I_0 a = IA \tag{3}$$

$$A = \frac{I_0}{I} a \tag{4}$$

Precision can be greatly increased under these conditions.

The sulfuric acid and phosphoric acid procedures have not proved satisfactory for assay of urinary corticosteroids because of the large number of steroid metabolites and other interfering substances which are excreted. The methods have been applied to other tissues. Sweat[24] used the method to follow the enzymatic conversion of 17-hydroxy-11-desoxycorticosterone to hydrocortisone. Other applications in enzymology should be possible.

B. Fluorescence Produced in Alkali

Bush and Sandberg[25] demonstrated that Δ^4-3-ketosteroids, on paper chromatograms, exhibit an orange fluorescence under ultraviolet light if the paper is sprayed with 10% NaOH and allowed to dry. They also utilized this reaction in the first rigorous identification of hydrocortisone in plasma. Since this observation there have been many applications of this, so-called, soda fluorescence, both qualitative and quantitative. Methods have been developed for fluorescence measurement on paper

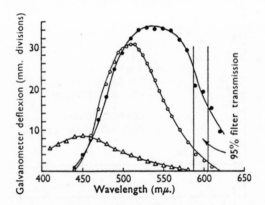

FIG. 7. Fluorescence spectrum of the soda fluorescence of aldosterone diacetate on a paper chromatogram. ○ — ○, spectrum as measured corrected for paper blank; ● — ●, spectrum corrected for the spectral sensitivity of the photomultiplier tube and for the paper blank; △ — △, paper blank (after Ayres et al.[26]).

chromatograms and in solution. The work of Ayres *et al.*[26, 27] is perhaps, best representative of the paper chromatographic applications of this procedure. On utilizing a fluorometer designed for measuring fluorescence directly on paper, they demonstrated that all Δ^4-3-ketosteroids yield comparable products, with excitation maxima near 365 mμ and fluorescence maxima between 550 and 600 mμ (Fig. 7).

The over-all procedure involves the use of C^{14}-labeled corticosterone, hydrocortisone, or aldosterone as indicators and to correct for recovery. One liter of urine is extracted continuously. The steroids or their acetylated derivatives are then chromatographed on paper, sprayed with NaOH, and the resulting fluorophors are measured in the fluorometer. Even with the isotope dilution step recoveries were reported to be about 70%. Urinary aldosterone was found to be 11 μg./day and hydrocortisone 35 μg./day.

The products formed with NaOH on paper do not fluoresce in aqueous solution, so that the procedure cannot be used directly in solution. However, Abelson and Bondy[28] were able to produce the same sort of fluorescence by carrying out the reaction in potassium *tert*-butoxide dissolved in *tert*-butyl alcohol.

For assay, 0.1 ml. of the solution (ethanol) is transferred to a fluorometer tube and evaporated to complete dryness. Following this, 0.5 ml. of butoxide reagent* is added, the tube is stoppered with an aluminum foil-covered cork, shaken, and allowed to stand for 1 hour at room temperature. The tube is covered to prevent absorption of CO_2 which produces turbidity. At the end of the 1-hour period, the tube is shaken again and its fluorescence measured.

All Δ^4-3-ketosteroids investigated yield products with an absorption maximum at about 385 mμ and a fluorescence maximum at about 580 mμ (Fig. 8). The development of this 580-mμ fluorescence is highly specific for the Δ^4-3-ketosteroids, including cortisone, hydrocortisone, corticosterone, progesterone, testosterone, and aldosterone. Most other types of steroids do not fluoresce in this reagent, a few exhibit slight blue or green fluorescence which is readily separable from the orange-yellow fluorescence of the Δ^4-3-ketosteroids. With filter instruments, the 365-mμ mercury line can be used with appropriate primary filter and a secondary filter which cuts off above 580 mμ. With many steroids as little as 0.02 μg. can be detected (if the reagent is properly prepared).

In subsequent publications Bondy *et al.*[29] and Bondy and Upton[30] reported the utilization of the potassium-*tert*-butoxide method along with

---

* The reagent is prepared by addition of potassium metal to *tert*-butanol and refluxing. The detailed procedure in the original article should be followed closely to insure a satisfactory reagent and to minimize the hazard involved in the use of potassium. The final concentration should be 0.3 N potassium *tert*-butoxide in *tert*-butanol.

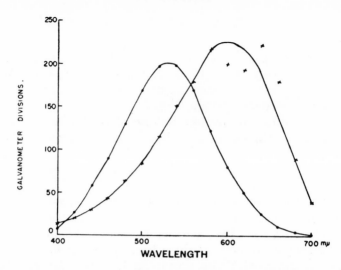

FIG. 8. Fluorescence spectrum of hydrocortisone product in *tert*-butoxide in *tert*-butyl alcohol. ● — ●, observed spectrum; ×—×, spectrum corrected for photomultiplier response and other instrumental variables (after Abelson and Bondy[28]).

solvent extraction and paper chromatography for estimation of hydrocortisone and corticosterone. In these procedures they corrected for incomplete recovery by utilizing C^{14} labeled steroids in the same manner as described by Peterson.[23] Comparison of the alkaline fluorometric-isotope procedure with other methods is shown in Table VII. The fluorometric assay yields values comparable to those obtained by the other methods. Staub and Dingman[32b] have utilized the *tert*-butoxide method along with chromatography on glass fiber filter paper to determine aldosterone in urine. The method is claimed to be accurate and relatively rapid.

TABLE VII

CORTICOSTERONE AND HYDROCORTISONE IN ADULT HUMAN PLASMA AS MEASURED BY
VARIOUS METHODS

Method	Corticosterone μg./100 ml.	Hydrocortisone μg./100 ml.
tert-Butoxide-isotope (30)	0–5	4–18
Double isotope (31)	—	5–20
Sulfuric-acid-isotope (23)	0–2	—
Colorimetric (32a)	—	6–25

A reaction which has not been applied analytically is one involving the condensation of salicylol hydrazide with ketonic steroids to yield fluorescent salicylol hydrazones.[33]

Another group of steroids which are of great interest are the $\Delta^{1,4}$-3-ketones, many of which are synthetically derived therapeutic reagents. Prednisone, the structure which is shown below, demonstrates the charac-

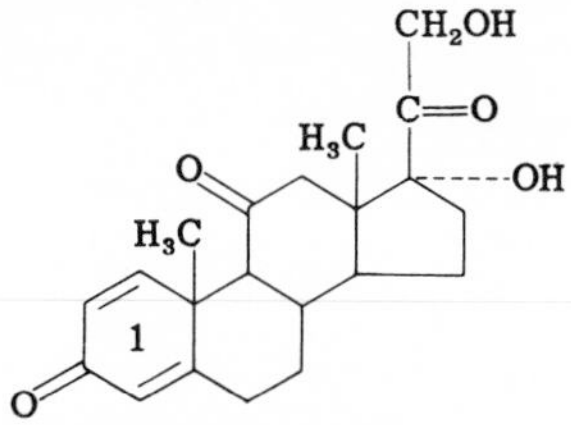

Prednisone

teristic structure of ring one. Prednisolone, triamcinolone, and their derivatives constitute a fair number of structurally related synthetic steroids. Smith and Foell[34] have described a procedure for distinguishing $\Delta^{1,4}$-3-ketones from Δ^4-3-ketones by a fluorescence procedure on paper chromatograms. Following development of the chromatograms, they are sprayed with a solution composed of 4 g. of isonicotinic acid hydrazide dissolved in 1 liter of methanol containing 5 ml. of concentrated HCl. Under these conditions, only the Δ^4-3-ketones appear as yellow fluorescent spots within 2 minutes. These are marked and the paper is set aside. After 3 to 16 hours at room temperature the $\Delta^{1,4}$-3-ketones also appear as yellow fluorescent spots and can now be distinguished from the rapidly appearing Δ^4-3-ketones. By using smaller amounts of acid and isonicotinic acid hydrazide, the reaction can be limited to the Δ^4-3-ketones and may be valuable for this group of compounds. However, there is need for a procedure to assay the $\Delta^{1,4}$-3-ketones in tissues. The findings of Smith and Foell[34] may prove applicable to quantitative application.

Although the fluorometric procedures are quite sensitive and specific for certain structures, their application is not always practical. The urine contains so many different steroid products and other materials which fluoresce or absorb light that many tedious purification procedures are required before material of reasonable homogeneity can be obtained. Fluorometry has therefore not proven of much use for urinary assays. Plasma is less complicated and relatively simple fluorescence assays are practical. For greater precision and specificity combined fluorescence-isotope dilution techniques are advisable.

V. BILE ACIDS

The bile acids, cholic, desoxycholic, chenodesoxycholic, and lithocholic, yield fluorescent products when heated in strong sulfuric acid solutions. According to Benard and Broer,[35] the products are maximally excited by the 436-mμ mercury line and the fluorescence maxima occur at 510 to 520 mμ (Fig. 9). These values are similar to those for the adrenal steroids. Levin et al.[36] have reported a spectrofluorometric procedure for the determination of total bile acids in bile.

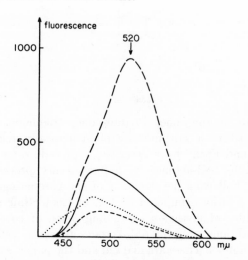

FIG. 9. Fluorescence spectra of bile acids in 12.95 M sulfuric acid. —, cholic acid; – – –, deoxycholic acid; —, chenodeoxycholic acid; $\cdots$, lithocholic acid (after Benard and Broer[35]).

REFERENCES

1. Kalant, H., *Biochem. J.* **69**, 79 (1958).
2. Braunsberg, H., and James, V. H. T., *Anal. Biochem.* **1**, 452 (1960).
3. Albers, R. W., and Lowry, O. H., *Anal. Chem.* **27**, 1829 (1955).
4. McDougal, D. B., Jr., and Farmer, H. S., *J. Lab. Clin. Med.* **50**, 485 (1957).
5. Carpenter, K. J., Gotsis, A., and Hegsted, D. M., *Clin. Chem.* **3**, 233 (1957).
6. Duggan, D. E., Bowman, R. L., Brodie, B. B., and Udenfriend, S., *Arch. Biochem. Biophys.* **68**, 1 (1957).
7. Bates, R. W., and Cohen, H., *Endocrinology* **47**, 166 (1950).
8a. Jailer, J. W., *J. Clin. Endocrinol.* **8**, 564 (1948).
8b. Finkelstein, M., Hestrin, S., and Koch, W., *Proc. Soc. Exptl. Biol. Med.* **64**, 64 (1947).
9. Bates, R. W., *Recent Progr. in Hormone Research* **9**, 95 (1954).
10. Bauld, W. S., Givner, M. L., Engel, and L. L., Goldzieher, J. W., *Can. J. Biochem. and Physiol.* **38**, 213 (1960).

11. Smith, E. P., Dickson, W. M., and Erb, R. E., *J. Dairy Sci.* **39,** 162 (1956).
12. Braunsberg, H., Stern, M. I., and Swyer, G. I. M., *J. Endocrinol.* **11,** 189 (1954).
13. Nakao, T., and Aizawa, Y., *Endocrinol. Japon* **3,** 92 (1956).
14. Oertel, G. W., West, C. D., and Eik-Nes, K. B., *J. Clin. Endocrinol.* **19,** 1619 (1959).
15. Strickler, H. S., Grauer, R. C., and Caughey, M. R., *Anal. Chem.* **28,** 1240 (1956).
16a. Eberlein, W. R., Bongiovanni, A. M., and Francis, C. M., *J. Clin. Endocrinol.* **18,** 1274 (1958).
16b. Finkelstein, M., Jewelewicz, R., Klein, O., Pfeiffer, V., and Sodmoriah-Beck, S., "Prenatal Care," p. 12. Noordhoff, Groningen, The Netherlands. 1960.
16c. Finkelstein, M., Forchialli, E., and Dorfman, R. I., *J. Clin. Endocrinol.* **21,** 98 (1961).
17. Sweat, M. L., *Anal. Chem.* **26,** 773 (1954).
18. Goldzieher, J. W., and Besch, P. H., *Anal. Chem.* **30,** 962 (1958).
19. Silber, R. H., Busch, R. D., and Oslapas, R., *Clin. Chem.* **4,** 278 (1958).
20. Zenker, N., and Bernstein, D. E., *J. Biol. Chem.* **231,** 695 (1958).
21. McLaughlin, J., Jr., Kaniecki, T. J., and Gray, I., *Anal. Chem.* **30,** 1517 (1958).
22. Sweat, M. L., *Anal. Chem.* **26,** 1964 (1954).
23. Peterson, R. E., *J. Biol. Chem.* **225,** 25 (1957).
24. Sweat, M. L., *J. Am. Chem. Soc.* **73,** 4056 (1951).
25. Bush, I. E., and Sandberg, A. A., *J. Biol. Chem.* **205,** 783 (1953).
26. Ayres, P. J., Garrod, O., Simpson, S. A., and Tait, J. F., *Biochem. J.* **65,** 639 (1957).
27. Ayres, P. J., Simpson, S. A., and Tait, J. F., *Biochem. J.* **65,** 647 (1957).
28. Abelson, D., and Bondy, P. K., *Arch. Biochem. Biophys.* **57,** 208 (1955).
29. Bondy, P. K., Abelson, D., Scheuer, J., Tseu, T. K. L., and Upton, V., *J. Biol. Chem.* **224,** 47 (1957).
30. Bondy, P. K., and Upton, V., *Proc. Soc. Exptl. Biol. Med.* **94,** 585 (1957).
31. Peterson, R. E., *in* "Lipids and the Steroid Hormones in Clinical Medicine" (F. W. Sunderman and F. W. Sunderman, Jr., eds.), p. 167. Lipincott, Philadelphia, Pennsylvania, 1960.
32a. Peterson, R. E., Karrer, A., and Guerra, S. L., *Anal. Chem.* **29,** 144 (1957).
32b. Staub, M. C., and Dingman, J. F., *J. Clin. Endocrinol.* **21,** 148 (1961).
33. Chen, P. S., Jr., *Anal. Chem.* **31,** 296 (1959).
34. Smith, L. L., and Foell, T., *Anal. Chem.* **31,** 102 (1959).
35. Benard, H., and Broer, Y., *Bull. soc. chim. biol.* **42,** 99 (1960).
36. Levin, S. J., Irvin, J. L., and Johnston, C. G., *Anal. Chem.* **33,** 856 (1961).

11

PLANTS

I. GENERAL REMARKS

In addition to many of the metabolites which are also found in animal tissues, there is a host of chemical substances which are unique to the plant kingdom. The chlorophylls, alkaloids, and flavonoids are three of the major categories in each of which there are, in turn, many examples. Because of their long interest in chlorophyll, which is highly fluorescent, plant scientists were among the first to be introduced to fluorometry and they have applied it widely for both qualitative and quantitative assay. An excellent survey of fluorescent compounds in plants and the application of fluorometry to their assay was presented by Goodwin.[1] It should be pointed out that many plant products are also important therapeutic agents. Many of these, including the alkaloids of cinchona, harmala, ergot, and rawolfia will be discussed in Chapter 13 on Drugs and Toxic Agents.

II. CHLOROPHYLL AND PHOTOSYNTHESIS

The chemical process which makes plant biochemistry unique is photosynthesis and the substance which makes photosynthesis possible is chlorophyll. Chlorophyll is a metalloporphyrin containing magnesium.

Chlorophyll *a*

Unlike iron and copper metalloporphyrins, chlorophyll is fluorescent both as the magnesium complex and as the metal-free porphyrin, pheophytin. There are actually several types of chlorophyll found in nature. Although the structural differences from one to another may not be very great, the different forms show characteristic fluorescent spectra. The spectra and fluorescence maxima of several forms of chlorophyll, shown in Fig. 1 and

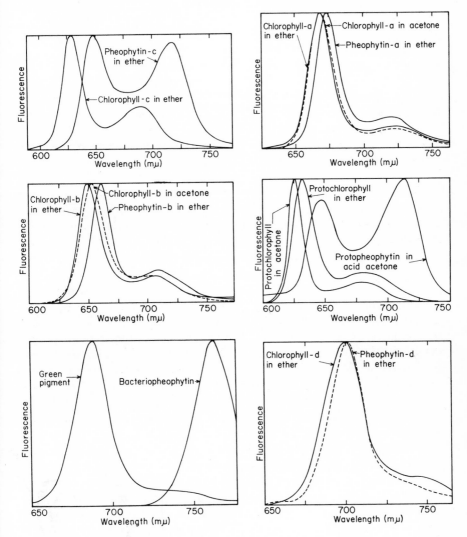

FIG. 1. Fluorescence spectra of various chlorophylls and their corresponding pheophytins (after French et al.[2a]).

TABLE I

Form of chlorophyll or pheophytin	Fluorescence maximum (mμ)	
	Chlorophyll	Pheophytin
a in ether	668	673
a in acetone	669	—
b in ether	648	661
b in acetone	652	—
c in ether	629	648
d in ether	699	701
Protochlorophyll in ether	631	—
Protochlorophyll in acetone	626	648–712
Bacteriochlorophyll ether	687	761

[a] After French et al.[2a]

Table I were obtained by French et al.[2a] It should be pointed out that the excitation wavelength used in these studies was generally 436 or 405 mμ. With the bacteriopheophytin, excitation was at 525 mμ. The emission in the red and near infrared necessitated the use of an RCA 6217 photomultiplier as a detector. However, it appears that the main peak of bacteriochlorophyll fluorescence is higher than 770 mμ, above the limits of detection used by French et al.[2a] The fluorescence spectra of chlorophyll a in ether, as obtained by Zscheile and Harris,[2b] are shown in Fig. 2.

In the case of the chlorophylls, light absorption and fluorescence are not merely useful as means for assay. They are actually intimately related to the mechanism for the conversion of light energy into chemical energy. For this reason fluorescence has been used by many investigators as a tool in studying the photosynthetic process; the literature on this subject is now immense. Some of the many reviews on the subject are listed in the bibliography.[3-5] An excellent survey concerning fluorescence spectrophotometry of photosynthetic pigments was presented by French.[6]

It should be pointed out that fluorescence occurs when a molecule absorbs radiation energy without using it. It is, therefore, difficult to ascribe any biological significance to this phenomenon. If the prime purpose of molecules in biological systems is to transmit energy, emission of fluorescence can signify only an inefficient utilization of absorbed radiation. As we shall see, chlorophyll in vivo shows very little fluorescence.

The interrelationship of light absorption, fluorescence, and energy utilization in photosynthesis may be formulated in the manner shown in

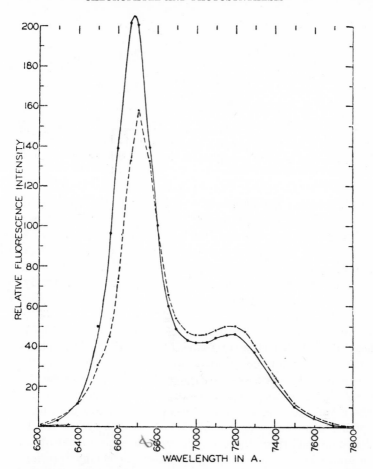

FIG. 2. Fluorescence spectra of chlorophyll *a* in ether. ○ — ○, thin layer of solution; ● – – – ●, thick layer of solution (after Zscheille and Harris[2b]).

Fig. 3. The highly absorbing chlorophyll molecule, on absorbing radiant energy, is converted to the excited electronic state designated as Chl*. The most recent findings[7, 8] suggest that the chlorophyll is brought into its triplet state, Chl***, and that it is in this "active" form that the chlorophyll passes its energy on to the acceptor, A_0, to initiate the chain of metabolic reactions. The Chl*** then returns to its ground state, Chl, and is capable of repeating the process. However, as in all phenomena of electronic excitation, alternative conversions are available. In addition to the transfer to an acceptor, the energy may be converted to heat or re-radiated as fluorescence, and there is a more or less reciprocal relationship among

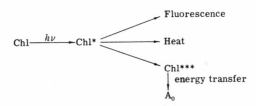

Fig. 3. Interrelationship of light absorption, fluorescence, and energy utilization in photosynthesis. Chl, chlorophyll in ground state; Chl*, chlorophyll in excited singlet state; Chl***, chlorophyll in excited triplet state; A_0, energy acceptor (substrate).

these phenomena. The fluorescence efficiency of chlorophyll a in live cells has been reported to be no higher than 2 to 2.8% and values as low as 0.15% have also been reported.[9] On the other hand, fluorescence efficiencies as high as 33% have been observed for chlorophyll a in solution. Brody and Rabinowitch[10] have presented additional data on excitation lifetimes and fluorescence yields of chlorophylls and related pigments. By contrast, the efficiency of photosynthesis has been reported to be as high as 90% by Warburg and Krippahl[11] and Burk[12] and about 25 to 30% by most other investigators. The balance of the absorbed energy, 10–75%, is lost as heat. Inhibitors such as urethane, decrease the yield of photosynthesis and produce a simultaneous, but not equivalent, increase in the yield of fluorescence.[4] However, the exact relationship between the intensity of chlorophyll fluorescence and the rate of photosynthesis in intact plant tissues is difficult to evaluate because both processes may be limited by competing processes such as interaction with oxygen, heat losses, etc. Under some conditions, however, fluorescence at any moment, may reflect in an approximate manner, the probability for energy transfer from the chlorophyll-protein complex.[4] Katz[13a] considers chlorophyll fluorescence as an energy flow meter for photosynthesis. Chlorophyll also undergoes photochemical bleaching which is associated with transient fluorescence changes.[13b] One example of the use of chlorophyll fluorescence in evaluating photosynthetic phenomena is in the explanation of action spectra of photosynthesis. In many instances, light, at wavelengths which are not maximally absorbed by chlorophyll, can evoke maximal rates of photosynthesis. It is now apparent that there are many other pigments in plants which can absorb light and fluoresce. In the intact plant many such pigments are effective in promoting photosynthesis. However, it can be shown that such pigments do not pass their excitation energy directly to the chemical metabolic mechanism but transfer it to chlorophyll first. The evidence for this comes from fluorescence studies on intact plants where chlorophyll fluorescence can be excited by irradiating with light which is

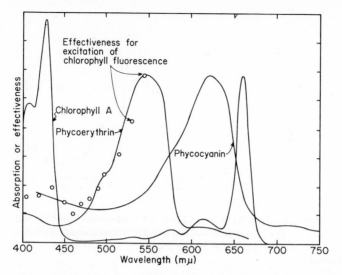

FIG. 4. Action spectrum for the excitation of chlorophyll a fluorescence in *Porphyridium* (open circles) compared with the absorption spectra of aqueous solutions of phycotrythrin and phycocyanin and an ether solution of chlorophyll *a* (after French and Young[14]).

optimal for excitation of the other pigments. French and Young[14] have shown that in *Porphyridium,* absorption of light in the region 500–600 mμ is effective for photosynthesis. In this region the pigments, phycoerythrin and phycocyanin, are responsible for most of the absorption. Examination of the action spectrum for the excitation of chlorophyll *a* fluorescence (685 mμ) in *Porphyridium* indicates that it coincides with the phycoerythrin absorption curve (Fig. 4). Since energy is transferred so effectively with respect to fluorescence, it is probable that the same energy transfer process occurs with respect to release of chemical energy in photosynthesis. In fact, chains of energy transferring pigments have been considered as ultimately leading to transfer to chlorophyll.[15] Fluorescence studies provide the evidence for such mechanisms. Bannister[16] has shown that even absorption by the tyrosine and tryptophan residues of a protein-pigment complex such as phycocyanin can yield the visible fluorescence of the pigment. The spectral characteristics of phycocyanin are shown in Fig. 5. Excitation at the 277-mμ band, in pure solutions of phycocyanin, is just as effective for fluorescence as is excitation in the region 300–400 mμ or higher.

Measurement of fluorescence spectra of intact plant cells as an index of photosynthesis and metabolism *in vivo* have been used widely not only with respect to chlorophyll and related pigments[17, 18] but also with respect

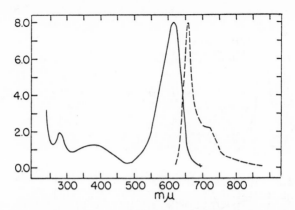

FIG. 5. Absorption spectrum (solid line) and fluorescence spectrum (broken line) of phycocyanin. The absorption bands at 615 and 350–400 mμ are attributed to the chromophor; the peak at 275 mμ is due to the protein. The fluorescence peak is at 655 mμ (after Bannister[16]).

to pyridine nucleotides.[19] Studies on pyridine nucleotide fluorescence in intact plant cells were discussed in Chapter 9.

An example of how valuable fluorescence spectroscopy can be in identifying newly isolated metabolites in plants is shown by the studies of Goodwin et al.[20] On using fluorescence microscopy they observed a pigment in the leaf of *Vicia* which differed from known plant pigments. It was possible to extract all the chlorophyll from portions of leaf epidermis with methanol.

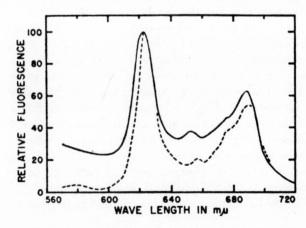

FIG. 6. Fluorescence spectra of methanol extracted epidermis of Vicia (solid line), and of uroporphyrin I octamethyl ester in methanol (dotted line) (after Goodwin et al.[20]).

The other pigment could not be extracted and the fluorescence spectrum of the extracted leaf epidermis was measured directly. This was found to compare almost exactly with that of uroporphyrin I (Fig. 6). The solubility of the plant pigment was so low that it was not possible to obtain an absorption spectrum. However, the excitation spectrum showed a peak at 408 mμ which is also characteristic of uroporphyrin I.

Chlorophyll derivatives and other porphyrins are also found in crude oil and in various other geological materials indicating the biological origin of these materials. Fluorescence is being used along with chromatography to identify and classify the various porphyrin substances[21, 22] in such geologic deposits. The application of fluorescence spectrometry in this area should prove extremely valuable.

It has been possible to touch only upon the uses of fluorescence in studies on plant pigments and photosynthesis. However, it should be obvious that fluorescence assay is an indispensable tool in this field. Most of the studies reported here and elsewhere have been done with individually designed or modified instruments. Corrections of the spectral output of the excitation source have generally been applied by workers in this field. Corrections of fluorescence spectra for the spectral sensitivity of the detector have also been applied by most investigators. It will be of interest to see whether commercially available fluorescence spectrometers will be able to provide data comparable to that obtained with the custom designed instruments. If this should prove to be the case it may lead to more intensive investigations in this area.

III. PLANT GROWTH SUBSTANCES, AUXINS

The discovery of auxins and other plant growth regulatory substances opened a new avenue of research in plant science. At the present time these agents are being intensively studied. Audus[23] has written an excellent monograph summarizing work in this field. Fluorescence assay has already proved valuable in the case of the two important classes of growth substances the indole acids and the gibberelins.

A. β-Indolylacetic (Heteroauxin)

β-Indolylacetic (heteroauxin, indoleacetic acid) acid is found in animal and plant tissues and is derived from tryptophan, at least in the former group. Being an indole it has the characteristic fluorescence of this class of compounds; excitation maximum at 278 mμ, fluorescence maximum at 348 mμ (corrected).[24] A method for assaying β-indolylacetic acid in animal tissues and urine was reported by Weissbach et al.[25] This procedure involved

$$CH_2COOH$$

β-Indolylacetic acid

extraction from the acidified tissue homogenate or urine into chloroform and reextraction from an aliquot of the chloroform into phosphate buffer, pH 7.0. In the assay, as reported, colorimetry was used. However, in many tissues it is possible to substitute fluorometry to measure the β-indolyl-acetic acid in the final extracts. In plant materials the simple extraction procedure may not be adequate for specific assay. It may, however, be possible to endow the procedure with greater specificity by proper choice of solvents, pH of extractions, pH of fluorescence assay, and excitation and fluorescence wavelengths.

Ebert[26a] has reported a more sensitive fluorometric method for β-indolyl-acetic acid. To 1 ml. of an aqueous extract containing 0.05–12.5 μg. of the acid, add 0.2 ml. of 0.05 M copper sulfate and then 0.5 ml. of concentrated sulfuric acid. The last addition is made with thorough mixing. The nature of the fluorescent product and its exact spectral characteristics have not been ascertained. Rakitin and Povolotskaya[26b] have modified and applied this procedure.

Mavrodineanu et al.[27] have reported using the native fluorescence of β-indolylacetic acid on paper in a quantitative assay procedure which they applied to plant extracts. This is rather surprising since it has been generally considered that indoles adsorbed on paper are not fluorescent.[28] It may be that β-indolylacetic acid behaves differently from other indoles in this respect. The procedure as described by Mavrodineanu et al.[27] would probably be best applied after an initial extraction from the plant tissue by a method similar to the one described by Weissbach et al.[25] or with ethanol as described in their paper.[27] The extract is concentrated and chromatographed on paper using isopropanol, water, and concentrated ammonium hydroxide (80:15:5). After 17 hours of development (ascending), the position of the β-indolylacetic acid can be determined by examination under an ultraviolet lamp. The fluorescence can be measured quantitatively by using a paper scanning device (see Chapter 3) and appropriate filters. In this study a General Electric 4-watt germicidal lamp was used for excitation with an ultraviolet filter (Schott U.G.2). The filtered exciting light contained a trace of the 297- and 303-mμ bands with the bulk of the radiation being at 313 and 366 mμ. The detector was a photomultiplier tube having a maximum in the visible region, probably a 1P28 tube. A filter having maximal transmission at 465 mμ was found to be most sensitive for this assay. As little as 0.5 μg. of β-indolylacetic acid per chromatographic spot can be estimated. An advantage of this procedure is that the compound is not destroyed in the analysis and can therefore be eluted and subjected to further tests for identification. Additional information concerning solvent systems for the development of paper chromatograms in the fluorescence assay of β-indolylacetic acid was presented by Pavillard and Beauchamp.[29] Audus has also presented data on various solvent systems for developing chromatograms for assay of heteroauxin and related indoles.

B. Gibberellins

The gibberellins are an important class of plant growth regulator substances which were isolated more recently than the indole auxins. Gibberellin X and some related products are shown in Fig. 7. Gibberellenic and gibberic acids are also found in plant extracts but are devoid of biological activity. Gibberellic acid itself is a colorless, polycyclic unsaturated compound which has no chemical properties that permit chemical assay. It was observed by Cross[30] that at high concentrations gibberellic acid yielded a red color when treated with concentrated sulfuric acid. What is of greater interest, this product was observed to be fluorescent. Kavanagh and Kuzel[31] have utilized this reaction in the development of a highly sensitive and specific method for the assay of gibberellic acid in extracts.

Transfer 1.00 ml. of extract or standard solution, containing 5–20 μg. of gibberellic acid, to a 50-ml. Erlenmeyer flask. Cool the flask in an ice bath

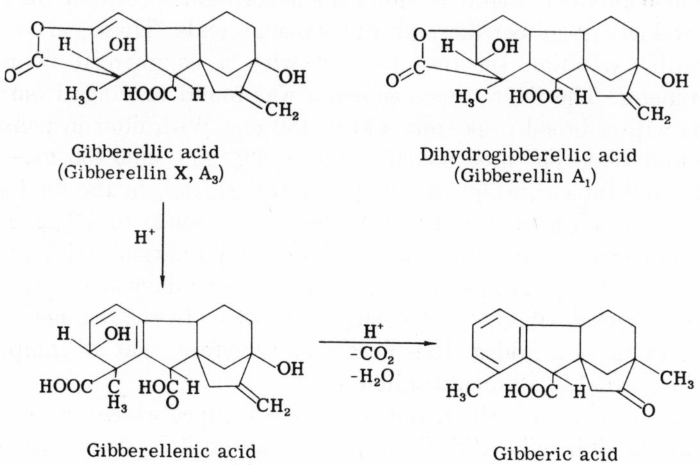

Gibberellic acid
(Gibberellin X, A$_3$)

Dihydrogibberellic acid
(Gibberellin A$_1$)

Gibberellenic acid

Gibberic acid

FIG. 7. Gibberellic acid and related products.

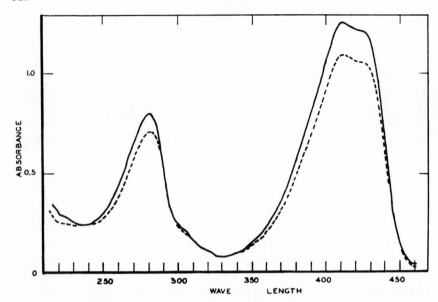

FIG. 8. Absorption spectra of gibberellic and gibberellenic acids in 85% sulfuric acid at a concentration of 0.02 mg./ml. ——, gibberellic acid; ------, gibberellenic acid (after Kavanagh and Kuzel[31]).

and add 25 ml. of chilled 85% sulfuric acid with continuous shaking of the flask. Allow the acid to flow as rapidly as it will from the pipet. Let the solution stand for at least 1 hour at room temperature to permit formation of the acid product. Figure 8 shows the absorption spectra of the sulfuric acid products of gibberellic and gibberellenic acids. These curves agreed well with excitation spectra observed with a commercial fluorescence spectrometer. The fluorescence emission was found to range from 440 to 550 mμ with a broad peak from 450 to 480 mμ. With filter instruments it was found that the three mercury lines—366, 405, and 436 mμ—served equally well for excitation. As little as 0.006 μg./ml. in the final sulfuric acid mixture can be detected and the response is linear to 3.2 μg./ml. The fluorescent product once formed is stable to illumination and fluorescence intensity remains constant over a period of several days at room temperature. The reproducibility of the assay is sensitive to the manner in which the sulfuric acid is added. It is necessary therefore, that all samples in a series be treated in an identical manner.

Gibberellenic acid is the major inactive substance which interferes with the assay of gibberellic acid. For rapid estimation it is possible to estimate the inactive compound by its characteristic absorption at 254 mμ and then subtract this value from the fluorometrically determined value. Gibberellic acid does not have an absorption peak at 254 mμ. For more precise measurements, gibberellic acid may be separated from gibberellenic acid by chromatography on a column of potassium bicarbonate-sodium sulfate (anhydrous) using acetonitrile (containing 1% water) as the eluant. Gibberellic acid is eluted with this solvent whereas gibberellenic acid and other contaminants can be eluted only with methanol.

C. Coumarins

Plants contain other substances which are considered by some as having auxin or antiauxin activity. The coumarins are widely distributed in plants. One coumarin derivative, scopoletin, is regarded by plant physiologist as

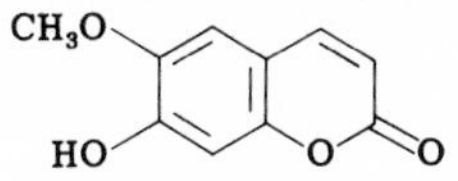

Scopoletin

an inhibitor of root growth,[32, 33] its effect being related to that of heteroauxin. Being a hydroxy, methoxy coumarin, scopoletin is highly fluorescent. Although its exact fluorescence characteristics have not been described, it yields a blue fluorescence in solutions above pH 7. It is also intensely fluorescent on paper chromatograms and directly in the plant tissues under ultraviolet illumination. Fluorescence methods have been used to show that scopoletin is increased in plant tissues in a variety of plant disorders including virus infections and boron deficiency.[23] Umbelliferone (7-hydroxycoumarin), one of the most intensely fluorescent compounds known, is also found in plants.

D. Caffeic and Chlorogenic Acids

The cinnamic acid derivatives caffeic acid and chlorogenic acid may also possess activity as plant growth regulatory substances. Both substances

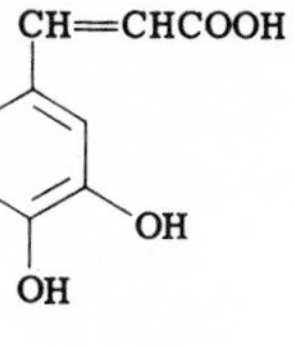

Caffeic acid

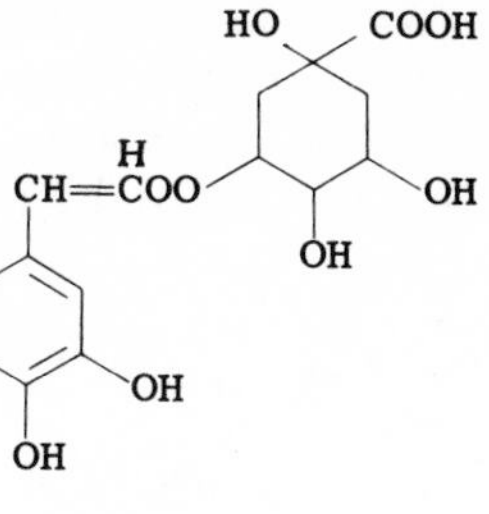

Chlorogenic acid

emit blue fluorescence on paper chromatograms and in plant tissues.[23] Perkins and Aronoff[34] have shown that the blue fluorescence formed around necrotic areas on boron deficient plants is due to the localization there of both chlorogenic and caffeic acids. In addition to their native fluorescence, these compounds should also yield highly fluorescent derivatives with ethylenediamine as described for catecholamines (Chapter 5). Such a reaction could prove to be highly specific for the two cinnamic acids derivatives.

IV. FLAVONES, ANTHOCYANINS, AND COUMARINS

The glycosides of various flavones and anthocyanins occur as pigments in flowers, seeds, and pollen. Rutin and quercetin are two polyhydroxylated flavone derivatives which are highly fluorescent. They have a therapeutic use in maintaining hemostasis and will therefore be discussed in Chapter 13 on Drugs and Toxic Agents. There are so many different flavones and anthocyanins that the mixture in a given plant is fairly characteristic. The presence of flavones in pollens has suggested their use in pollen detection and classification. Inglett et al.[35] have applied fluorescence methods for this purpose. They report that hydroxyflavone metal complexes with Al^{3+}, Be^{++}, and Zr^{4+} are most sensitive for this purpose and use filter paper impregnated with these metals to distinguish different compounds. Ragweed pollen is rich in isorhamnetin. With the Al^{3+} impregnated paper,

Isorhamnetin

as little as 0.001 μg. of isorhamnetin can be detected. The procedure, because of its sensitivity, may prove extremely useful for quantitating ragweed pollen in the atmosphere. Another sensitive fluorometric reagent for hydroxyflavones is tetraphenylborate.[36] This reagent is discussed in the section relating to rutin and quercetin in Chapter 13.

V. TOXIC PRODUCTS

An important agricultural problem concerns the presence of toxic agents in plants which are fed to animals or humans. A hemorrhagic disease of cattle was found to be caused by a toxic agent in spoiled sweet clover. This substance was later identified as bishydroxycoumarin (dicumarol)

Dicumarol
(bishydroxycoumarin)

formed by bacterial action on coumarin, which is the sweet-smelling substance of clover. In man, dicumarol produces a decrease of plasma prothrombin and is used in anticoagulant therapy. Although there have been no reports concerning the fluorescence of this agent, it appears to be a structure that should lend itself to fluorometric assay.

Another toxic agent in plants which is related to the flavones is coumestrol. This compound was isolated from ladino clover, but is found in a

Coumestrol

variety of leguminous plants used for forage.[37-39] The compound possesses estrogenic activity which can be beneficial with relation to growth and milk production. However, it can also lead to undesirable reproductive disturbances. Livingston et al.[40] have developed a method for the assay of coumestrol utilizing the native fluorescence of the compound on paper chromatograms. For quantitative assay extracts, prepared as described by Lyman et al.,[38] are applied as 2- to 10-μl. aliquots on marks 5 cm. apart along a line 11 cm. from the bottom edge of a sheet of Whatman No. 1 paper (40 $\times$ 57 cm.). The chromatograms are developed with glacial acetic acid and water (1:1) for 16 hours at room temperature and are then dried in air for 24 hours. Coumestrol appears as a fluorescent spot at R_f 0.5. The intensity of fluorescence is measured with an attachment for measuring fluorescence directly on paper and is compared with appropriate standards. It is suggested that a wide aperture be used, one which is larger than the spot. The galvanometer deflection is then maximal and directly proportional to the coumestrol concentration.

Coumestrol exhibits maximal absorption at 303 and 343 mμ and emits blue fluorescence. The fluorescence on paper is directly proportional to

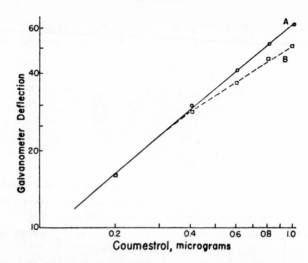

FIG. 9. Fluorescence intensity of coumestrol on paper chromatograms as a function of the quantity in the developed spot. Curve A, with wide apertures; curve B, with a narrow aperture (after Livingston et al.[40]).

coumestrol concentration (Fig. 9) and is stable over a period of days. The specificity of this method is largely dependent on the method of isolation which is used prior to the paper chromatography.

GENERAL REFERENCES

Brugsch, J., "Porphyrine." J. A. Barth, Leipzig, 1952.
Bladergroen, W., "Problems in Photosynthesis." C. C Thomas, Springfield, Illinois, 1960.
Duysens, L. N. M., "Transfer of Excitation Energy in Photosynthesis." Kemink en Zoon, Utrecht, 1952.
Franck, J., and Loomis, W. E. (eds.), "Photosynthesis in Plants." Iowa State College Press, Ames, Iowa, 1949.
Gaffron, H., Brown, A. H., French, C. S., Livingston, R., Rabinowitch, E. I., Strehler, B. L., and Tolbert, N. E. (eds.), "Research in Photosynthesis." Interscience, New York, 1957.
Johnson, F. H. (ed.), "The Luminescence of Biological Systems." Am. Assoc. Advanc. Sci., Washington, D. C., 1955.
The Photochemical Apparatus, Its Structure and Function. Brookhaven Symposia in Biol., No. 11, 1958.

REFERENCES

1. Goodwin, R. H., Ann. Rev. Plant Physiol. 4, 283 (1953).
2a. French, C. S., Smith, J. H. C., Virgin, H. I., and Airth, R. L., Plant Physiol. 31, 369 (1956).
2b. Zscheille, F. P., and Harris, D. G., J. Phys. Chem. 47, 623 (1943).
3. Rabinowitch, E., "Photosynthesis and Related Processes," Vol. II, p. 740. Interscience, New York, 1951.
4. Wassink, E. C., Advances in Enzymol. 11, 91 (1951).

5. Gaffron, H., *et al.* (eds.), "Research in Photosynthesis." Interscience, New York, 1957.

6. French, C. S., *in* "The Luminescence of Biological Systems" (F. H. Johnson, ed.), p. 51. Am. Assoc. Advance. Sci., Washington, D. C., 1955.

7. Commoner, B., Heise, J. J., Lippincott, B. B., Norberg, R. E., Passonneau, J. V., and Townsend, J., *Science* **126**, 57 (1957).

8. Calvin, M., and Sogo, P. B., *Science* **125**, 499 (1957).

9. Latimer, P., Bannister, T. T., and Rabinowitch, E., *in* "Research in Photosynthesis" (H. Gaffron *et al.*, eds.), p. 110. Interscience, New York, 1957.

10. Brody, S. S., and Rabinowitch, E., *Science* **125**, 555 (1957).

11. Warburg, O., and Krippahl, G., *Angew. Chem.* **66**, 493 (1954).

12. Burk, D., *Federation Proc.* **12**, 611 (1953).

13a. Katz, E., *in* "Photosynthesis in Plants" (J. Franck and W. E. Loomis, eds.), p. 287. Iowa State College Press, Ames, Iowa, 1949.

13b. Olson, R. A., *Rev. Sci. Instr.* **31**, 844 (1960).

14. French, C. S., and Young, V. K., *J. Gen. Physiol.* **35**, 873 (1952).

15. Duysens, L. N. M., *Nature* **168**, 548 (1951).

16. Bannister, T. T., *Arch. Biochem. Biophys.* **49**, 222 (1954).

17. Virgin, H. I., *Physiol. Plantarum* **9**, 674 (1956).

18. Epstein, H. T., de la Tour, B. E., and Schiff, J. S., *Nature* **185**, 825 (1960).

19. Duysens, L. N. M., and Sweep, G., *Biochim. et Biophys. Acta* **25**, 13 (1957).

20. Goodwin, R. H., Koski, V. M., and Owens, O. V. H., *Am. J. Botany* **38**, 629 (1951).

21. Blumer, M., *Anal. Chem.* **28**, 1640 (1956).

22. Dunning, H. N., and Carlton, J. K., *Anal. Chem.* **28**, 1362 (1956).

23. Audus, L. J., "Plant Growth Substances." Leonard Hill, London, 1959.

24. Teale, F. W. J., and Weber, G., *Biochem. J.* **65**, 476 (1957).

25. Weissbach, H., King, W., Sjoerdsma, A., and Udenfriend, S., *J. Biol. Chem.* **234**, 81 (1959).

26a. Ebert, V. A., *Phytopathol. Z.* **24**, 216 (1955).

26b. Rakitin, Yu. V., and Povolotskaya, K. L., *Fiziol. Rastenii Akad. Nauk S.S.S.R.* **4**, 285 (1957).

27. Mavrodineanu, R., Sanford, W. W., and Hitchcock, A. E., *Contrib. Boyce Thompson Inst.* **18**, 167 (1955).

28. Smith, I., "Chromatographic Techniques," p. 128. William Heinemann Medical Books, London, 1958.

29. Pavillard, J., and Beauchamp, C., *Compt. rend. acad. Sci.* **244**, 678 (1957).

30. Cross, B. E., *J. Chem. Soc.* p. 4670 (1954).

31. Kavanagh, F., and Kuzel, N. R., *J. Agr. Food Chem.* **6**, 459 (1958).

32. Thimann, K. V., and Bonner, W. D., Jr., *Proc. Natl. Acad. Sci. U.S.* **35**, 272 (1949).

33. Andreae, W. A., *Nature* **170**, 83 (1952).

34. Perkins, H. J., and Aronoff, S., *Arch. Biochem. Biophys.* **64**, 506 (1956).

35. Inglett, G. E., Miller, R. R., and Lodge, J. P., *Mikrochim. Acta No.* **1**, 95 (1959).

36. Neu, R., *Mikrochim. Acta Nos.* **7–8** 74 (1956).

37. Bickoff, E. M., Booth, A. N., Lyman, R. L., Livingston, A. L., Thompson, C. R., and DeEds, F., *Science* **126**, 969 (1957).

38. Lyman, R. L., Bickoff, E. M., Booth, A. N., and Livingston, A. L., *Arch. Biochem. Biophys.* **80**, 61 (1957).

39. Bickoff, E. M., Lyman, R. L., Livingston, A. L., and Booth, A. N., *J. Am. Chem. Soc.* **80**, 3969 (1958).

40. Livingston, A. L., Bickoff, E. M., Guggolz, J., and Thompson, C. R., *Anal. Chem.* **32**, 1620 (1960).

12

ANALYSIS OF INORGANIC CONSTITUENTS

I. INTRODUCTION

MANY inorganic constituents, both normally occurring and present as toxic constituents, can be assayed by fluorescence procedures. These inorganic constituents include metal ions, anions, oxygen, and ozone.

Not many elements possess native fluorescence in aqueous solution. Uranium salts, such as potassium uranyl sulfate or acetate, fluoresce only slightly in aqueous solution. The intensity increases markedly with increasing amounts of sulfuric acid and with lowering of the temperature. Uranium in solid form is highly fluorescent. Thallium chloride emits a characteristic native fluorescence when dissolved in water or dilute salt solutions. As little as 5 μg./ml. yields appreciable fluorescence. Although none of the metal ions which are of interest to the biologist possess native fluorescence, many of them form fluorescent complexes with organic re-

TABLE I

REAGENTS FOR INORGANIC FLUOROMETRIC ANALYSIS[a]

Element	Reagent	Excitation maximum mμ	Fluorescence maximum mμ	Sensitivity μg.
Al	Pontochrome BBR	470, 580	630	0.2
Al	Morin	430	500	0.0005
Al	Acid alizarin garnet R	470	590	—
Be	Morin	470	555–585	0.004
B	Benzoin	370	480	0.2
F⁻	Al—alizarin garnet R	—	550–600	0.1
F⁻	Mg—oxine	420	530	—
Li	8-Hydroxyquinoline (oxine)	370	580	3.0
Mg	Bissalicylidene ethylenediamine	355	439	0.0001
Mg[b]	8-Hydroxyquinoline (oxine)	420	530	0.5

[a] After White[1] and White and Cutitta.[3]
[b] After Schachter.[4]

agents which can be utilized in the design of specific analytical procedures. Inorganic analyses based on fluorometry and utilizing specific fluorescence yielding chelating agents are now widely used. Various publications by White[1] and his colleagues[2] summarize recent advances in this field. Some typical reagents and the properties of the fluorescent chelates are shown in Table I. Specific applications of fluorescent complexes are presented in the following sections.

II. MAGNESIUM IN SERUM AND URINE[4]

Ethanolic solutions of Mg^{++}-8-hydroxyquinoline exhibit characteristic fluorescence which is maximal at pH 6.5 and almost negligible at pH 3.5. Under appropriate conditions the increment in fluorescence between these two pH values may be used as an index of Mg^{++} concentration.

The following two reagents are used for serum. Reagent 1 (pH 3.5) comprises 2 volumes of 2 M acetate buffer, pH 3.5, 2 volumes of 5% 8-hydroxyquinoline in ethanol, 30 volumes of ethanol, and 5 volumes of deionized water. Reagent 2 (pH 6.5) is prepared in the same way except that 2 M acetate buffer, pH 6.5, is used instead of the pH 3.5 buffer. To two separate glass-stoppered centrifuge tubes are added 3.9 ml. of reagents 1 and 2. A 0.1-ml. aliquot of serum is added to each tube, they are stoppered, shaken by hand for 2 minutes, and centrifuged. The clear supernatant solutions are transferred to cuvettes and fluorescence is measured; excitation maximum, 420 mμ; fluorescence maximum, 530 mμ (uncorrected). Reagent blanks, substituting water for serum, and standard solutions (0.1 μM of Mg^{++} in 0.1 ml.) are prepared in the same manner. Urine samples, about 0.04 ml., are treated in a similar manner except that the buffering capacity of the two reagents is increased 2.5-fold.

TABLE II

COMPARISON OF NORMAL HUMAN SERUM MAGNESIUM LEVELS OBTAINED BY VARIOUS PROCEDURES[a]

Method	Serum magnesium[b] mole equivalent/liter
Fluorometric	2.05 ± 0.22
Flame photometer	2.05 ± 0.18
Titan yellow	1.87 ± 0.12
Molybdivanadate	1.95 ± 0.21

[a] After Schachter.[4]

[b] Mean ± standard deviation; 14–45 subjects used with each method.

The only normally occurring ion in animal tissues which yields appreciable fluorescence with the reagent is zinc. However, the zinc contents of human serum and urine are only 2–3% that of magnesium. Since the zinc fluorescence is also somewhat lower, the error due to zinc should be less than 1.5%. A further indication of specificity is provided with the chelating agent ethylenediamine tetraacetate which quenches the magnesium fluorescence completely. In serum, there is no fluorescence in the presence of the chelating agent. In urine, 10% of the fluorescence persists even after addition of the chelating agent. Such a correction is applied to urine measurements. Recoveries of magnesium from serum and urine range from 95 to 105% and values obtained with this procedure compare well with other methods including flame photometry (Table II).

The use of other dye reagents, such as bissalicylidene ethylenediamine, suggested by White and Cutitta,[3] may make possible even more sensitive fluorescence assays for magnesium.

III. ALUMINUM

A fluorometric procedure for the quantitative assay of aluminum in beer, based on the reaction with 8-hydroxyquinoline, was reported by Tullo et al.[5] in 1949. A review of fluorometric methods for aluminum using 8-hydroxyquinoline was published by Goon et al.[6] More recently, Rubins and Hagstrom[7] have reported a modification of this procedure which is readily applicable to plant tissues.

The plant tissue is carried through a wet digestion using nitric and perchloric acids. The digest is then made up to volume and an aliquot, containing no more than 10 μg. of aluminum, is transferred to a separatory funnel. Iron, which would otherwise interfere, is removed by adding 4,7-diphenyl-1,10 phenanthroline in the presence of hydroxylamine. The ferrous phenanthroline complex is removed by adjusting the pH to about 3.5 and diluting and extracting with a mixture of chloroform and isoamyl alcohol (450 + 50 v/v). To the residual aqueous phase (about 120 ml.) are added 2 ml. of 8-hydroxyquinoline solution (10 g. of 8-hydroxyquinoline in 500 ml. of water containing 30 ml. of glacial acetic acid) and sufficient ammonium hydroxide (about 5 ml.) to raise the pH to 5.5. After 10 minutes, two extractions with 15-ml. portions of chloroform are carried out and the combined chloroform extracts are made up to volume. Standards and reagent blanks are carried through the entire procedure.

The chloroform extracts are transferred to cuvettes and fluorescence is measured. With filter instruments, excitation with the 360-mμ mercury line is used (with fluorescence spectrometers this should be checked).

Fluorescence occurs as a broad band from 450 to 610 mμ.[1] Recoveries are quantitative and, when properly carried out, other metal ions such as copper, zinc, and manganese do not interfere. The magnesium complex does not extract into chloroform at the low pH values used. Aluminum contents of some plants, obtained by the fluorometric method, are (μg./g.) : alfalfa, 47.2; red clover, 38.4; quack grass, 11.5; timothy, 10.9; and tobacco, 1225.

IV. BERYLLIUM

Beryllium salts are now being used industrially on a wide scale, in fluorescent lamps, as X-ray tube windows, etc. Those working with beryllium frequently develop chronic pneumonitis with coughing, dyspnea, and extreme loss of weight. Berylliosis, the term used for beryllium toxicosis, is due to beryllium dust in the air. It can even occur in individuals living in the vicinity of a beryllium producing plant, but not employed there. In such areas the beryllium concentration in air must be monitored and kept to a minimum (less than 0.01 μg./cu. m.).[8] Methods for its detection in air-dust, smear samples, urine, and other biological material are important in maintaining proper monitoring procedures and a number of methods have been reported. The one by Sill and Willis[9] is described in some detail.

All samples are digested with nitric and perchloric acid. However, specific modifications in the digestion procedure are required for each material being assayed. Following the digestion and removal of nitric acid, ethylenediamine tetraacetate is added to complex heavy metlas and the solution is partly neutralized with ammonium hydroxide (with cooling) using phenol red. The solution, which may be about 75 to 100 ml. at this point, is transferred to a separatory funnel, about 10 drops of acetylacetone are added, and additional ammonium hydroxide is added to the red alkaline form of phenol red indicator. The beryllium acetylacetonate is extracted with two 10-ml. portions of chloroform, vigorous shaking being used. The chloroform is evaporated over a mixture of nitric and perchloric acid and wet ashing is repeated.* Following digestion and evaporation, beryllium from the unknown sample is obtained in about 0.5 ml. of 72% perchloric acid. To this are added 3.0 ml. of 8.2 M NaOH and 3 drops of 0.01% quinine sulfate as an indicator. The mixture is neutralized by dropwise addition of 72% perchloric acid (to blue fluorescence) and made definitely acidic. It is then transferred to a 25-ml. volumetric flask to which

---

* The reaction between acetylacetone and nitric acid is vigorous and the precautions in reference 9 should be followed closely.

the following additions are made with careful mixing between each: 0.5 ml. of triethanolamine-EDTA solution (5 g. disodium ethylenediamine tetraacetate and 3 ml. of triethanolamine in 95 ml. of water); 1 N NaOH, dropwise until quinine fluorescence is quenched, then 2 more drops; 0.5 ml. of freshly prepared alkaline stannite solution†; and 5.0 ml. of piperidine buffer. Make up to volume, then add 1.0 ml. of morin (2′,4′,3,5,7-penta-hydroxyflavine) solution, mix, adjust to constant temperature and transfer to a cuvette for fluorometric assay. Blanks and standards, 0.02–2.5 μg., are carried through the entire procedure.

Sill and Willis[9] used a mercury lamp and excited with a filter for the 360-mμ band. The secondary filter was one which cut off at 460 mμ. According to White and Cutitta,[3] the beryllium morin complex has maximal excitation at 470 mμ and maximal fluorescence at 555 to 585 mμ. As little as 0.001 μg. can be detected and 0.5 μg. can be recovered to the extent of about 95%. The method of isolation separates the beryllium from most interfering substances.

V. SELENIUM

Selenium can replace the sulfur in cystine and methionine in many plants. As a result, selenium poisoning constitutes a public health and agricultural problem in those areas which have a high selenium content in the soil. However, selenium is now of greater interest than as a toxicological problem since it appears to be an essential trace element[10] in many animal species. Its addition to the diet can prevent "white muscle" disease in sheep[11] and other disorders which are also associated with vitamin E. Obviously, sensitive and specific methods for selenium assay can be useful to ascertain the biological role of this element.

Methods for the assay of selenium in biological material, based on the fluorescence of the complex with 3,3′-diaminobenzidine, have recently been published by Cousins[12] and by Watkinson.[13a] Some details of the latter procedure are presented here.

Plant material, containing 0.02–2 μg. of selenium, is dried and the

---

† Beryllium reagents—*Alkaline stannite*: Dissolve 1.5 g. stannous chloride dihydrate in 2 ml. of water. Add 25 ml. of cold water and pour the milky suspension into 10 ml. of 6 N NaOH and dilute to 50 ml. Prepare immediately before use.

Piperidine buffer: Weigh 10.0 g. of hydrazine sulfate and 5.0 g. of ethylenediamine tetraacetate into a large beaker. Add 30 ml. of water and 50 ml. of redistilled piperdine and dissolve. Dilute to 500 ml.

Morin 0.0078%: Dissolve 7.8 mg. of morin in 25 ml. of 95% ethanol. Mix with 25 ml. of 8 M sodium perchlorate solution (iron-free) and 2 drops of 10% ethylenediamine tetraacetate and make up to 100-ml. volume with water.

pulverized material is transferred to a borosilicate test tube. Digestion in nitric acid and perchloric acid is carried out as described and, after boiling off the nitric acid and removing precipitated potassium perchlorate, the sample is adjusted to a volume of 35 ml. To this are added 50 ml. of concentrated HCl and 4 ml. of a 1% suspension of the zinc complex of toluene-3,4-dithiol (dithiol) in 95% ethanol. After 15 minutes the selenium-dithiol complex is extracted into 10 ml. of a 50% mixture of ethylene chloride and carbon tetrachloride. The organic solvent is collected in a 6-× 1-inch borosilicate tube, and another 5 ml. of organic solvent is used for extraction and added to the first extract. To the tube are then added 1 ml. of 72% perchloric acid, 10 drops of concentrated nitric acid, and a glass ball to prevent bumping. The organic solvents are boiled off and the residue is digested until the appearance of perchloric acid fumes. At this point, water is added and boiling is continued to remove traces of nitric acid. Cool the mixture and add about 20 ml. of water, 2 ml. of 2.5 M formic acid, 1 drop of m-cresol purple indicator, and sufficient 7 M ammonium hydroxide until the indicator turns yellow. At this point, 2 ml. of freshly prepared 0.5% aqueous solution of 3,3'-diaminobenzidene are added, the solution is diluted to 45 ml. and permitted to stand for 30 to 40 minutes. Following this, 7 M ammonium hydroxide is added until the indicator turns purple and the volume is adjusted to 50 ml. The selenium-diaminobenzidene complex is then extracted into 10 ml. of toluene, the solvent is dried by shaking with anhydrous calcium chloride and transferred to a cuvette for fluorometric assay. Blanks and standards are carried through the entire procedure. The fluorescent properties of the selenium-diaminobenzidine complex have recently been studied in detail by Parker and Harvey.[13b]

Both Cousins[12] and Watkinson[13a] found that excitation must be carried out above 420 mμ to minimize blank fluorescence. The fluorescence maximum was reported as being approximately 580 mμ. A fluorescence spectrometer was not used for these measurements. Recoveries were from 85 to 100%. Various grasses, fruits, and bulbs were found to contain from 0.01 to 0.08 μg. of selenium per gram. Soil content (New Zealand) was reported as 1.4 to 1.5 μg./g.

VI. PHOSPHATE AND PYROPHOSPHATE

Colorimetric methods for phosphate in solution are quite sensitive and adequate. The colorimetric reagents are, however, difficult to use on paper either with chromatography or electrophoresis and do not provide adequate sensitivity. Rorem[14] reported a procedure for the detection of phosphate esters on paper chromatograms by a fluorescence procedure. This same procedure can also be used for both phosphate and pyrophosphate. The dried chromatogram is immersed in a solution of 0.5% quinine sulfate in ethanol. After drying at room temperature, the phosphate spots can be detected under an ultraviolet lamp (either 254- or 360-mμ excitation). They show up as light areas against an intense gray-blue fluorescent background. As little as 0.7 μg. of either phosphate or pyrophosphate can be detected. Organic phosphate esters are less fluorescent but must be separated on the chromatogram. Amino acids and sugars interfere when present in amounts 50 times as great. Inorganic salts in the developing solvents must be avoided.

VII. BORATE

Borate is of interest for toxicological reasons. It has also been implicated as a plant nutrient. With benzoin, α-hydroxy-α-phenyl-acetophenone, borate yields a highly fluorescent and characteristic derivative. As little as 0.1 μg. can be detected. White[1] and Parker and Barnes[15a, 15b] have discussed this reaction and the fluorescence characteristics of the complex. It should be possible to utilize this reagent for the assay of borate in biological material.

VIII. FLUORIDE

Fluoride has been implicated in dental health as an agent to prevent formation of caries. It is also used as a metabolic inhibitor. Although no fluorometric method for its assay in biological materials has been developed, Powell and Saylor[16] have shown that the aluminum-alizarin garnet complex is an excellent quantitative reagent for fluoride. The decrease in fluorescence of the complex provides a quantitative measure of the amount of fluoride present. According to White,[1] other fluorescent complexes have been used in the same way. These may be applicable to biological materials. Detection of fluoride in air is of great importance in the steel industry where high levels of fluoride are released into the air in the steel-processing. Monitoring is so important that kits for assaying fluoride in air are commercially available (see Chapter 14). These utilize magnesium oxinate impregnated paper strips; when metered volumes of air are drawn through the paper, the decrease in fluorescence is proportional to the fluoride content.

IX. CYANIDE

Cyanide can be made to yield a fluorescent derivative by conversion to cyanogen chloride and subsequent reaction with nicotinamide. A procedure based on this reaction was reported by Hanker et al.[17] To 0.1 ml. of solution, containing 0.03–0.6 μg. of cyanide in a 5-ml. graduated cylinder, are added 0.4 ml. of 0.5 N potassium hydroxide, 0.4 ml. of 2 N potassium bicarbonate, 0.2 ml. of 25% nicotinamide solution, and 0.1 ml. of 10% chloramine T in water. The volume is made up to 2.0 ml. and, after mixing, the solution is transferred to a cuvette. Exactly 3 minutes later, 0.4 ml. of 6 N potassium hydroxide are added and the reading is taken by following the fluorescence increase on the galvanometer to a maximum, usually within 45 seconds after adding the 6 N alkali. A reagent blank is made up by substituting 0.1 ml. of water for standards or extracts. The fluorophor is most likely a pyridinium product having fluorescent properties comparable to N'-methylnicotinamide (Chapter 7). Fluorescence is not exactly proportional to cyanide concentration over the entire range (Fig. 1).

Another highly sensitive method for cyanide has been developed by the same group.[18] The method depends upon the demasking of the nonfluorescent palladium complex of 8-hydroxyquinoline-5-sulfonate and the subse-

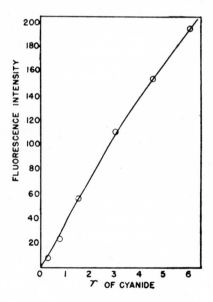

FIG. 1. Standard curve for the fluorometric estimation of cyanide by the nicotin amide-chloramine-T procedure (after Hanker et al.[17]).

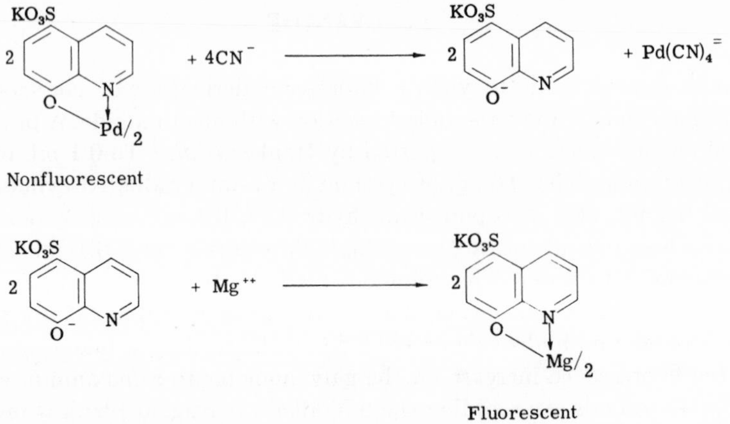

FIG. 2. Demasking of the palladium-8-hydroxyquinoline-sulfonate complex by cyanide and the subsequent conversion of the free reagent to the fluorescent Mg^{++} complex (after Hanker et al.[18]).

quent addition of magnesium ion to form the highly fluorescent magnesium complex with the released reagent (Fig. 2). The resulting fluorescence is directly proportional to the amount of cyanide added. As little as 0.02 μg. of cyanide can be assayed with this procedure. Thiocyanate, sulfide, thiols, and certain disulfides also give this reaction, and it may be that sensitive methods for these compounds can also be devised utilizing this demasking reaction following appropriate isolations.

X. OXYGEN

Biologists have not made much use of fluorometric methods for detection and assay of oxygen. However, fluorescence can be extremely sensitive to oxygen and this sensitivity can be used for analytical purposes. For example, Parker and Barnes[15a] have shown that the borate-benzoin complex in alochol is quite sensitive to oxygen. As shown in Fig. 3, the fluorescence remains fairly constant for almost 2 hours under nitrogen, but each time a mixture of 0.9% oxygen is passed through there is a fall in fluorescence which is partly reversible. In this instance, oxygen produces two effects, a rapid reversible quenching and a slower irreversible reaction. While this is not a too sensitive system, it can be used for quantitative purposes, as shown in Fig. 4. The values F_0/F represent the amount of quenching. In this instance, 0.1% oxygen in the gas stream produces 10% quenching of the fluorescence. Borate-benzoin is not the most sensitive system with

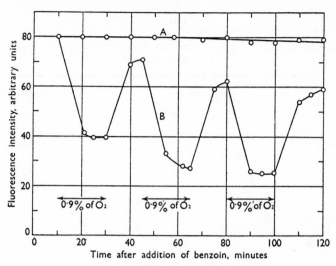

FIG. 3. Quenching of the fluorescence of the borate-benzoin complex by oxygen (0.018 μg. of boron per ml. of aqueous ethanol). Curve A, pure nitrogen passing through the solution; curve B, pure nitrogen passing through, but with 0.9% (v/v) of oxygen periodically injected into the gas stream (after Parker and Barnes[15a]).

respect to oxygen. Konstantinova-Schlezinger[19] utilized dihydroacridine to measure oxygen fluorometrically. In the presence of oxygen this compound is oxidized to acridine. The extremely high intensity of the dihydroacridine fluorescence and the rapidity of its reaction with oxygen make it possible to determine as little as 1 μg. of oxygen in a 5-ml. volume of gas, with a precision of ±15%. The method has actually been applied. It is much

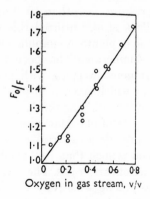

FIG. 4. Relationship between oxygen tension and fluorescence of the borate-benzoin complex; F_0 = fluorescence in the absence of oxygen, F = fluorescence in the presence of oxygen (after Parker and Barnes[15a]).

TABLE III

DETECTION OF OXYGEN ON A TWO-GRAM COLUMN OF TRYPAFLAVINE-SILICA
GEL ADSORBATE[a,b]

Oxygen, moles	Phosphorescence
1.7×10^{-10}	Strong
1.9×10^{-11}	Moderate
9.5×10^{-12}	Weak
4.7×10^{-12}	Still visible
1.2×10^{-12}	Not visible

[a] 5×10^{-6} mole of dye.
[b] After Kautsky and Müller.[22]

more sensitive to ozone and can be modified to determine 1 μg. of ozone in 100 ml. of air to within ±5%. On using this fluorometric procedure, Konstantinova-Schlezinger found that the ozone content of the air in Moscow was 0.02 μg./liter at a height of 100 meters.

A much more sensitive method for detecting oxygen, utilizing phosphorescence of adsorbed trypaflavine (acriflavine), was first suggested by Kautsky and Hirsch[20] and later studied by Franck and Pringsheim[21] and by Kautsky and Müller.[22] They found that the dye trypaflavine, adsorbed on silica gel (or other solids), has a greater yield of both fluorescence and phosphorescence than it does in solution. The green phosphorescence of trypaflavine is extremely sensitive to oxygen and is half-quenched by oxygen at 5×10^{-5}-mm. pressure. Less than 10^{-11} moles of oxygen can be detected (Table III); Pollack et al.[23] utilized this in a micromethod for measuring oxygen production in plants. A simplified version of this method, based on a modification by Zimmerman,[24] has been described by Tolmach.[25] The method involves continuous sweeping with pure nitrogen (or other inert gas) of the sample in which oxygen is produced. The carrier gas transports the oxygen to a tube containing the dye trypaflavine adsorbed on silica gel. Variations in intensity of phosphorescence with varying amounts of oxygen are measured to provide a calibration curve. The sensitivity of this phosphorescence method is such that 10^{-9} atmosphere of oxygen can be detected in an atmosphere of carrier gas. The changes in oxygen tension can be recorded continuously. The sensitivity is so great that samples of leaf juice or chloroplast suspensions in a volume contained in a 2-mm.-diameter loop are sufficient for studying oxygen production during metabolism.

GENERAL REFERENCES

Stevens, H. M., The effect of the electronic structure of the cation upon fluorescence in metal–8–hydroxyquinoline complexes. *Anal. Chim. Acta.* **20**, 389 (1959).

REFERENCES

1. White, C. E., *in* "Fluorescence Analysis in 'Trace Analysis'" (J. H. Yoe, ed.), p. 211. Wiley, New York, 1957.
2. White, C. E., Hoffman, D. E., and Magee, J. S., Jr., *Spectrochim. Acta* **9**, 105 (1957).
3. White, C. E., and Cutitta, F., *Anal. Chem.* **31**, 2083 (1959).
4. Schachter, D., *J. Lab. Clin. Med.* **54**, 763 (1959).
5. Tullo, J. W., Stringer, W. J., and Harrison, G. A. F., *Analyst* **74**, 296 (1949).
6. Goon, E., Petley, J. E., McMullen, W. H., and Wiberley, S. E., *Anal. Chem.* **25**, 608, 1216 (1953).
7. Rubins, E. J., and Hagstrom, G. R., *J. Agr. Food Chem.* **7**, 722 (1959).
8. Eisenbid, M., Wanta, R. C., Dustan, D., Steadman, L. T., Harris, W. B., and Wolf, B. S., *J. Ind. Hyg. Toxicol.* **31**, 282 (1949).
9. Sill, C. W., and Willis, C. P., *Anal. Chem.* **31**, 598 (1959).
10. Schwarz, K., *Am. Inst. Biol. Sci. Publ. No.* **4**, 509 (1958).
11. Hartley, W. J., *New Zealand J. Agr.* **99**, 259 (1959).
12. Cousins, F. B., *Australian J. Exptl. Biol. Med. Sci.* **38**, 11 (1960).
13a. Watkinson, J. H., *Anal. Chem.* **32**, 981 (1960).
13b. Parker, C. A., and Harvey, L. G., *Analyst* **86**, 54 (1961).
14. Rorem, E. S., *Nature* **183**, 1739 (1959).
15a. Parker, C. A., and Barnes, W. J., *Analyst* **82**, 606 (1957).
15b. Parker, C. A., and Barnes, W. J., *Analyst* **85**, 828 (1960).
16. Powell, W. A., and Saylor, J. H., *Anal. Chem.* **25**, 960 (1953).
17. Hanker, J. S., Gamson, R. M., and Klapper, H., *Anal. Chem.* **29**, 879 (1957).
18. Hanker, J. S., Gelberg, A., and Witten, B., *Anal. Chem.* **30**, 93 (1958).
19. Konstantinova-Schlezinger, M. A., *Trudy Fiz. Inst. Akad. Nauk S.S.S.R. Fiz. Inst. imi. P. N. Lebedeva* **2**, 7 (1942).
20. Kautsky, H., and Hirsch, A., *Z. anorg. u. allgem. Chem.* **222**, 126 (1935).
21. Franck, J., and Pringsheim, P., *J. Chem. Phys.* **11**, 21 (1943).
22. Kautsky, H., and Müller, G. O., *Z. Naturforsc.* **2a**, 167 (1947).
23. Pollack, M., Pringsheim, P., and Terwoord, D., *J. Chem. Phys.* **12**, 295 (1944).
24. Zimmerman, G. L., Ph.D. thesis, University of Chicago, Chicago, Illinois, 1949.
25. Tolmach, L. J., *Arch. Biochem. Biophys.* **33**, 120 (1951).

13

Drugs and
TOXIC AGENTS

I. GENERAL REMARKS

PHARMACOLOGY and toxicology are particularly in need of sensitive and specific analytical methods. Newer and more potent drugs are administered at such small dosages that detection in the tissues requires the most sensitive procedures. Fluorescence can attain the sensitivities required. For example, lysergic acid diethylamide (L.S.D.) is administered in doses of about 100 µg./day, yet even with this small dosage its absorption and excretion can be followed by fluorometric assay, as we shall see. Another important feature, in many cases, is that the drug itself is highly fluorescent so that no chemical changes are required. This permits the development of relatively simple assay methods involving only an isolation procedure (extraction or adsorption) followed by measurement of fluorescence. Such methods are actually physical rather than chemical.

A most important consideration in analytical methods designed for pharmacological studies, and particularly for toxicological analysis, is the need to distinguish between the administered drug and metabolic products of the drug. The former are generally the active agents and are metabolized by various enzymes to inactive metabolic products. Though inactive, they may be changed so slightly that they still resemble the parent drug very closely. Occasionally, the metabolite may be the active agent. This again requires distinguishing the active from the inactive form. The sensitivity of fluorescence to slight changes in chemical structure permits the evaluation of specificity of a fluorescence procedure by relatively simple techniques. The shape of excitation and fluorescence spectra, and the influence of pH on them, are sensitive criteria of specificity. Distribution coefficients (determined fluorometrically) and their change with pH are extremely sensitive to changes in chemical structure. The effects of metabolism most frequently lead to compounds which are less readily extracted into organic solvents from water, although the reverse can also occur. Aromatic compounds are hydroxylated to phenols; aliphatic compounds to alcohols and acids. Conjugation with glucuronic and sulfuric acids, a frequent occurrence, markedly changes the properties of the compound. All such

400

metabolic changes are not only detectable by combinations of extraction and fluorometry, but information gathered regarding the nature of the contaminant can be utilized in the design of a modification which will result in a more specific method. This can be arranged by modifying the extracting conditions by changing pH, salt concentration, or solvent. Chromatography, utilizing differential adsorption or partition, may prove useful. If complete separation cannot be achieved in the isolation, it may be accomplished by the proper choice of wavelengths for excitation and for measuring fluorescence, by modifying the pH of the final solution, or by the use of chemical agents which can selectively influence fluorescence.

The drugs and methods listed in this chapter are many in number but are not all inclusive. They were selected somewhat arbitrarily in an attempt to present examples of different procedures in as many areas as possible. Steroid drugs, such as prednisone and prednisolone, were discussed in Chapter 10.

II. CHEMOTHERAPY OF INFECTIOUS DISEASES

A. Antimalarials

It is perhaps fitting that the first class of drugs to be discussed are the antimalarials and that the first of these be quinine. Quinine is not only one of the oldest drugs, but is perhaps the first known to fluoresce. In fact, the use of quinine as a reference standard for fluorescence assay is almost as old as fluorometry itself. As a result of this widespread use of quinine in fluorometry, a great deal is known about its fluorescent properties. Quinine

Quinine

is an alkaloid which can be extracted into organic solvents as the free base and returned to an aqueous phase as the salt. It fluoresces maximally in acid media and is, in fact, an excellent pH indicator with a sharp end point (disappearance of fluorescence) in the range pH 5.9–6.1.[1] The fluorescence is extremely sensitive to halide ions so that in 0.1 N HCl quinine is essentially nonfluorescent, yet it is maximally fluorescent in 0.1 N H_2SO_4. Quinine has an absorption spectrum characterized by a major peak at 250 mμ, a smaller one at 350 mμ, and a third peak at about 315 mμ which is barely resolved from the one at 350 mμ^2 (see Chapter 3). With fluorescence spectrometers which use a xenon arc and fixed slits, one observes two excitation bands corresponding to the 250- and 350-mμ absorption peaks.[3] However, unless corrections are applied for the smaller light output of the xenon arc at 250 mμ, the 350-mμ peak appears at the major one. By using a linear energy spectrophotofluorometer, Slavin[4] was able to obtain an excitation spectrum for quinine which coincided fairly well with the absorption spectrum, including the third band at about 315 mμ. However, for practical purposes, excitation at 350 mμ yields more measurable fluorescence on most instruments. The fluorescence maximum has been reported as 450 mμ[2, 3] and as 475 mμ by Slavin.[4] Quinine in dilute sulfuric acid has a fluorescence efficiency of about 0.3.[5] This, combined with a relatively high absorption coefficient, makes quinine a most intense fluorophor. With appropriate instrumentation, less than 0.001 μg./ml. can be detected. All these properties would indicate that quinine should lend itself well to the development of simple, sensitive, and specific fluorometric assay procedures. This is true, and a number of methods for measuring quinine in tissues, based on the foregoing properties, have appeared.

A simple and rapid method for measuring quinine in plasma was published by Brodie and Udenfriend in 1943.[6] It involves precipitation of the proteins with metaphosphoric acid in aqueous solution at a high dilution. The quinine in the filtrate is estimated fluorometrically directly in the filtrate.

To 1 ml. of plasma are added 39 ml. of water and, after mixing, 10 ml. of 20% metaphosphoric acid are added with vigorous shaking. After 15 minutes the mixture is centrifuged at high speed for 10 minutes and a portion of the clear supernatant fluid is transferred to a cuvette. Standards are prepared by dilutions made up in quinine-free plasma filtrates (0.5 ml of concentrated standard plus 10 ml. of filtrate). This is done because quinine fluorescence is less in plasma filtrates than in aqueous controls owing to the chloride content of the former. The quenching due to chloride, about 17%, is fairly constant from one filtrate to another. Recoveries of quinine added to plasma are quantitative. However, the method also measures metabolic products of quinine. In plasma, this results in values which are on the average 20% too high.

There have been many procedures for estimating quinine in biological systems, which utilize solvent extraction procedures. One which has yielded excellent data was devised by Kelsey and Geiling.[7] This procedure and others like it utilize large volumes of solvent and multiple extractions and are therefore cumbersome. The following extraction procedure is both simple and highly specific.[8]

Add 1–10 ml. of plasma, urine, or tissue homogenate (containing up to

1 μg. of quinine) and an equal volume of 0.1 N sodium hydroxide to 30 ml. of benzene in a 60-ml. glass-stoppered bottle. Shake for 30 minutes on a shaking apparatus, then centrifuge. Add 1 ml. of isoamyl alcohol to the supernatant benzene layer* mixing carefully so as not to disturb the aqueous phase. Transfer 20 ml. of the benzene to a 30-ml. glass-stoppered conical tube, containing 3 ml. of 0.1 N H_2SO_4. Shake for 3 minutes, centrifuge, and remove the supernatant benzene phase by aspiration. Transfer the aqueous phase to a cuvette and measure fluorescence. Reagent and normal tissue blanks, as well as standards, are carried through the entire extraction procedure. The procedure is not only sensitive but is specific for quinine in the presence of its known metabolites. A simpler version of this method is described in the same paper.

Quinine is only one of several pharmacologically active alkaloids isolated from cinchona bark. Quinidine, the dextra-rotatory stereoisomer of quinine, is also an active antimalarial. However, its most important use is in treating cardiac arrhythmias (see below). It can be assayed by exactly the same procedures as are used for quinine. Cinchonine and cinchonidine are another pair of isomeric cinchona alkaloids possessing appreciable anti-malarial activity. They differ from quinine and quinidine in not having a methoxy group at the 6′ position on the quinoline ring. As a result, the excitation and fluorescence maxima are shifted to shorter wavelengths (315–320 mμ/420 mμ, uncorrected). Measurement of this fluorescence requires the use of an instrument with quartz optics. The same type of extraction procedure can be used for cinchonine and cinchonidine assay as is used for quinine.

Quinacrine (atabrine) was for a long time the most widely used syn-

$$CH_3$$
$$CH(CH_2)_3\,N(C_2H_5)_2$$
HN

CH$_3$O

N   Cl

Quinacrine

thetic antimalarial agent. Today it is used in the therapy of other parasitic diseases as well. Being an acridine, it is highly fluorescent and exhibits maximal fluorescence in alkaline solution at pH 11 with two excitation peaks, one at 285 mμ and one at 420 mμ; the fluorescence peak is at 500 mμ (uncorrected). Many extraction procedures for the assay of quinacrine by fluorometric methods have been reported.[9, 10] The following is a simple

---

* To minimize adsorption onto glass surfaces.

and relatively specific method for its assay.[8] Following the extraction of the compound (0.05–1 µg.) from alkalinized biological material into heptane, a small amount of isoamyl alcohol is added to the organic solvent. After careful mixing, an aliquot of the heptane is transferred to a cuvette containing trichloroacetic acid in ethylene dichloride. Fluorescence is measured in the acidified organic solvent mixture. For some purposes, re-extraction from the heptane into aqueous acid may be required to remove interfering substances. With the usual regimes of quinacrine therapy, plasma concentrations range from 5 to 50 µg./liter.

Pamaquine (plasmochin) is a synthetic antimalarial which has marked

Pamaquine

activity against the gametocyte or extraerythrocyte forms of the disease. Its structure is such that one would expect it to fluoresce in aqueous solution, and in fact a report on the fluorescence characteristics of pamaquine has appeared.[3] However, the findings are no doubt in error because measurements were made on the naphthoate salt of pamaquine, an aromatic anion which is itself fluorescent. No doubt the compound does have useful fluorescence. If so, it should be possible to utilize the same solvent extraction procedure as devised by Brodie et al.[11] and measure the pamaquine in the final extract fluorometrically, using an instrument with a photomultiplier detector. Such a procedure would be much simpler and more sensitive than the original colorimetric one.

Irvin and Irvin[12] devised a fluorometric procedure for pamaquine and related compounds based on the violet-blue fluorescence emitted by these compounds in 18 M sulfuric acid. The absorption peak of the acid fluorophore is at 350 mµ. Plasmochin is isolated from biological material by extraction with hexane. After two such extractions the hexane is evaporated and the residue is taken up in 18 M sulfuric acid for fluorometric assay.

Another important quinoline drug is chloroquine. Although it was developed originally as an antimalarial, it is now used in treating other parasitic disorders. Chloroquine, a substituted aminoquinoline, fluoresces maximally at 400 mµ, at pH 11, with excitation at 335 mµ. Again, because the intensity is not as great as for quinine or quinacrine, and the excitation maximum does not coincide with the intense 365-mµ mercury line, its

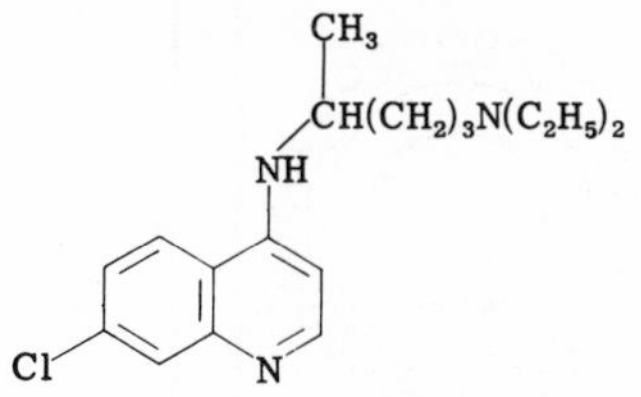

$$\text{CH}_3$$
$$\text{CH(CH}_2)_3\text{N(C}_2\text{H}_5)_2$$

Chloroquine

native fluorescence could not be utilized until instruments utilizing the xenon arc and selective monochromators were made available. Brodie *et al.*[13] devised a photochemical procedure which converts chloroquine and related compounds to, as yet, unidentified highly fluorescent derivatives with the absorption maximum shifted to coincide more closely to the 365-mμ mercury line. The procedure, as originally described, is as follows:

Add 1–10 ml. of plasma or tissue homogenate (containing up to 1 μg. of chloroquine) and an equal volume of 0.1 N NaOH to 30 ml. of heptane in a 60-ml. glass-stoppered bottle. Shake for 30 minutes on a shaking apparatus, then centrifuge the mixture, if necessary, to separate the phases. Add 8 drops of ethanol and mix with the heptane so as not to disturb the aqueous phase. Then transfer as much of the heptane as possible to a 125-ml. glass-stoppered bottle, add about twice the volume of 0.1 N NaOH, and shake for 5 minutes. After the two phases have separated, add 8 drops of ethanol to the heptane and mix as before. Remove the aqueous phase by aspiration, add the same volume of fresh 0.1 N NaOH, and shake for 5 minutes. The two washings with 0.1 N NaOH remove a metabolic product of chloroquine which interferes in the assay. After the second wash, add 5 drops of ethanol to the heptane layer, mix, and transfer 20 ml. of the organic phase to a 60-ml. glass-stoppered centrifuge tube containing 6 ml. of 0.1 N HCl. Shake for 3 minutes, centrifuge, and remove the heptane by aspiration. Transfer 5 ml. of the acid to a fluorometer tube containing 1 ml. of 0.5 N NaOH and 1.5 ml. of 0.6 M borate buffer, pH 9.5. Add 0.5 ml. of neutralized 5% cysteine (prepared fresh daily) and mix thoroughly. Reagent blanks, run through the entire procedure, and standards prepared in 5 ml. of 0.1 N HCl, are treated with alkali, buffer, and cysteine in the same way. After 30 minutes, to allow for reaction between cysteine and oxygen, all tubes are irradiated with ultraviolet light in the irradiation apparatus shown in Fig. 1.

The irradiator is made up of an H-4 (see Chapter 3) mercury arc lamp, with appropriate transformer, set up in the center of a circular rack. Cooling is effected by passing air around the lamp and tubes. The solutions must be maintained below 35° during irradiation to minimize side reac-

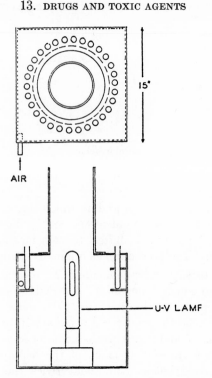

FIG. 1. Schematic diagram of an apparatus for the irradiation of solutions with ultraviolet light for photochemical reactions (after Brodie et al.[13]).

tions. With the H-4 lamp and the dimensions used, 3 hours are required to achieve maximal fluorescence. The time may be reduced by using more intense light sources or changing the distance from the lamp. Exact conditions must be worked out in each laboratory. However, with intense light sources, precautions must be taken to avoid the damaging effects of ultraviolet light on the eyes. Irradiators of this type can be used for other photochemical reactions such as in the assay for 3,4-dihydroxyphenylethylamine (Chapter 5).

Fluorescence of the chloroquine photochemical product can be measured by exciting at 360 mμ and measuring the fluorescence emitted above 410 mμ. However, the precise excitation and fluorescence maxima of the compound(s) have not yet been reported. The sensitivity and precision of the method are adequate for amounts as low as 0.1 μg.

Although the photochemical method is an interesting one and was absolutely necessary with the photofluorometers available 15 years ago, it should now be possible to utilize the native fluorescence of chloroquine for fluorometric assay. A simplified version of the foregoing assay could be devised which utilizes the same extraction procedure to isolate the compound. The final extract could be adjusted to pH 11 and fluorescence measured at 400 mμ with excitation at 335 mμ. Although such a modification has not yet been tested, it is likely that an assay can be developed with the omission of the time consuming irradiation step.

Oxychloroquine, an analog of chloroquine containing a hydroxyl group in the side chain, is also used as an antimalarial. This compound exhibits excitation and fluorescence maxima at 335 and 380 mμ, respectively (uncorrected), at pH 11.[3]

B. Antibacterial Agents

Fluorometric procedures have been reported for most of the better known antibiotics and for other antibacterial agents. It is anticipated that procedures employing the inherent fluorescence of sulfonamides will be devised since they are aniline derivatives and as such possess native fluorescence.

The penicillins are a group of antibiotics having the structure shown in Fig. 2 where R may be either an aliphatic or aromatic group. Only one of these structures (penicillin X or III) could be expected to possess some degree of fluorescence in the ultraviolet region since R is a hydroxybenzyl group. This has not yet been ascertained. None of the penicillins exhibit fluorescence in the visible region. A fluorometric method for the determination of the penicillins has, however, been developed. Scudi[14] demonstrated the reaction shown in Fig. 2. Amines can be made to react with the penicillins to yield acidic amides in almost quantitative yield; the latter can be separated from the unreacted amine reagent. In the first report Scudi[14] utilized an amine form of an azo dye and measured the product colorimetrically. Subsequently, Scudi and Jelinek[15] introduced a flavine amine, 2-methoxy-6-chloro-9-(β-aminoethyl) aminoacridine (Fig. 3) and

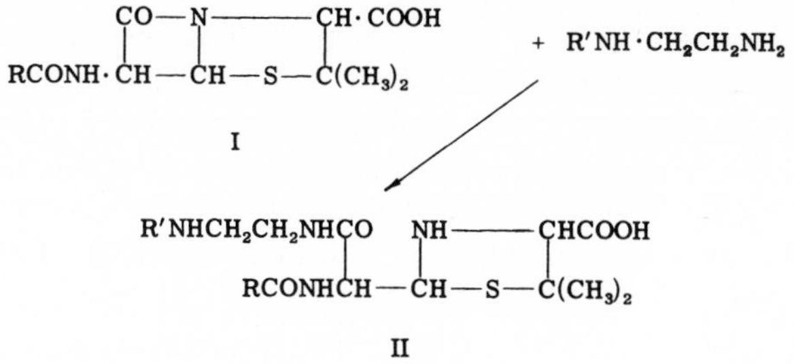

Fig. 2. Reaction of penicillins (I) with amines to form acidic amides (II) (after Scudi[14]).

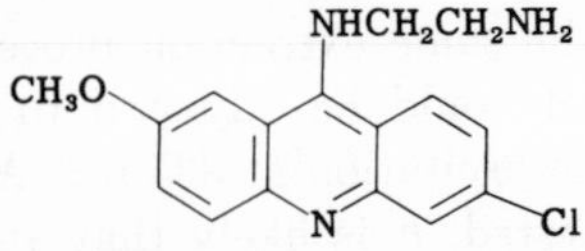

FIG. 3. 2-Methoxy-6-chloro-9-(β-aminoethyl) acridine, the amine used for the fluorometric assay of penicillin by the procedure of Scudi and Jelinek.[15]

utilized fluorometry for the estimation of the resulting amide derivative. Their procedure for fluorometric assay of penicillin in blood and urine involves multiple extractions in order to separate the penicillin from normally occurring compounds, metabolic products of the drug, and excess amine reagent.

Urine is diluted 1 to 50 before assay. Whole blood and plasma require the preparation of a protein free filtrate by the Haden modification[16] of the Folin-Wu method. The latter procedure results in a loss of 20% by sorption on the precipitated proteins. To 8 ml. of diluted urine or protein-free filtrate (in an ice bath) containing 0.5–5 μg. of penicillin there are successively added 12 ml. of chloroform and 2 ml. of Sorenson's glycine buffer, pH 2.0.* The mixture is shaken vigorously for 30 seconds and the phases are then permitted to separate. The chloroform is transferred to a chilled glass-stoppered graduate and rapidly dried with 1 to 2 g. of anhydrous sodium sulfate. A 10-ml. aliquot of the dried chloroform is transferred to a separatory funnel containing 5 ml. of a benzene solution of aminoacridine (10 mg.), 2 ml. of acetone, and 5 ml. of glacial acetic acid in benzene (10 plus 1000 v/v). The reaction mixture is allowed to stand in the closed funnel at room temperature, for 1 hour, in the absence of light.†
At the end of an hour, 10 ml. of 0.5 N NaOH are added and the mixture is shaken. The alkaline phase, containing the penicillin-acridine derivative, is washed twice with 5-ml. portions of chloroform. After the last wash the chloroform is discarded and 1 ml. of glacial acetic acid is added to the alkaline solution. The condensation product is then extracted into 15 ml. of butanol-benzene (1:2 v/v) and after discarding the aqueous layer the organic phase is washed once with 10 ml. of 5% aqueous acetic acid. After discarding the acetic acid, 50 ml. of chloroform and 15 ml. of 0.5 N NaOH are shaken with the organic phase and the layers are permitted to separate. The lower, organic, phase is discarded and 1 ml. of concentrated hydrochloric acid is added to the alkaline solution. A portion of the acidified

* In order to extract the penicillin into chloroform it is necessary that the pH of the aqueous phase be 2.0. However, the penicillin is unstable in acid and the solution must be neutralized as soon as possible. The cooling also minimizes destruction.

† All subsequent reactions should be carried out with minimal exposure to light and as rapidly as possible.

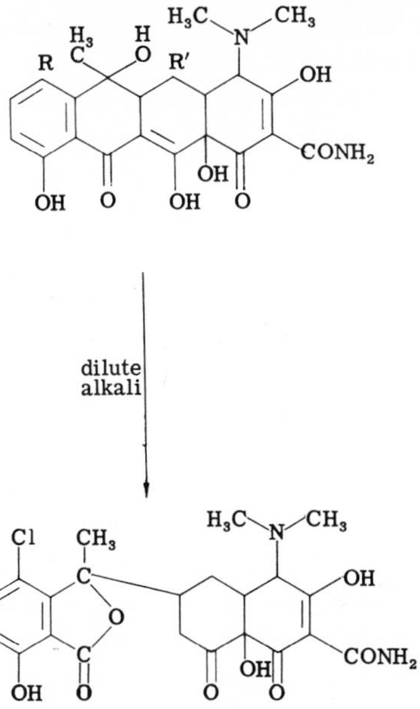

	R	R'
Tetracycline –	H	H
Chlortetracycline –	Cl	H
Oxytetracycline –	H	OH

Isochlortetracycline

FIG. 4. Tetracyclines and isochlortetracycline, the blue fluorophor formed by treating chlortetracycline with base.

solution is transferred to a cuvette and its fluorescence measured. The fluorescence resembles that of quinacrine, excitation maximum 420 mμ, fluorescence maximum 500 mμ (uncorrected). Standards and reagent blanks must, of course, be carried through the entire procedure. It should be possible to reduce all volumes tenfold and measure the final fluorescence in 1.5 ml. instead of 15 ml. This would permit assay on 1 ml. of plasma or less.

The tetracyclines are another class of antibiotics. The three shown in Fig. 4 are in general use. According to Udenfriend et al.,[3] all three compounds are fluorescent. Tetracycline and oxytetracycline have excitation maxima at 390 mμ and fluorescence maxima at 515 to 520 mμ at pH 11. The fluorescence characteristics of chlortetracycline (aureomycin) were reported as 355-mμ excitation and 445-mμ fluorescence, maximal fluorescence occurring at pH 11. It appears that the fluorescence reported for chlortetracycline is, in part, due to a derived product, isochlortetracycline (Fig. 4) to which the antibiotic is rapidly converted in basic solution[17, 18] and even at neutral pH.[19] According to Feldman et al.,[19] the derived product has an absorption maximum at 350 mμ as compared to about 385 mμ for chlortetracycline. These authors reported the fluorescence maximum to be about 425 mμ (blue). All the native tetracyclines fluoresce at about 515 to 520 mμ (golden yellow). Since the excitation peak is at 390 mμ, the major spectral line of mercury, which is at 360 mμ, is not efficient for evoking the native fluorescence.

Several fluorometric methods for measuring chlortetracycline in pharmaceutical preparations have been reported.[17-19] The following procedure for the determination of chlortetracycline in blood and urine, by Saltzman,[20a] may be found applicable with slight modifications, to all the tetracyclines.

Measure exactly 1 ml. of plasma or 0.1 ml. of urine, containing 0.5–10 μg. of chlortetracycline, into a test tube and dilute to about 5 ml. with water. Transfer the entire solution to a column of Decalso (9 $\times$ 70 mm.) with 2 ml. of water to effect quantitative transfer. The column is washed twice with water—15 ml., then 5 ml. Suction is then applied and the column is washed with 35 ml. of absolute ethanol. The suction is maintained until all traces of moisture have disappeared from the sides of the column and is then discontinued. The chlortetracycline is eluted with five 3-ml. portions of 5% sodium carbonate warmed to 60°. The first 0.5 ml. is discarded and slight suction (or pressure) is used to complete the elution. The treatment with warm sodium carbonate is sufficient to bring about the conversion to isochlortetracycline with its characteristic blue fluorescence. Reagent blanks and standards are also run through the procedure. Serum levels observed in patients on therapy range from about 0.8 to 7.0 μg./ml. Comparisons of fluorometric assay with bioassay were good.

More recently Kohn[20b] has described the formation of highly fluorescent mixed complexes formed through the interaction of tetracyclines, calcium, and barbiturates. The complexes are readily extractable into organic solvents permitting their separation from biological material. Procedures for the assay of tetracyclines (with the exception of oxytetracycline) in animal tissues are described.

An interesting property of all the tetracyclines is their ability to be visualized *in vivo* in newly proliferated normal bone tissue[21, 22] where they remain for long periods of time. The compounds are detected by their golden-yellow fluorescence under long-wavelength ultraviolet light. In animals the uptake in bone is rapid and persists long after fluorescence has disappeared from all other tissues; it is still apparent at the end of a 10-week period of observation. The bone fluorescence following tetracycline is due to some, as yet, unexplained binding process between drug and bone, possibly with calcium ion. The biological half-life of the fluorescent complex is quite long, but a precise measurement requires methods for quantitative assay of the bone fluorescence. Tetracycline fluorescence also persists in certain tumors.[23]

Another interesting localization of tetracyclines *in vivo* has been observed recently.[24] When administered to patients with filariasis, the tetracyclines concentrate in the filarial worms to such an extent that the parasites can be detected beneath the skin by their bright golden-yellow color under ultraviolet light. Such a procedure may be of diagnostic value.

DuBuy and Showacre[25] have demonstrated a highly selective localization of tetracycline in the mitochondria of living cells. Their observations were made by means of fluorescence microscopy. The selective fluorescence labeling of mitochondria may prove to be a useful tool in enzyme studies. In bacteria, too, tetracycline binds to the mitochondrial material and endows it with fluorescence. In this case the localization may help in elucidating the mechanism of the antibacterial action.

The sulfonamides are structurally related to *p*-aminobenzoic which is

p-Aminobenzoic
acid

Sulfonamide

highly fluorescent in solution at pH 11; the excitation maximum is 295 mμ and the fluorescence maximum is 345 mμ (uncorrected).[26] Colorimetric procedures for sulfonamide drugs are adequate for routine blood and urine assays. However, if the need for more sensitive methodology were to arise, it should not be difficult to devise simple fluorometric procedures for sulfonamide assay. It may be pointed out, however, that although the sulfonamide structure is inherently one which should yield fluorescence, some of the substituent groups may interact with it to lower the fluorescence efficiency below useful levels. A detailed study of sulfonamide fluorescence would be required before generalizations can be made.

C. Antitubercular Drugs

There are a large number of drugs which have been and are used in the treatment of tuberculosis, but few have been studied fluorometrically. The sulfones, as exemplified by 4,4'-diaminodiphenylsulfone, are structurally related to *p*-aminobenzoic acid and the sulfonamides and should therefore be fluorescent.

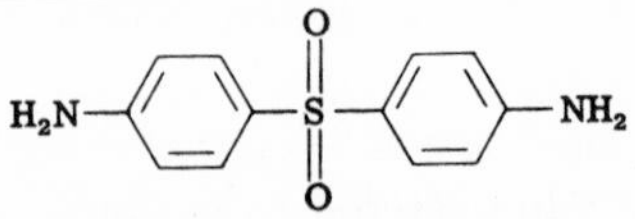

$$H_2N - \langle\bigcirc\rangle - \overset{\overset{O}{\|}}{\underset{\underset{O}{\|}}{S}} - \langle\bigcirc\rangle - NH_2$$

4,4'-Diaminodiphenylsulfone

p-Aminosalicylic acid is highly fluorescent in aqueous solution at pH 11; excitation maximum at 300 mμ, fluorescence maximum at 405 mμ (uncorrected).[3] It should be relatively simple to devise sensitive fluorometric methods for assaying this compound in tissues.

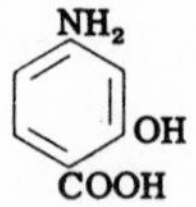

NH$_2$
OH
COOH

p-Aminosalicylic acid

Auerbach and Angell[27] reported that a number of substituted 2-thiohydantoins, which possessed antitubercular activity, combined with 2:6-dichloroquinone chloroimide at pH 10–10.5 to yield a highly fluorescent product. A procedure for estimating 5-*n*-heptyl-2-thiohydantoin in plasma and urine was developed which is sensitive to less than 1 μg. of the compound. Unsubstituted thiohydantoin, hydantoin, and thiouracil do not yield fluorescent derivatives with the reagent.

D. Antifungal Agents

One of the more interesting antifungal agents is the antibiotic griseofulvin. Bedford *et al.*[28] have found that in aqueous solution, between pH

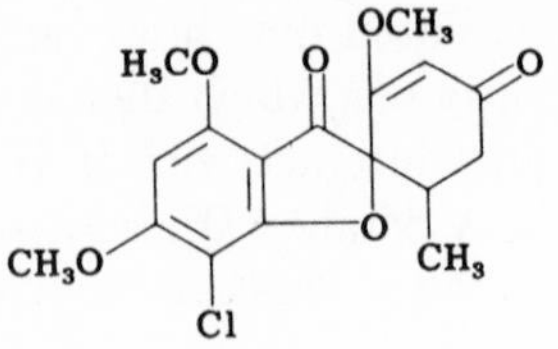

Griseofulvin

3 and 10, and in ethanol, griseofulvin fluoresces strongly (Fig. 5). The excitation spectrum contains two well-defined peaks, one at 295 mμ and another at 335 mμ. The fluorescence maximum is at 450 mμ (uncorrected values). The following procedure has been applied for its assay in biological materials.

One milliliter or less of blood, plasma, or serum, containing 0.1–1.0 μg.

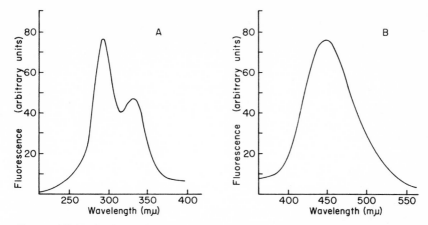

Fig. 5. (*A*) Excitation and (*B*) fluorescence spectra of griseofulvin in ethanol (after Bedford *et al.*[28]).

of griseofulvin, is transferred to a glass-stoppered centrifuge tube containing 1 ml. of 1% ethanol and 10 ml. of ether. Urine is adjusted to pH 10 and treated in the same manner. The tube is shaken for about 15 seconds and, after separation of the phases (by centrifugation if necessary), 8 ml. of the ether is transferred to a large test tube and evaporated to dry-

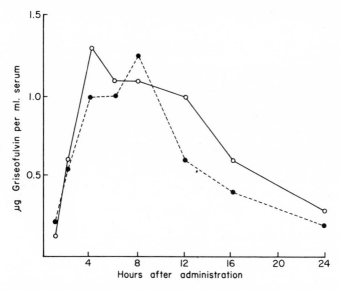

Fig. 6. Serum levels of griseofulvin in man following a single oral dose of 0.5 g. Each point represents the mean for three individuals. o — o, fluorometric assay; ● — ●, biological assay (after Bedford *et al.*[28]).

TABLE I

Fluorescence Characteristics of Some Carcinostatic Agents

Compound	Structure	Excitation maximum mμ	Fluorescence maximum mμ	Optimal pH	Reference
Methotrexate (amethopterin)	 R = CH₃	280, 370	450		(29)
Aminopterin	Structure is as above, with R = H	280, 370	460	7	(3)
8-Azaguanine (guanazolo)		285	405	7	(3)

TABLE I (continued)

Compound	Structure	Excitation maximum mμ	Fluorescence maximum mμ	Optimal pH	Reference
Podophyllo-toxin		280	325	11	(3)
Sarcolysine		260	365	7	(30)

ness. The residue is dissolved in 10 ml. of ethanol and fluorescence is measured and compared to reagent blanks, normal tissue blanks, and standards taken through the procedure. The 295-mμ peak is preferable for excitation. Recoveries are quantitative and the blood levels obtained with the fluorometric assay in man are comparable to those obtained by the longer and more tedious bioassay, as shown in Fig. 6.

III. CARCINOSTATIC AND CARCINOGENIC AGENTS

A. Carcinostatic Drugs

Fluorescence assay has been widely used in cancer research. Until recently, however, this was mainly concerned with carcinogenic agents, particularly the polycyclic hydrocarbons. More recently, fluorometry has been applied to the assay of anticancer agents both in experimental and in therapeutic use. The high toxicity of these agents makes necessary close chemical supervision during therapy and this requires specific and simple analytical methods. The drugs methotrexate, azaguanine, podophyllotoxin, and sarcolysine are all highly fluorescent. Their fluorescence characteristics are shown in Table I.

Aminopterin and methotrexate are both structurally related to folic acid and resemble it in their fluorescence characteristics. Both aminopterin and methotrexate can also be oxidized to the much more fluorescent pteroyl-6-carboxylic acid. Zakrzewski and Nichol[31] utilized the oxidation procedure along with paper chromatography to measure both aminopterin and methotrexate and pteroyl glutamic acid in mixtures. Freeman[32] has also utilized the oxidation procedure to develop a fluorometric method for the assay of methotrexate in biological tissues and has applied it to studies on the absorption, distribution, and excretion of the drug in man. To 1 ml. of plasma, containing 0.06–1 μg. of methotrexate, add 7 ml. of water and mix. Add 2 ml. of trichloroacetic acid (15%), mix, and centrifuge. To 5 ml. of the supernatant fluid, add 1.0 ml. of 3.24% sodium hydroxide then 0.1 ml. of 5 M acetate buffer, pH 5. After mixing, measure the fluorescence (A). Following this, add 0.05 ml. of 4% potassium permanganate and, after 5 minutes, destroy the excess by adding 0.1 ml. of 3% hydrogen peroxide. After another wait of 3 minutes fluorescence is measured once more at the same wavelength settings (B). The increment in fluorescence due to oxidation ($B - A$) is proportional to the methotrexate concentration. As little as 0.005 μg./ml. in the final dilution can be detected, and blood levels in patients during therapy are readily measured (Fig. 7).

A similar procedure would probably be applicable to aminopterin and related folic acid antagonists.

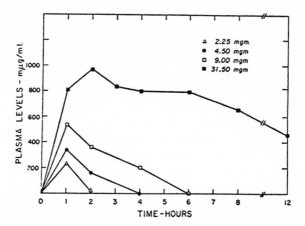

FIG. 7. Plasma levels of methotrexate after oral administration of a single dose of the drug (after Freeman[32]).

It should be noted (Table I) that azaguanine fluorescence is at the borderline of the visible range. In addition to its experimental use as a carcinostatic agent, 8-azaguanine has also been used as an antiviral agent in plants.[33] The compound has been assayed fluorometrically following isolation from plant material and separation on paper chromatograms according to the procedure of Markham and Smith.[34] Similar procedures are probably applicable to extracts from animal tissues.

A procedure for extracting podophyllotoxin for fluorometric assay was presented by Udenfriend et al.[3] The compound can be extracted with several volumes of n-butanol from aqueous solutions at pH 5–6. It can be returned to an aqueous phase by adding an equivalent amount of heptane to an aliquot of the n-butanol and extracting with 0.1 N NaOH. The final concentration should be about 0.4 μg./ml. While this procedure has not actually been applied to biological material, it should be possible to use it for such purposes with, perhaps, minor modifications.

Sarcolysine is a most interesting nitrogen mustard derivative since it is also an aromatic α-amino acid. Chirigos and Mead[30] have devised a fluorometric assay for sarcolysine in animal tissues. Tissues are homogenized in 6% trichloroacetic acid; 100 mg. of tissue in 10 ml. of acid. Plasma proteins are precipitated by addition of 0.1 ml. of plasma to 9.9 ml. of 6% trichloroacetic acid. After centrifugation, a portion of the protein free solution is extracted twice with several volumes of ether to remove the trichloroacetic acid. The ether is discarded and residual solvent is removed by warming the solution at about 55° for 5 minutes. After cooling, 1 ml. of the solution is transferred to a cuvette containing 2 ml. of 0.5 M

phosphate buffer, pH 7.0. Standards and blanks are carried through the procedures. Recoveries of known amounts of sarcolysine added to tissue are quantitative and as little as 0.02 μg. can be detected. Other aromatic nitrogen mustards may also have useful fluorescence.

B. Carcinogenic Agents

Based on results of animal experimentation and on implications from the prevalence of human carcinoma in certain industrial areas, it is now generally recognized that many polycyclic hydrocarbons are carcinogenic.[35, 36] Benzo[a]pyrene is, perhaps, the best studied example of these

Benzo[a]pyrene
(3,4-benzpyrene)

compounds. However, the direct implication of these carcinogenic agents in the environment with human carcinoma has not yet been solidly established. One of the difficulties in such investigations has been the lack of simple, specific, and sensitive methods for detecting and quantifying these chemicals both in the environment and in tissues. Actually, the polycyclic hydrocarbons are among the most intensely fluorescent compounds known, many of them having fluorescence efficiencies of 0.5 to 1.0. Detection is not much of a problem for this reason. However, large numbers of related hydrocarbons are known, few of which have been demonstrated to be carcinogenic in animals. It is, therefore, necessary that the methods be endowed with specificity. Many reports concerning the application of fluorometry to the detection and determination of hydrocarbons have appeared. Of these, the recent studies of Sawicki and his colleagues[37] are the most noteworthy. In a series of papers they have presented absorption and fluorescent spectra and isolation procedures for many of the important carcinogens and for related noncarcinogenic hydrocarbons. These studies were directed mainly to the problem of air pollution and the details will be presented in Chapter 14. However, similar procedures can be applied to studies of the carcinogens in animal tissues, food, food containers and other materials.

IV. ANALGESICS, SEDATIVES, AND MUSCLE RELAXANTS

Many drugs in this category possess native fluorescence (Table II) or can be converted to fluorescent derivatives for analytical purposes. Spe-

TABLE II

Analgesics, Sedatives, and Muscle Relaxants

Compound	Structure	Excitation maximum $m\mu$	Fluorescence maximum $m\mu$	Optimal pH	Reference
Procaine		275	345	11	(3)
Tolserol (mephenesin)		280	315	1	(3)
Flexin (zoxazol- amine)		280	320	11	(3)

TABLE II (continued)

Compound	Structure	Excitation maximum mμ	Fluorescence maximum mμ	Optimal pH	Reference
N-allylnor-morphine		285	355	1	(3)
Dromoran		275	320	1	(3)
Eserine (physostig-mine)		300	360	1	(3)

TABLE II (*continued*)

Compound	Structure	Excitation maximum mμ	Fluorescence maximum mμ	Optimal pH	Reference
Presidon		370^a	460^a	13	(38)
Salicylic acid		310	435	11	(3)
Neocincophen		275, 345	455	7	(3)

a Estimated from optical filters suggested for assay.

cific procedures based on this native fluorescence have, thus far, been reported for relatively few of these compounds.

Udenfriend et al.[3] reported that dromoran could be extracted from basic solution, pH 11, into benzene and then re-extracted into 0.1 N HCl. As little as 0.2 μg. can be determined. This procedure has not yet been used for biological material.

DeRitter et al.[38] have modified previously reported procedures for presidon in urine. Ether extraction from salt saturated urine, buffered to pH 9 to 9.5, is used to isolate the compound. The ether is evaporated (in the presence of ascorbic acid to prevent oxidative destruction), the presidon is taken up in water diluted with alcoholic sodium hydroxide, and fluorescence is assayed directly (Table II). Treatment with hydroxylamine destroys the fluorescence and yields a blank. As little as 1 μg. in 5 ml. of urine can be determined.

It is likely that morphine exhibits the same fluorescence characteristics as does N-allylnormorphine (see Table II), although data for the former

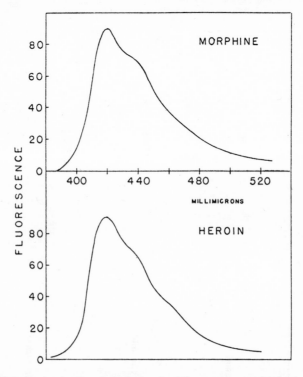

FIG. 8. Fluorescence spectra, in isobutanol, of the reaction products of morphine and heroin formed in the procedure of Nadeau and Sobolewski.[39] Excitation was with the 365-mμ mercury line.

compound have not been reported. The intensity of morphine's native fluorescence is probably not sufficient for measurement of blood and urine levels in man, nor have such applications been reported. Nadeau and Sobolewski[39] have found that when morphine is treated with concentrated sulfuric acid and the reaction mixture is then made alkaline with ammonium hydroxide, an intensely blue fluorescence appears. Although several fluorescent products are formed, the bulk of the fluorescence is due to only one of them. This derivative can be extracted into isobutanol where it fluoresces at 420 mμ when excited at 365 mμ (Fig. 8). For assay, the morphine solution, in a test tube, is evaporated to dryness (all traces of moisture must be removed for uniform results), 0.5 ml. of concentrated sulfuric acid is added, and the tube is heated at 50° for 8 minutes. Following this, 5 ml. of water and 6 ml. of concentrated ammonium hydroxide are added and the tube is heated at 50° for 2 hours. It is then cooled, 10 ml. of isobutanol are added, and the tube is shaken to extract the fluorophor. Blanks and standards are carried through in the same manner and measured fluorometrically. As little as 0.02 μg. of morphine can be assayed in this manner. The procedure has been applied mainly to assays of morphine in opium extracts. According to Nadeau and Sobolewski,[39] it is so specific that it permits the estimation of morphine in opium in the presence of other constituents without prior separation. With appropriate isolation procedures it may be possible to utilize this technique for assay of morphine in animal tissues. Heroin (diacetylmorphine) gives the same fluorescence as morphine. Probably N-allylnormorphine and other related products will yield similar fluorescent derivatives.

Tabun (ethyl dimethylphosphoramidocyanidate) is a potent cholinesterase inhibitor which has been developed as a nerve gas for chemical warfare purposes. While it is not a fluorophor, it is readily hydrolyzed to yield an equivalent of cyanide which can then be assayed fluorometrically as described in Chapter 12.[40]

Salicylic acid is, of course, the most widely used of all drugs; either as such or in its acetylated form. Although it is one of the oldest of the synthetic drugs, its mechanism of action still remains an enigma and is still being actively investigated in the laboratory. Measurement of salicylate in tissues, blood, and urine is also important in toxicology and in conducting human experiments with other drugs, where the improper or unauthorized use of aspirin by patients is a problem requiring continuous and rigid detection.

Simple and satisfactory colorimetric procedures for measuring salicylate in plasma and urine are available. However, the amounts of salicylate found in tissues are not sufficient to yield appreciable color intensities with these procedures. The intense fluorescence of salicylate has made it possible to develop a sensitive and specific procedure for its assay in tissues.[41] This involves extraction from the tissue by ether, return to borate buffer, and subsequent measurement of the fluorescence in a fluorescence spectrometer.

Tissues (1 g.) are homogenized in 0.01 N HCl (3 ml.), centrifuged at 10,000 × g and the supernatants are taken for analysis. Plasma is diluted by adding 1 ml. of plasma to 3 ml. of 0.01 N HCl. To a 30-ml. glass-stoppered centrifuge tube, add 0.5 ml. of tissue supernatant or diluted plasma, 1 ml. of water, 1 ml. of 6 N HCl, and mix. Add exactly 20 ml. of washed ether to the mixture, shake vigorously, and centrifuge at moderate speed for 5 minutes. Transfer exactly 10 ml. of the ether layer to another 30-ml. shaking tube containing 3 ml. of 0.5 M borate buffer, pH 10.0, and shake vigorously. Centrifuge for 5 minutes at moderate speed, then remove the ether layer by aspiration. Place the tubes in a water bath at 55° for 5 minutes to evaporate any residual ether and cool to room temperature. Transfer 1 ml. of the borate buffer to a cuvette and measure the fluorescence with excitation set at 310 mμ and fluorescence at 400 mμ. Standards are prepared by carrying known amounts of salicylate through the entire extraction procedure.

Specificity. Excitation and fluorescence spectra determined on salicylate standards, reagent blanks, and experimental tissue extracts are shown in Fig. 9. The absence of detectable blanks in salicylate-free tissues indicate that the constituents of normal tissues do not interfere with the fluorescence assay. Of the metabolic products of salicylic acid, gentisic acid also exhibits weak fluorescence. However, this occurs at 440 mμ when excited at 315 mμ. A solution containing 2 μg. each of salicylic acid and gentisic acid in borate buffer exhibits exactly the same fluorescence, qualitatively and quantitatively, as that of pure salicylic acid with excitation set at 310 mμ and fluorescence at 400 mμ.

Salicylic acid added to tissues may be recovered with satisfactory precision. The procedure has been shown to be sufficiently sensitive to detect

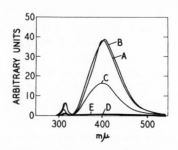

FIG. 9. Fluorescence spectra of salicylate standard (curve A), brain extracts from animals given the drug (curves B and C), control brain extracts (curve D), and reagent blank (curve E) (after Chirigos and Udenfriend[41]).

salicylate in rat brain as long as 24 hours following intraperitoneal injection of the drug. Being a physical method, without chemical reactions, the assay is certainly the simplest available. The fluorescence occurs just in the visible region and is therefore amenable to detection through glass optics and detecting devices of the usual fluorophotometers. However, the excitation at 310 mμ requires quartz optics and special exciting sources (either a xenon arc or a quartz-jacketed mercury lamp) for maximum sensitivity. Obviously, all these requirements are met in fluorescence spectrometers.

Aspirin can be assayed by measuring the liberated salicylate following hydrolysis. Gentisic acid, a major metabolite of salicylate, also fluoresces (excitation, 315 mμ; fluorescence, 440 mμ; uncorrected). The fluorescence is not as intense as that of salicylate. Being more water soluble than salicylate, it can be separated from the latter by extraction and then assayed fluorometrically. Gentisic acid can also be assayed by fluorescence on paper chromatograms[42] (see Chapter 3).

The extraction procedure for salicylate, previously reported by Saltzman,[43] would also be improved by using a fluorescence spectrometer with appropriate wavelength settings instead of a filter instrument, as reported.

Neocincophen can be extracted from alkaline solution (pH 13) with heptane and re-extracted from the heptane with 0.1 N HCl. For fluorescence measurement, the pH of the final extract should be adjusted to 7.[3]

V. ANESTHETICS

As shown in Table III, most of the barbiturate anesthetics are fluorescent in the native state. Fluorometric assay of barbiturates was first described by Pesez.[44] Udenfriend et al.[3] described extraction procedures for pentobarbital and pentothal which are applicable to fluorometric assay. The compounds, in aqueous solution, are buffered to pH 5.5 and are extracted with heptane containing 1.5% isoamyl alcohol. An aliquot of the organic solvent is then shaken with either 1 N NaOH (pentothal) or buffer pH 11 (pentobarbital) to return the compounds to an aqueous phase. Fluorescence is measured at the appropriate wavelengths (Table III) with optimal concentrations, in the final extract, of about 1 μg./ml. for pentothal and 0.5 μg./ml. for pentobarbital. The intensity of barbiturate fluorescence is not very great so that the concentrations in biological material may become limiting. However, the methods are so simple and specific that, if necessary, this disadvantage can be overcome by using more material so that higher concentrations may be obtained in the final extracts.

TABLE III

Anesthetics

Compound	Structure	Excitation maximum mμ	Fluorescence maximum mμ	Optimal pH	Reference
Phenobarbital	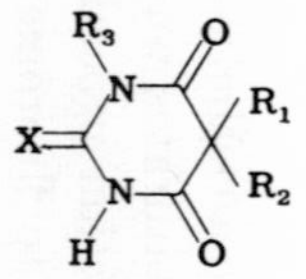 X = O R_1 = C_2H_5 R_2 = phenyl R_3 = H	265	440	13	(3)
Pentobarbital	Structure is as above, with X = O R_1 = C_2H_5 R_2 = 1-methylbutyl R_3 = H	265	440	13	(3)
Amytal (amobarbital)	Structure is as above, with X = O R_1 = C_2H_5 R_2 = isoamyl R_3 = H	265	410	14	(3)
Pentothal (thiopental)	Structure is as above, with X = S R_1 = C_2H_5 R_2 = 1-methylbutyl R_3 = H	315	530	13–14	(3)

TABLE III *(continued)*

Compound	Structure	Excitation maximum mμ	Fluorescence maximum mμ	Optimal pH	Reference
Surital (Thiamylal)	Structure is as above, with X = S R₁ = allyl R₂ = 1-methylbutyl R₃ = H	310	530	13	(3)

TABLE IV

Drugs Influencing the Cardiovascular System and Hemostasis

Compound	Structure	Excitation maximum mμ	Fluorescence maximum mμ	Optimal pH	Reference
Quinidine	Same as quinine (see Section II. A)	250, 350	450	1	(3)
Procaine amide		295	385	11	(3)
Paredrine (p-hydroxy-amphetamine)		275	300	1	(3)
Neosynephrine		270	305	1	(3)

TABLE IV (continued)

Compound	Structure	Excitation maximum mμ	Fluorescence maximum mμ	Optimal pH	Reference
Piperoxan		290	325	7	(3)
Yohimbine		270	360	1	(3)
α-Methyl 3, 4-dihydroxy-phenyl-alanine		285	325	1-7	(3)

TABLE IV (*continued*)

Compound	Structure	Excitation maximum mμ	Fluorescence maximum mμ	Optimal pH	Reference
Menadione		335	480	Alcohol	(3)

VI. CARDIOVASCULAR SYSTEM AND HEMOSTASIS

Table IV lists some drugs in this category which fluoresce in the native state. Quinidine has the same physical properties as quinine and can be assayed by exactly the same procedures.[6, 8] Edgar and Sokolow[45] have also reported a method for measuring quinidine in blood.

Yohimbine can be extracted from aqueous solutions, made alkaline by addition of ammonium hydroxide to 0.1 N, by shaking with ethylene dichloride. An aliquot of the solvent is shaken with 0.1 N HCl to return the compound to an aqueous phase for fluorometric assay. The final concentration should be about 0.2 μg./ml. Related ergot alkaloids may be assayed with a similar procedure.

The catecholamine derivative, α-methyl 3,4-dihydroxyphenylalanine, is a drug which is being used experimentally in the treatment of hypertension.[46] It possesses the native fluorescence of catecholamines and can also be assayed by other methods devised for epinephrine and norepinephrine (Chapter 5).

Procedures for the separation and fluorometric assay of the various cardiac glycosides in digitalis plant samples have been reported. Jensen[47–49]

Digitoxigenin

has utilized paper chromatography for separating the various genins and their glycosides. Following chromatography, the paper is dried and sprayed with a freshly prepared 25% solution of trichloroacetic acid in chloroform containing 1% ethanol. The spray reagent can also be prepared with the further addition of 1 or 2 drops of 30% hydrogen peroxide to 10 ml. of reagent. Instead of hydrogen peroxide, chloramine T or benzoyl peroxide can also be used. The various compounds exhibit characteristic fluorescence when examined under an ultraviolet lamp (365 mμ) (Table V). Estimations of amounts present can be made directly on paper by comparison with control spots. More precise estimations are made by eluting the spots with alcohol, filtering off particles of filter paper, and diluting to a convenient volume. An aliquot containing 1–10 μg. of the glycoside is transferred to a test tube and the solvent is evaporated on the water bath. After cooling the tube, the residue is dissolved in 10 ml. of a mixture of equal volumes of glycerol and concentrated hydrochloric acid. The solu-

TABLE V

FLUORESCENCE OF DIGITALIS PRODUCTS ON PAPER CHROMATOGRAMS[a]

	Fluorescence color with	
Glycoside	Trichloroacetic acid alone	Trichloroacetic acid plus H_2O_2
Lanatoside A _____	No fluorescence	Yellow
Digitoxin _____	No fluorescence	Yellow-orange
Lanatoside B _____	Bright blue	Grey-blue
Gitoxin _____	Bright blue	Greenish-blue
Lanatoside C _____	Bright gray-blue	Blue
Digoxin _____	No fluorescence	Blue

[a] After Silberman and Thorp.[50]

tion is shaken and, after 20 minutes, fluorescence is measured. The various steroid glycosides yield fluorescent products having different spectral characteristics. The products of digitoxigenin and related glycosides are maximally excited at 430 mμ and the fluorescence maxima are above 525 mμ. Gitoxigenin and related glycosides yield products which absorb maximally at 350 mμ and fluoresce maximally at about 465 mμ. Recoveries are quantitative and submicrogram quantities can be assayed. Silberman and Thorp[50] have utilized methods similar to those of Jensen. Pesez[51] has made use of fluorescence in strong phosphoric acid to distinguish the cardiac glycosides.

The methylated purine, theobromine, is not itself fluorescent. However, according to Florentin and Heros,[52] treatment with benzyliodide yields the N-benzyl derivative which is highly fluorescent.

Hydrastine, an important hemostatic agent, is found in the plant *Hydrastis canadensis*. Brochmann-Hanssen and Evers[53] have assayed hydrastine by oxidation in strong nitric acid solution which yields the highly fluorescent product, hydrastinine. In crude plant extracts, other alkaloids do not interfere. It should be pointed out that hydrastinine is

Hydrastinine

itself used as a myocardial stimulant and as a hemostatic agent.

Coumarin is the parent substance of many compounds which are active in promoting and retarding hemostasis. It is found in plants and is present in large amount in sweet clover. During spoilage of the clover, the coumarin

Coumarin

is apparently converted by a fermentation process to the important anti-coagulant dicumarol (bishydroxy coumarin). Coumarin itself does not fluoresce appreciably, but it develops a green fluorescence (maximum 510 mμ) on irradiation in 0.01 N NaOH solution with ultraviolet light.[54, 55] Most hydroxylated coumarins are highly fluorescent.[55-57] Dicumarol, which is also a hydroxylated coumarin derivative, probably possesses native fluorescence, although this has not yet been reported.

Coumarin is also the parent substance of the flavonoid hemostatic agents. It is likely that rutin and its aglycone, quercetin, two important flavonoids, possess native fluorescence. However, this has not yet been reported. Glazko et al.[58] demonstrated that the boric acid complexes of the flavonoids are highly fluorescent. The boric acid complex of rutin has an excitation maximum at about 430 to 440 mμ and a fluorescence above 520 mμ. As little as 0.1 μg. of rutin or quercitin can be detected. Tetraphenyl-boroxide complexes of the flavones can also be used for fluorometric measurement.[59]

Menadione is a synthetic analog of vitamin K which is just as active as the natural vitamin in promoting blood clotting. It is fluorescent in alcohol with an excitation maximum at 335 mμ and a fluorescence maximum at 480 mμ.

VII. CENTRAL NERVOUS SYSTEM

Centrally active agents are under intensive study both experimentally and clinically. Methods for their assay are widely used to detect and quantify their presence in brain. As can be seen from Table VI, many of the more interesting agents possess native fluorescence.

A. Lysergic Acid Diethylamide (L.S.D.)

Lysergic acid diethylamide (L.S.D.) is a partially synthetic hallucino-genic compound and is probably one of the most potent centrally active agents yet to be described. As little as 50 μg. when administered to a man produces its typical actions. Experimental studies therefore require the most sensitive methods of assay. Fortunately, L.S.D. is a highly fluores-cent compound permitting the development of methods with the required sensitivity. Axelrod et al.[60] have devised a method for measuring L.S.D. in

TABLE VI

Drugs Active on the Central Nervous System

Compound	Structure	Excitation maximum mμ	Fluorescence maximum mμ	Optimal pH	Reference
Lysergic acid diethylamide (L.S.D.)	CON(C_2H_5)$_2$, NCH$_3$, R, N–H, R = H	325 325	465 445	7 —	(3) (60)
2-Brom-lysergic acid diethylamide (brom L.S.D.)	Structure is as above, with R = Br	315	460	1	(3)
Chlorpromazine	Cl, S, N, CH$_2$CH$_2$CH$_2$N(CH$_3$)$_2$	350	480	11	(3)

TABLE VI (*continued*)

Compound	Structure	Excitation maximum mμ	Fluorescent maximum mμ	Optimal pH	Reference
Chlorpromazine sulfoxide		335	400	7	(3)
Harmine		300, 365	400	1	(3)
Reserpine		300	375	1	(3)

TABLE VI (*continued*)

Compound	Structure	Excitation maximum mμ	Fluorescence maximum mμ	Optimal pH	Reference
Rescinnamine	A, B, C, D, and E rings as in reserpine; bridge between E and F different	310	400	1	(3)
11-Desmethoxy-reserpine	Same as reserpine but no methoxy group on the A ring	280	365	1-2	(61)

tissues which can determine as little as 0.003 μg. of the compound in biological materials.

Homogenate, plasma, or urine (up to 5 ml.) is added to a 60-ml. glass-stoppered bottle containing 25 ml. of n-heptane-isoamylalcohol (98:2 v/v), 0.5 ml. of 1 N sodium hydroxide, and sufficient sodium chloride to saturate (about 3 g.). The mixture is shaken on an automatic shaker for 15 minutes and then centrifuged. Twenty milliliters of the heptane is transferred to a conical glass-stoppered centrifuge tube and shaken with 3 ml. of 0.004 N HCl, for 10 minutes. The organic solvent is removed by aspiration and the fluorescence of an aliquot of the acid phase is measured directly in a cuvette. Standards and reagent blanks are carried through the entire procedure. Recoveries from tissue are essentially quantitative. The method is highly specific and metabolites of L.S.D. formed in mammalian tissues do not interfere. The plasma level of L.S.D. in the monkey, 60 minutes after administering 0.2 mg./kg. of the compound, was 0.13 μg./ml. Most of the compound was metabolized, less than 1% of the unchanged drug appearing in the urine and feces.

B. Chlorpromazine

Chlorpromazine, one of the better-known tranquilizing agents, is fluorescent. The compound can be extracted from biological material, at pH 11, into heptane and then returned to 0.1 N HCl for fluorometric assay. Unfortunately, the fluorescence intensity of chlorpromazine is so weak that fluorometry can have only limited application in its study. It is, however, useful for purposes of identification and comparison. One of the metabolic products, which clorpromazine forms in man and experimental animals, is the oxidation product, chlorpromazine sulfoxide. This product is more highly fluorescent than the parent drug. Since its fluorescence also differs from that of chlorpromazine it is possible to follow the appearance of the metabolite during chlorpromazine administration.

C. Harmala Alkaloids

Harmine is one of a number of related compounds, the harmala alkaloids. Most of these compounds are potent inhibitors of the enzyme monoamine oxidase. They are now used mainly for experimental studies in animals, although they had been used clinically in the past. The harmala alkaloids are highly fluorescent and can be extracted from tissue extracts by procedures comparable to those described for quinine.

D. Reserpine

Reserpine is another important tranquilizer and antihypertensive agent. Although it has sufficient native fluorescence to permit direct assay,[62a] the

methods of analysis which have been developed utilize oxidative procedures to convert reserpine to a product which is both more highly fluorescent and which is excited by radiation to which glass is transparent (320
mμ). The oxidation product also emits fluorescence in the visible region.
Oxidation procedures were probably designed for the purpose of using
filter instruments. However, it should be pointed out that the native
fluorescence, excitation at 300 mμ and fluorescence at 375 mμ, would
probably overlap that due to normally occurring indoles which are excited
at about 285 mμ and fluoresce at 350 mμ. Oxidation procedures eliminate
such interferences.

A number of assays for reserpine have appeared. Poet and Kelly[62b] published a procedure for tissues which was later modified by Hess et al.[63]
Transfer 1–4 ml. of tissue homogenate, plasma, or urine (containing 0.1–1.0
μg. of reserpine) to a 60-ml. glass-stoppered bottle. Adjust the pH to
about 8.5 with a few drops of 0.5 N NaOH, then add 1 ml. of borate buffer,
pH 8.5. Add 25 ml. of heptane containing 1.5% isoamyl alcohol and shake
for 20 minutes. Centrifuge and transfer 20 ml. of the heptane to another
60-ml. glass-stoppered bottle containing 6 ml. of 1.5 N sulfuric acid. Shake,
centrifuge, and transfer 5 ml. of the acid layer to a test tube containing
1 ml. of 10% selenious acid in water, 2 ml. of 1.5 N sulfuric acid, and 0.1
ml. of isoamyl alcohol. Cover the tube with a glass marble and heat in a
boiling water bath for 20 minutes. Cool the tube and measure the fluorescence. The excitation and fluorescence maxima were not presented, but the
same filters as are used for quinine fluorescence, 365-mμ excitation, 420-mμ
fluorescence, are satisfactory. The method is apparently specific and recoveries from most tissues are above 90%; recoveries from brain, however, may be as low as 70%.

TABLE VII

FLUORESCENCE OF 11-DESMETHOXYRESERPINE AND ITS OXIDATION PRODUCT[a]

Compound	Excitation maximum[b] mμ	Fluorescence maximum[b] mμ	Relative fluorescence intensity[c]
11-Desmethoxyreserpine_____	280	360	6
Oxidation product_____	315	450	70

[a] After Gordon and Campbell.[61]

[b] Uncorrected data.

[c] Obtained with Aminco-Bowman spectrophotofluorometer. Corrections were not
made for instrumental differences in excitation energy and detection at the different
wavelengths.

Dechene[64] has used hydrogen peroxide to oxidize the reserpine to the visibly fluorescing derivative. The characteristic fluorescence of reserpine and related alkaloids on paper chromatograms has been utilized in the development of a paper chromatographic procedure for determining the alkaloid composition of various *Rawolfia* species.[65] Gordon and Campbell[61] have described, in detail, the oxidation of 11-desmethoxyreserpine to a more highly fluorescent product. On using 0.001 N ceric sulfate in 0.1 N sulfuric acid, they were able to obtain maximal yields of the fluorescent product. A comparison of the fluorescent properties of the parent compound with that of the oxidized product is shown in Table VII. The method was found to be applicable to reserpine with only slightly lower sensitivity. Less than 0.5 μg. can be measured.

VIII. MISCELLANEOUS DRUGS

A. Alloxan

Alloxan is a drug which has been of great interest in studies on experimental diabetes. Archibald[66, 67] described a number of methods for determining alloxan, one of them was a fluorometric procedure based on the condensation with *o*-phenylenediamine. Alloxan and compounds which dissociate into alloxan yield a green fluorescence with this reagent. Ninhydrin and ascorbic acid also react to give compounds with green and blue fluorescence, respectively (Fig. 10).

Alloxan is rapidly destroyed in biological material and specific isolation procedures must be used. These are discussed in detail by Archibald.[67] Once alloxan is obtained in an acidified protein-free filtrate, the assay can be carried out very simply. To 5 ml. of solution, containing about 1 μg. of alloxan, add 0.5 ml. of *o*-phenylenediamine reagent (5 mg. of *o*-phenylenediamine dissolved in 10 ml. of glycerol and diluted with 100 ml. of 1 M sodium dihydrogen phosphate). Set the tubes in a dark place and after 1 hour compare in the fluorometer against standards and blanks. The filters

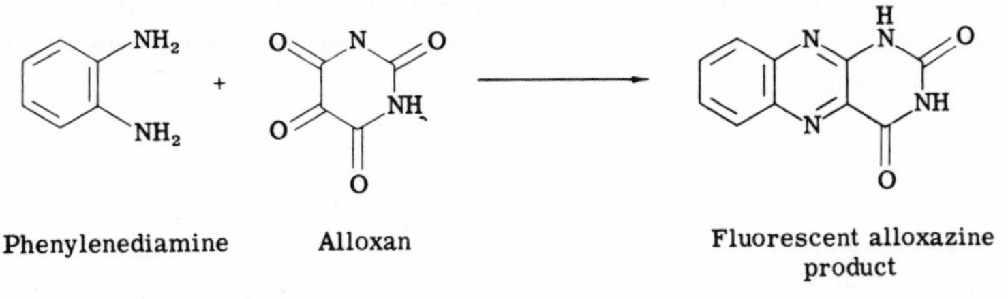

o-Phenylenediamine   Alloxan   Fluorescent alloxazine
                                                                product

FIG. 10. Condensation of *o*-phenylenediamine with alloxan to yield the fluorescent alloxazine (according to Archibald[67]).

used for excitation and fluorescence are the same as those used for ribo-flavin assay, excitation, 405 mμ; fluorescence, 520 mμ.

There has been some doubt concerning the occurrence of alloxan in normal animal tissues. On using the highly sensitive fluorometric procedure, Archibald[67] found that, if it is present at all, it must occur in amounts below 20 μg./100 ml. of blood or urine. More recently, Loubatières and Bouyard,[68] using the procedure of Archibald,[67] have made observations which support the natural occurrence of alloxan. They have also reported that alloxan, when heated in water at elevated temperatures (80°), gives rise to a characteristic fluorescent product which can be used for its identification and assay.

B. Stilbamidine

Stilbamidine has been used as a therapeutic agent in various tropical infectious diseases, in arthritis, and in multiple myeloma. The drug, when

Stilbamidine

dissolved in acid solution, exhibits a brilliant blue fluorescence upon excitation with ultraviolet light (365 mμ). A procedure devised by Saltzman[69] involves separation from plasma or urine on a Decalso column, washing of the column with hot water, and subsequent elution with 0.2 N HCl in 50% ethanol. Fluorescence is measured after the eluates are made more acid by addition of a small amount of concentrated hydrochloric acid. Reagent blanks, tissue blanks, and standards are carried through the entire procedure.

Both stilbamidine and 2-hydroxystilbamidine localize in myeloma cells. The former combines with ribonucleic acid only, the latter with both ribonucleic and deoxyribonucleic acids. The nucleic acid complexes emit intense and characteristic fluorescence which is visible in tissue sections, under the microscope, and can be used to localize the drugs and to study their action.

C. Diethylstilbestrol

Diethylstilbestrol is a synthetic compound used for replacement therapy in estrogen deficiency. It is also widely used in poultry and cattle breeding. Sensitive methods for its assay are required not only for research purposes but for detection of residues in meat and poultry products

Diethylstilbestrol

as well. Diethylstilbestrol is a phenol and probably fluoresces in the native state. However, no data is available concerning this. In any event, such fluorescence would not be expected to be very specific since it is characteristic of all phenols. Recently, Goodyear and Jenkins[70, 71] have described a procedure for diethylstilbestrol assay which involves photochemical conversion to a derivative which is fluorescent in the visible region. The product, although not identified, is maximally excited with light at 410 mμ and exhibits a fluorescence maximum above 510 mμ. Following extraction of the diethylstilbestrol from the tissue homogenate, the organic solvent is evaporated and the residue is dissolved in 5 ml. of ethanol. A 2.5-ml. aliquot of this solution is transferred to a quartz tube for irradiation along with standards and blanks. An apparatus for irradiation such as described in the procedure for chloroquine may be used. The sample is irradiated for an optimum time determined by previous studies on standards, and is then transferred to a cuvette in which its fluorescence is measured along with that of an unirradiated aliquot. Following these measurements, 0.05 ml. of 1 N potassium hydroxide are added to each tube and the fluorescence is measured again. The diethylstilbestrol is thus distinguished from extraneous tissue fluorescence in two ways: (a) by the increment in fluorescence upon irradiation and (b) by the characteristic shift in the fluorescence spectrum of the photochemical product following addition of alkali.

References

1. *Org. Chem. Bull.* **29**, No. 4 (1957).
2. Parker, C. A., and Rees, W. T., *Analyst* **85**, 587 (1960).
3. Udenfriend, S., Duggan, D. E., Vasta, B. M., and Brodie, B. B., *J. Pharmacol. Exptl. Therap.* **120**, 26 (1957).
4. Slavin, W., Pittsburgh Conference on Analytical Chemistry and Applied Spectroscopy, p. 34, March 1958.
5. Bowen, E. J., and Wokes, F., "Fluorescence of Solutions." Longmans, Green, London, 1953, p. 22.
6. Brodie, B. B., and Udenfriend, S., *J. Pharmacol. Exptl. Therap.* **78**, 154 (1943).
7. Kelsey, F. I., and Geiling, E. M. K., *J. Pharmacol. Exptl. Therap.* **75**, 183 (1942).
8. Brodie, B. B., Udenfriend, S., Dill, W., and Downing, G., *J. Biol. Chem.* **168**, 311 (1947).

9. Brodie, B. B., and Udenfriend, S., *J. Biol. Chem.* **151,** 299 (1943).
10. Craig, L. C., *J. Biol. Chem.* **150,** 33 (1943).
11. Brodie, B. B., Udenfriend, S., and Taggart, J. V., *J. Biol. Chem.* **168,** 327 (1947).
12. Irvin, J. L., and Irvin, E. M., *J. Biol. Chem.* **174,** 589 (1948).
13. Brodie, B. B., Udenfriend, S., Dill, W., and Chenkin, T., *J. Biol. Chem.* **168,** 319 (1947).
14. Scudi, J. V., *J. Biol. Chem.* **164,** 183 (1946).
15. Scudi, J. V., and Jelinek, V. C., *J. Biol. Chem.* **164,** 195 (1946).
16. Haden, R. L., *J. Biol. Chem.* **56,** 469 (1923).
17. Levine, J., Garlock, E. A., Jr., and Fischbach, H., *J. Am. Pharm. Assoc. Sci. Ed.* **38,** 473 (1949).
18. Chiccarelli, F. S., Van Gieson, P., and Woolford, M. H., Jr., *J. Am. Pharm. Assoc. Sci. Ed.* **45,** 418 (1956).
19. Feldman, D. H., Kelsey, H. S., and Cavagnol, J. C., *Anal. Chem.* **29,** 1697 (1957).
20a. Saltzman, A., *J. Lab. Clin. Med.* **35,** 123 (1950).
20b. Kohn, K. W., *Anal. Chem.* **33,** 862 (1961).
21. Milch, R. A., Rall, D. P., and Tobie, J. E., *J. Natl. Cancer Inst.* **19,** 87 (1957).
22. Milch, R. A., Rall, D. P., and Tobie, J. E., *J. Bone and Joint Surg.* **40A,** 897 (1958).
23. Rall, D. P., Loo, T. L., Lane, M., and Kelly, M. G., *J. Natl. Cancer Inst.* **19,** 79 (1957).
24. Tobie, J. E., and Beye, H. K., *Proc. Soc. Exptl. Biol. Med.* **104,** 137 (1960).
25. DuBuy, H. G., and Showacre, J. L., *Science* **133,** 196 (1961).
26. Duggan, D. E., Bowman, R. L., Brodie, B. B., and Udenfriend, S., *Arch. Biochem. Biophys.* **68,** 1 (1957).
27. Auerbach, M. E., and Angell, E., *J. Pharm. and Pharmacol.* **10,** 776 (1958).
28. Bedford, C., Child, K. J., and Tomich, E. G., *Nature* **184,** 364 (1959).
29. Freeman, M. V., *J. Pharmacol. Exptl. Therap.* **120,** 1 (1957).
30. Chirigos, M. A., and Mead, J. A. R., to be published.
31. Zakrzewski, S. F., and Nichol, C. A., *J. Biol. Chem.* **205,** 361 (1953).
32. Freeman, M. V., *J. Pharmacol. Exptl. Therap.* **122,** 154 (1958).
33. Mathews, R. E. F., *Nature* **171,** 1065 (1953).
34. Markham, R., and Smith, J. D., *Biochem. J.* **52,** 552 (1952).
35. Haddow, A., and Kon, G. A. R., *Brit. Med. Bull.* **4,** 314 (1947).
36. Hartwell, J. L., *U. S. Public Health Serv. Publ. No.* **149,** 1951.
37. Sawicki, E., Hauser, T. R., and Stanley, T. W., *Intern. J. Air Pollution* **2,** 253 (1960).
38. DeRitter, E., Jahns, F. W., and Rubin, S. H., *J. Am. Pharm. Assoc. Sci. Ed.* **38,** 319 (1949).
39. Nadeau, G., and Sobolewski, G., *Can. J. Biochem. Physiol.* **36,** 625 (1958).
40. Hanker, J. S., Gamson, R. M., and Klapper, H., *Anal. Chem.* **29,** 879 (1957).
41. Chirigos, M., and Udenfriend, S., *J. Lab. Clin. Med.* **54,** 769 (1959).
42. Nanninga, L., and Bink, B., *Nature* **168,** 389 (1951).
43. Saltzman, A., *J. Biol. Chem.* **174,** 399 (1948).
44. Pesez, M., *J. pharm. chim.* **27,** 247 (1938).
45. Edgar, A. L., and Sokolow, M., *J. Lab. Clin. Med.* **36,** 478 (1950).
46. Oates, J. A., Jr., Gillespie, L., Udenfriend, S., and Sjoerdsma, A., *Science* **131,** 1890 (1960).
47. Jensen, K. B., *Acta Pharmacol. Toxicol.* **8,** 101 (1952).
48. Jensen, K. B., *Acta Pharmacol. Toxicol.* **9,** 66 (1953).
49. Jensen, K. B., *Acta Pharmacol. Toxicol.* **12,** 27 (1956).

50. Silberman, H., and Thorp, R. H., *J. Pharm. and Pharmacol.* **6,** 546 (1954).
51. Pesez, M., *Ann. pharm. franç,* **8,** 746 (1950).
52. Florentin, D., and Heros, M., *Bull. soc. chim. france* p. 90 (1947).
53. Brochmann-Hanssen, E., and Evers, J. A., *J. Am. Pharm. Assoc. Sci. Ed.* **40,** 620 (1951).
54. Goodwin, R. H., and Kavanagh, F., *Arch. Biochem. Biophys.* **36,** 442 (1952).
55. Mattoo, B. N., *Trans. Faraday Soc.* **52,** 1184 (1956).
56. Williams, R. T., *J. Roy. Inst. Chem.* **83,** 611 (1959).
57. Wheelock, C. E., *J. Am. Chem. Soc.* **81,** 1348 (1959).
58. Glazko, A. J., Adair, F., Papageorge, E., and Lewis, G. T., *Science* **105,** 48 (1947).
59. Neu, R., *Z. anal. Chem.* **142,** 335 (1954).
60. Axelrod, J., Brady, R. O., Witkop, B., and Evarts, E. V., *Ann. N. Y. Acad. Sci.* **66,** 435 (1957).
61. Gordon, J. A., and Campbell, D. J., *Anal. Chem.* **29,** 488 (1957).
62a. Haycock, R. P., Sheth, P. B., and Mader, W. J., *J. Am. Pharm. Assoc. Sci. Ed.* **48,** 479 (1959).
62b. Poet, R. B., and Kelly, J. M., abstracts of papers presented at 126*th Meeting Am. Chem. Soc., 1954,* New York, p. 83C.
63. Hess, S., Shore, P. A., and Brodie, B. B., *J. Pharmacol. Exptl. Therap.* **118,** 84 (1956).
64. Dechene, E. B., *J. Am. Pharm. Assoc. Sci. Ed.* **44,** 657 (1957).
65. Korzun, B. P., St. André, A. F., and Ulshafer, P. R., *J. Am. Pharm. Assoc. Sci. Ed.* **46,** 720 (1957).
66. Archibald, R. M., *J. Biol. Chem.* **157,** 507 (1945).
67. Archibald, R. M., *J. Biol. Chem.* **158,** 347 (1945).
68. Loubatières, A., and Bouyard, P., *J. physiol.* **46,** 437 (1954).
69. Saltzman, A., *J. Biol. Chem.* **168,** 699 (1947).
70. Goodyear, J. M., and Jenkins, N. R., *Anal. Chem.* **32,** 1203 (1960).
71. Goodyear, J. M., and Jenkins, N. R., *Anal. Chem.* **33,** 853 (1961).

14

PUBLIC HEALTH
AND SANITATION

I. GENERAL REMARKS

FLUORESCENCE, as a qualitative and quantitative tool, has been used effectively in areas which can be referred to as Public Health and Sanitation. Under this heading we find many important fluorescence applications to the problems of air and water pollution. Food inspection and analysis is an extremely large field in which fluorescence has been used to evaluate quality, adulteration, nutritional value, sanitation, and the presence of chemical residues. In the latter category there is a great need for ultrasensitive and specific methods for detecting, in foods, traces of chemical preservatives, antioxidants, drugs, hormones, pesticides, flavorings, and colorings. The final section of this chapter contains several uses of fluorescence in medicine other than the metabolic applications reported in previous chapters.

II. AIR AND WATER POLLUTION

The problems of air and water pollution are many and include chemical agents produced by industrial and nonindustrial activities of man and products of bacterial decay and plant growth. The agents used in chemical warfare are also best handled under this category.

A. Dust

According to Przibram,[1] there is an atmospheric dust, in most areas of the world, much of which is organic in nature and fluoresces. This dust apparently endows with fluorescence many otherwise nonfluorescent chemicals and accounts for the bluish fluorescence of "pure" water and of snow. This atmospheric dust must be composed of a host of organic compounds and no doubt varies in composition from one area to another and from one season to another. During certain seasons pollen dust is a most serious health hazard, rural air containing more than urban air. As was pointed out in Chapter 11, pollens contain, among other things, flavonoids which can yield intensely fluorescent metal complexes.[2] The flavonoid composition differs from one type of pollen to another and ragweed pollen contains the compound isorhamnetin which yields such a highly fluorescent aluminum complex that as little as 0.0001 μg. of this flavonoid can be detected. The procedure for detecting and quantifying different pollen flavonoids by fluorescence on paper[2] may prove to be a most useful adjunct to the present-day pollen counting procedures since it will permit discrimination of the allergenic pollens in the over-all count.

B. Fluoride in Air

Industrial plants release many substances into the atmosphere which are of concern from a health standpoint. In many steel mills sufficient amounts of fluoride are released in the smelting processes so as to create a real hazard.[3] Measurement of fluoride in air is now carried out in many steel mills to permit effective regulation. The procedure used for assaying fluoride in air makes use of the highly fluorescent magnesium-oxine complex[4] (see Chapter 12). A definite volume of air is drawn through a piece of filter paper impregnated with the magnesium oxine complex. Fluoride combines with the magnesium dissociating the fluorescent complex. Diminution of fluorescence is proportional to the amount of fluoride. An adapter for measuring fluorescence on paper provides a rapid and precise means of assay. The excitation maximum of the magnesium-oxine complex is at 420 mμ, fluorescence maximum at 530 mμ (uncorrected).[5] Interest in fluoride control is so great that a special fluoride measuring kit has been made commercially available for just this purpose.* It, too, makes use of the diminution of magnesium-oxine fluorescence on paper.

C. Oil Mists

Low-pressure air produced by compressors contains appreciable amounts of entrained oil mist. Such a contamination can be objectionable for many reasons including its presenting a health hazard. Parker and Barnes[6] examined a number of procedures for detecting oil in air, including ultraviolet and infrared spectroscopy and fluorometry. Only the latter was sufficiently sensitive for the purpose. They found that all commercial oils are highly fluorescent, yielding an emission band with a maximum in the region 357 mμ. Excitation is maximal at 248 mμ. To determine oil in air, an apparatus is used for sampling the air through small disks of filter paper. After collection from a suitable volume of air the filter paper is eluted with hexane and the fluorescence of the solution is measured in a

---

* Fluoride Measuring Kit, Engineering Specialties Company, Saxonville, Massachusetts.

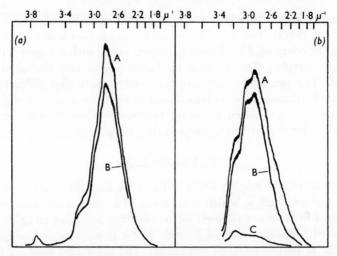

Fig. 1. Fluorescence spectra of compressor lubricant with excitation at (a) 4.03 μ^{-1} (248 mμ) and (b) 3.5 μ^{-1} (286 mμ). Curves A represent oil from the sump of the compressor; curves B oil from the air; curve C is the filter paper blank (after Parker and Barnes[6]).

fluorescence spectrometer and compared with appropriate standards. Figure 1 compares fluorescence spectra obtained on a sample of oil from the sump of a compressor and from the surrounding air. By drawing 50 liters of air through the paper it was possible to detect as little as 0.2 to 0.3 μg. of oil per liter of air. It was found that the air in the vicinity of many compressors contained from 0.5 to 1 μg. of oil per liter.

TABLE I

COMPARISON OF FLUORESCENCE INTENSITIES OF POLYNUCLEAR HYDROCARBONS IN PENTANE[a]

| Compound | Wavelength maxima | | K_Q[c] |
	Excitation	Fluorescence	
Quinine			1.0
Anthanthrene	420	430	45
Naphtho[2,3-a]pyrene	457	458	~33
Perylene	430	438	32
Dibenzo[b,k]chrysene	308	428	17[b]
7-Methyldibenzo[a,h]pyrene	460	467	13[b]
Benzo[k]fluoranthene	302	400	13
Benzo[a]pyrene	381	403	6
1,4-Diphenylbutadiene	328	370	6
p-Terphenyl	284	338	6
Benzo[k,l]xanthene	363	418	5

TABLE I (*Continued*)

Compound	Wavelength maxima		$K_Q{}^c$
	Excitation	Fluorescence	
11H-Benzo[a]fluorene	317	340	4
11H-Benzo[b]fluorene	312	340	4
7H-Benzo[c]fluorene	334	337	4
Benzo[b]naphtho[2,3]furan	320	350	4
Anthracene	350	398	3
Benzo[b]chrysene	283	398	2
9-Methylanthracene	382	410	2
3-Methylcholanthrene	297	392	2
Benzo[b]fluoranthene	300	428	2
Dibenzo[a,e]pyrene	370	401	2
Tribenzo[a,e,i]pyrene	384	448	2
Dibenz[a,h]anthracene	292	394	1
Fluorene	300	321	1
Benz[a]anthracene	284	382	0.9
7,12-Dimethylbenz[a]anthracene	293	427	0.8
Fluoranthene	354	464	0.8
Dibenz[a,j]anthracene	300	410	0.7
Dibenz[a,c]anthracene	280	381	0.7
Picene	281	398	0.7
4-Methylpyrene	338	386	0.7
o-Phenylenepyrene	360	506	0.6
1-Methylpyrene	336	394	0.6
Chrysene	264	381	0.6
Coronene	337	450	0.6
Pyrene	330	382	0.5
3-Methylphenanthrene	292	368	0.5
Benzo[e]pyrene	329	389	0.4
Benzo[g,h,i]perylene	280	419	0.4
Acenaphthene	291	341	0.4
Phenanthrene	252	362	0.2
2-Methylphenanthrene	257	357	0.2
4-Cyclopenta[d,e,f]phenanthrene	294	362	0.2
Triphenylene	288	357	0.1
5,12-Dihydronaphthacene	282	340	0.1

[a] Measurements were made with an Aminco-Bowman spectrophotofluorometer and are reported as uncorrected instrumental readings (after Sawicki *et al.*[9]).

[b] In chloroform.

[c] K_Q = Fluorescence relative to quinine.

TABLE II

Excitation and Fluorescence Spectra in Sulfuric Acid

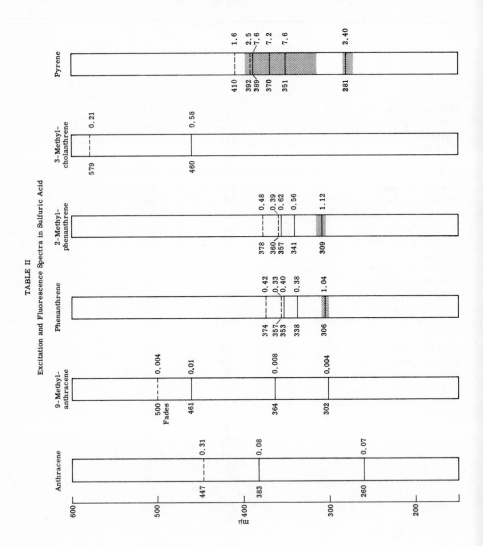

TABLE II

Excitation and Fluorescence Spectra in Sulfuric Acid (continued)

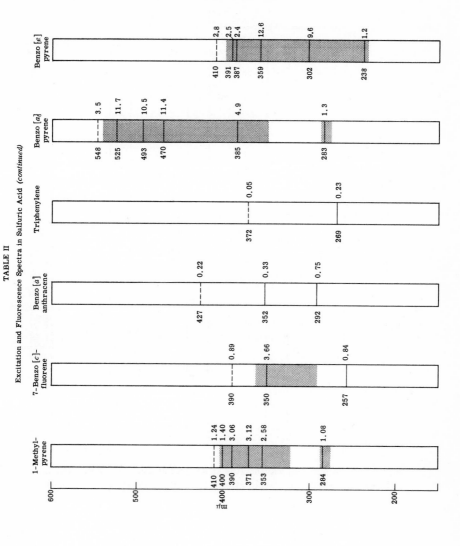

TABLE II

Excitation and Fluorescence Spectra in Sulfuric Acid *(continued)*

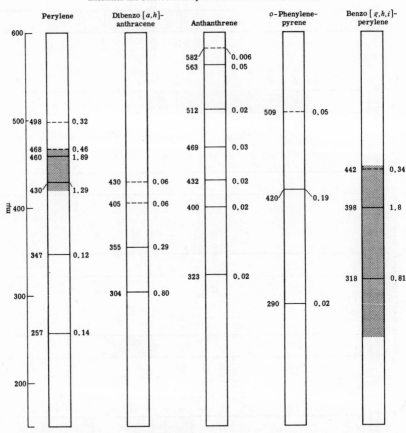

—— Excitation maxima; wavelengths at left and arbitrary units at right.
---- Fluorescence maxima; wavelengths at left and K_Q values at right. (Sawicki *et al.*[9])

D. Carcinogenic Hydrocarbons

A most serious problem concerns the presence in urban atmospheres of hydrocarbons, many of which are characterized as carcinogenic. A number of reports concerning the fluorescence of carcinogenic hydrocarbons have appeared.[7,8] However, the recent studies of Sawicki and his colleagues at the Robert A. Taft Sanitary Engineering Center of the U. S. Public Health Service represent the most thorough work done in this field. In one report[9] they presented absorption spectra and excitation and fluorescence maxima for a large number of hydrocarbons (including carcinogens) which are found in urban air. The fluorescence characteristics in pentane are shown in Table I. Similar data are presented for solutions in concentrated sulfuric acid, Table II. The value K_Q, in Tables I and II, represents the fluorescence intensity compared to that of an equivalent concentration of quinine. The fluorescence intensities of most of these compounds are so high that less than 0.001 μg./ml. can be readily detected. With the fluorescence spectrometer used, fairly good agreement was obtained between the absorption and excitation spectra of a given compound, as shown in Fig. 2. The studies revealed some remarkably specific characteristics of these closely related compounds which made possible the development of specific assay pro-

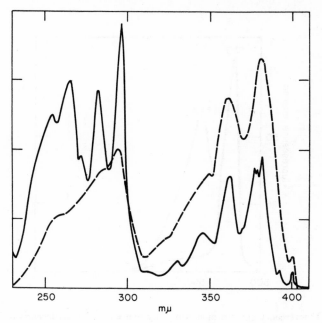

Fig. 2. Benzo[*a*]pyrene in pentane. —, absorption spectrum; - - -, excitation spectrum, fluorescence measured at 403 mμ (after Sawicki *et al.*[9]).

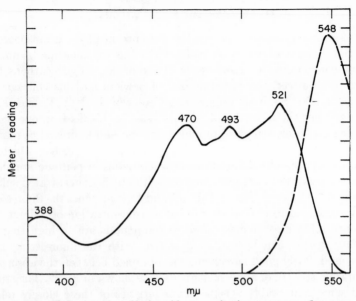

FIG. 3. Benzo[a]pyrene in sulfuric acid. —, excitation spectrum, fluorescence measured at 545 mμ; – – –, fluorescence spectrum, excitation set at 520 mμ (after Sawicki et al.[9]).

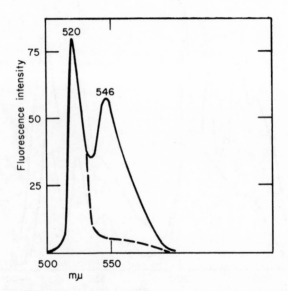

FIG. 4. Fluorescence spectra of benzo[a]pyrene and other hydrocarbons in sulfuric acid; excitation wavelength at 520 mμ. – – –, mixture of 50 hydrocarbons, each at 10^{-6} M; —, same solution plus 5×10^{-7} benzo[a]pyrene, BaP (after Sawicki et al.[9]).

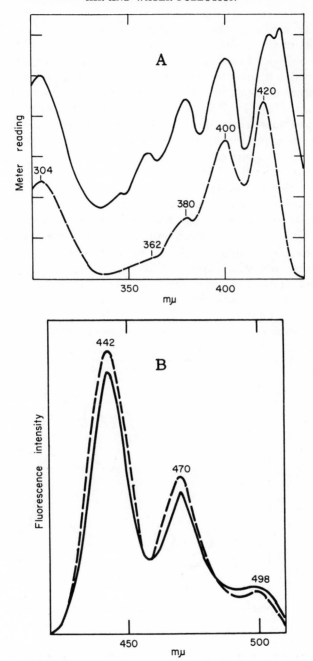

FIG. 5. *A*. Excitation spectra in pentane of authentic anthanthralene, – – –, and corresponding fraction isolated from air, —; fluorescence measured at 430 mμ (after Sawicki *et al.*[10]). *B*. Fluorescence spectra of authentic coronene, – – –, and of the coronene fraction isolated from air, —; excitation wavelength 327 mμ, solvent sulfuric acid.

cedures. For instance, when the fluorescence spectra, in concentrated sulfuric acid, of all the different hydrocarbons are compared, benzo[a]pyrene is the only one found to have a maximum at 548 mµ (Fig. 3). This specificity makes it possible to detect as little as 0.04 µg. of benzo[a]pyrene in the presence of a mixture of 1 µg. each (total 50 µg.) of all the other hydrocarbons (Fig. 4). The combined use of absorption spectra and excitation and fluorescence spectra in the two solvents is so highly specific a procedure that traces of hydrocarbon impurities can be detected in what are considered to be pure samples of a given hydrocarbon. As little as 0.1% perylene were detected in "pure" benzo[e]pyrene.

The absorption and fluorescence data obtained in the foregoing study[9] were combined with a chromatographic procedure to divide the hydrocarbons in air into several distinct fractions in order to simplify identification.[10] With such a combined procedure it was possible to identify many hydrocarbons from samples of urban air by their characteristic excitation and fluorescence spectra (Fig. 5).

A detailed study of benzo[a]pyrene in urban airborne particulates was carried out.[11] First it was shown that the hydrocarbon mixture eluted from the chromatograms along with benzo[a]pyrene were comparable in composition from city to city. In concentrated sulfuric acid the charac-

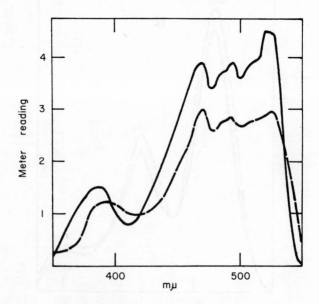

Fig. 6. Excitation spectra in sulfuric acid of authentic benzo[a]pyrene, —, and of the benzopyrene fraction from urban air, – – –, fluorescence measured at 545 mµ (after Sawicki et al.[11]).

TABLE III

BENZO[a]PYRENE IN THE AROMATIC FRACTION OF AIR PARTICULATES OF SEVERAL
AMERICAN CITIES[a]

City	mg./g. of aromatic fraction
A	0.15
B	0.64
C	0.75
D	0.87
E	2.4
F	3.5
G	4.6
H	5.9
I	7.7

[a] November, 1958. After Sawicki et al.[11]

teristic 545- to 548-mμ fluorescence peak indicated the presence of benzo[a]-pyrene. Further identification was made by comparing excitation spectra of the isolated and authentic material using the 545- to 548-mμ fluorescence peak (Fig. 6). These and other studies established the specificity of the method. On using this procedure they determined the concentrations of benzo[a]pyrene in the air of nine American cities (Table III). Other data indicated that an increase in the concentration of benzo[a]pyrene denotes an increase in the concentration of many of the other polynuclear hydrocarbons.

There have, of course, been prior studies of benzo[a]pyrene in air, many of them also utilizing fluorometry. Goulden and Tipler[12] identified it in domestic soot by its fluorescence spectrum and Waller[7] used similar techniques for measuring the benzo[a]pyrene content of air samples from many English towns and cities. Measurement of specific carcinogenic hydrocarbons in air as a monitoring device appears to be a procedure which will be used routinely by more and more public health agencies.

A most controversial subject at the moment is the relationship between cigarette smoking and lung cancer. Carcinogens have been reported in the inhaled cigarette smoke by a number of investigators. However, before conclusions can be drawn it is necessary to obtain quantitative data. Van Duuren[8, 13, 14] has developed procedures for detecting and quantifying hydrocarbons in cigarette smoke utilizing fluorescence spectrometry in conjunction with chromatography (many of the procedures of Sawicki are based on Van Duuren's work). He was able to identify benzo[a]pyrene in cigarette smoke by its characteristic fluorescence spectrum (Fig. 7). This

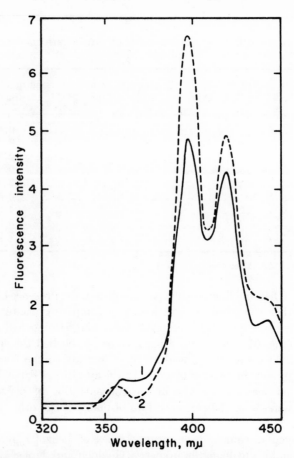

FIG. 7. Fluorescence spectra of material from cigarette smoke, curve 1, —; and of authentic benzo[*a*]pyrene, curve 2, – – –; excitation at 375 mμ, in cyclohexane (after Van Duuren[13]).

compound and dibenz[*a*, *h*]anthracene, both strong carcinogens, were found in amounts of 0.5 and 0.05 μg. per 100 cigarettes. However, Van Duuren does not think that these amounts are large enough to account for the observed carcinogenic activity in experimental animals. It is hoped that quantitative and specific methods will permit the resolution of many problems in this controversial field.

E. Chemical Warfare

Although not exactly in the area of public health, detection of chemical warfare agents, such as nerve gas vapors, presents a medical problem of interest to the military and to civil defense agencies. Nerve gases give no sensory warning when lethal concentrations are present in the atmosphere. Procedures for their detection must therefore be sensitive, rapid, and continuous. Nerve gases with reactive P-F or P-CN linkages such as methane phosphonyl fluoride methylester and dimethylaminocyanophosphoric acid ethyl ester, in the presence of peroxides, bring about the oxidation of indole (I), which is not fluorescent in the visible region, to the blue-green fluorescent indoxyl (II) and diindoxyl (indigo white) (III) as shown in Fig. 8. This reaction was developed as a laboratory assay for nerve gases by Gehauf and Goldenson.[15] More recently, Cherry *et al.*[16] have utilized the reaction in an automatic nerve gas alarm device. A

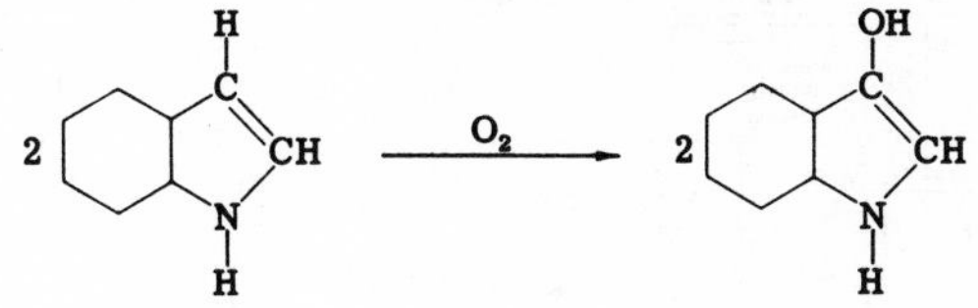

Indole
I

Indoxyl
(highly fluorescent)
II

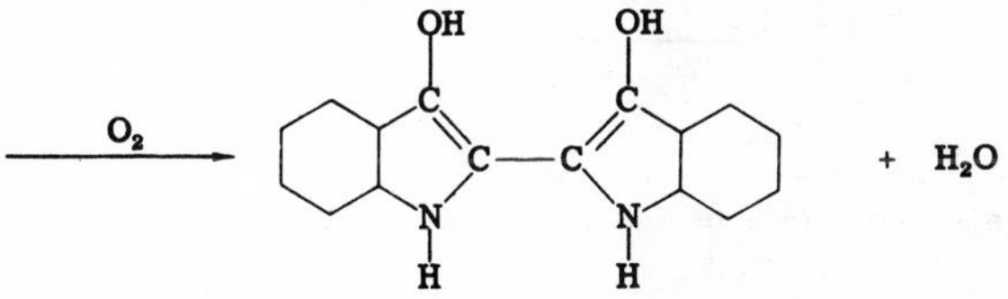

Indigo white
(highly fluorescent)
III

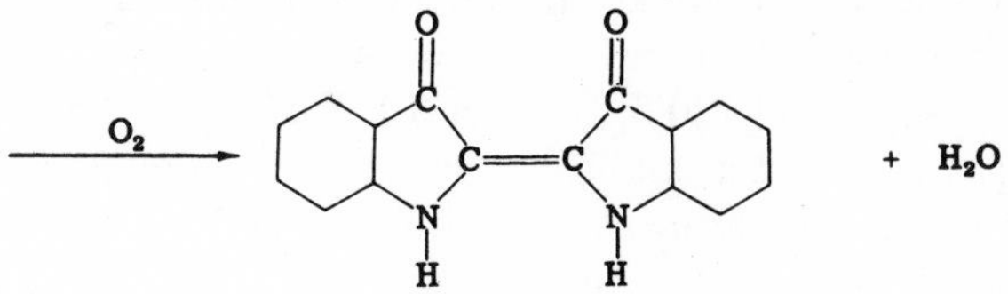

Indigo
(nonfluorescent)
IV

FIG. 8. Oxidation of indole to visibly fluorescing products. The six-member ring is unsaturated.

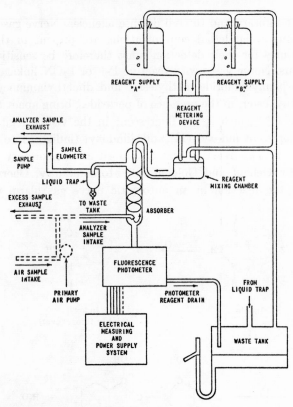

Fig. 9. Schematic diagram of the fluorometric nerve gas analyzer as devised by Cherry *et al.*[16] (see text for details).

schematic diagram of the nerve gas alarm and analyzer is shown in Fig. 9. Reagent A is prepared by mixing 1000 ml. of water, 150 ml. of 2-propanol, 100 ml. of acetone, 2 g. of indole, and 4 g. of carbonate-bicarbonate to bring the pH to 10.7. Reagent B contains 1000 ml. of water, 250 ml. of 2-propanol, and 3.5 ml. of Superoxol (30% H_2O_2 in 0.1 N H_2SO_4). The two solutions mix, pass over the absorber, and then through the fluorescence photometer. Fluorescence is continually measured with respect to a quinine standard. When the apparatus is appropriately calibrated, continuous quantitative data is available. The sensitivity is such that less than 1 µg. of nerve gas per liter of air gives a full scale deflection on the meter. Concentrations less than 0.1 µg./liter are apparently detectable. An alarm device can be set to go off when the fluorescence reaches a certain level. Nerve gases can also be detected by chemiluminescence produced with luminol and alkaline peroxide.[17] This procedures is less sensitive than the fluorometric method.

F. Stream and Harbor Pollution

Pollution of streams and water supplies represents a serious public health problem. Fluorometry can certainly be used to detect traces of specific organic and inorganic materials. Carpenter[18] has reported a most interesting use of fluorescence in tracing circulation and mixing in natural waters. In experiments in Baltimore Harbor, solutions of the highly fluorescent dye, Rhodamine B, have been released at specific sites. A null-point fluorometer with a continuous flow cell was placed aboard a small vessel to permit continuous analysis as the boat moved about the harbor. Water from selected depths was then pumped through the instrument yielding a continuous flow of data showing the concentration of dye in the sampled water. From this data it was possible to reconstruct the detailed circulation of water in Baltimore Harbor. This is one situation where fluorescence is a much better indicator than radioactive isotopes.

III. FOOD INSPECTION AND ANALYSIS

The present large-scale use of commercially prepared foods and the introduction of new procedures for food preservation, flavoring, and packaging makes inspection and analysis a necessity for manufacturers and public health authorities. Inspection is required to determine the quality of food and to detect spoilage and adulterants. The effects of various procedures on vitamins and other nutrients must be checked. Sanitation in the handling and processing of food must also be monitored. A host of chemical agents find their way into foods as pesticides, drugs, and antibiotics used on livestock, antioxidants, preservatives, flavorings, and colorings. Some of these are acceptable, others are toxic. Only by inspection and analysis can the consumer be safeguarded. Manufacturers, local and state authorities, and federal agencies such as the Food and Drug Administration and the Department of Agriculture maintain field stations and laboratories which are equipped to carry out a multitude of analyses on various food products. Fluorescence assay plays an important role in this field since sensitivity is extremely important. Specificity and simplicity can also be attained in fluorometric assays.

A. Decay and Spoilage

Inspection under ultraviolet light is routinely used to detect decay and spoilage. In some foods spoilage is indicated by the appearance of fluorescence. Orange rot appears as a light fluorescence on a dark background of the fruit. Pseudomonas infected eggs can be detected by the brilliantly green fluorescent pigment, pyoverdine, which is elaborated by the growing bacteria. This is so intense that it can be seen within the egg by using an ultraviolet candling device. Other foods possess characteristic fluorescence which is either altered or disappears with spoilage. Traces of organic matter and detergents in food containers can be detected by fluorescence under ultraviolet light. Fluorescence characteristics are an important test for quality of dried egg powder.[19] Browning in milk and milk products, which is due to interaction of amino acids and sugars, is detectable in very early stages (before discoloration) by fluorescence under ultraviolet light.[20, 21] The latter group investigated the nature of the fluorescent product. Simonson and Tarassuk[22] utilized quantitative fluorometric measurement of the amino acid-sugar products to determine optimal conditions for preparation and storage of evaporated milk. Another important index with regard to storage and handling of fresh milk concerns its native fluorescence. Fresh milk and cream emit greenish-yellow fluorescence which is due mainly to riboflavin. When exposed to daylight for appreciable periods of time or irradiated in any other way, the fluorescence turns from yellow to blue. The latter is characteristic of lumichrome.[23] These chemical changes in riboflavin are discussed in Chapter 7. The effects of food preparation on the distribution and content of vitamins is of great interest to nutritionists and modifications of the methods described in Chapter 7 are used.

B. Adulteration

An interesting example of the use of fluorescence assay in detecting and quantifying adulteration in lemon oil was reported by Vannier and Stanley.[24a] It appears that grapefruit oil, which is cheaper, is frequently found as an adulterant in samples of lemon oil. Chemical studies on these oils revealed the presence of distinctive coumarins in each. Grapefruit oil contains an ether of the highly fluorescent 7-hydroxycoumarin (umbelliferone) which is released on acid hydrolysis. Lemon oil contains other hydroxycoumarins which are not fluorescent. The umbelliferone in the grapefruit oil made possible the development of methods which can detect and measure as little as 0.5% of grapefruit oil ín lemon oil.

Ciusa and Nebbiea[24b] have measured excitation and fluorescence spectra of a variety of olive oils.

C. Flavorings

Various synthetic sweetening agents are used and there may be need of methods for their detection. Saccharin and dulcin are two commonly used sweetening agents. According to Radley,[25] when saccharin is treated with resorcinol in strong sulfuric acid at elevated temperatures, it gives

Saccharin   Dulcin

rise to sulfofluorescin. This product, in alkaline solution, yields an intensely green fluorescence. Dulcin also yields a fluorescent condensation product with resorcinol in strong sulfuric acid. Hydrolysis of the amide bond in dulcin should release the product p-ethoxy aniline which would be expected to possess appreciable native fluorescence in the ultraviolet region.

Glycerine is occasionally used as a sweetening agent; more frequently as a moistening agent and as a plasticizer. It is possible to oxidize glycerol and condense it with a variety of agents in strong sulfuric acid.[25] Glycerine can be extracted from the food material with ethanol, which is then evaporated. The residue is extracted with dilute aqueous acid. About 1 ml. of the extract may be added to a tube containing 10 ml. of freshly prepared 0.5% bromine water. After heating on a boiling water bath for 20 minutes, an alcoholic solution of β-naphthol and concentrated sulfuric acid are added. The mixture is heated for 2 minutes yielding a product with yellow-green fluorescence. Glycerine also yields fluorescent condensation products in sulfuric acid with 2,7-dihydroxy-naphthalene and anthrone. According to Radley,[25] the latter reaction is the most sensitive. The presence of glycerine is also an indication of rancidity since the compound is released during the enzymatic breakdown of fat.

D. Chemical Residues

The following chemical residues in food can be detected fluorometrically. Ammonium persulfate is used as a flour maturing agent. Auerbach et al.[26] developed a fluorometric procedure for the quantitative determination of as little as 5 to 10 parts of ammonium persulfate per million parts of flour or dough. Persulfate, which is an oxidant reoxidizes leuco-fluorescein to fluorescein. Under appropriate conditions the developed fluorescence is proportional to the persulfate content. Leuco-fluorescein can be used as a reagent for determining traces of other oxidizing agents as well,[27] including hydrogen peroxide.

The compound 6-ethoxy-1,2-dihydro-2,2,4-trimethylquinoline is used as an antioxidant for carotene in alfalfa meal. It is used in cattle and poultry feeds. A fluorometric method for detecting traces of this antioxidant in animal tissues, eggs, milk, and other products was developed by

TABLE IV

Excitation and Fluorescence Wavelengths for Several Pesticides [a]

Compound	Structure	Wavelength, mμ		Solvent
		Excitation	Fluorescence, λmax.	
		λmax. λ50[b]		
Guthion[c]		312 298, 327 250	380	Water (pH 11)
Potasan		320 292, 345	385	Methanol
Warfarin		320 290, 342	385	Methanol

TABLE IV (continued)

Compound	Structure	Wavelength, mμ			Solvent
		Excitation		Fluorescence, λmax.	
		λmax.	λ50		
Piperonyl butoxide		292 248	282, 302	318	Methanol
n-Propyl isome		292 248	280, 305	326	Methanol
Indoleacetic acid		285	265, 295	345	Water (pH 7)

TABLE IV (continued)

Compound	Structure	Wavelength, mμ			Solvent
		Excitation		Fluorescence,	
		λmax.	λ50	λmax.	
Naphthalene-acetic acid	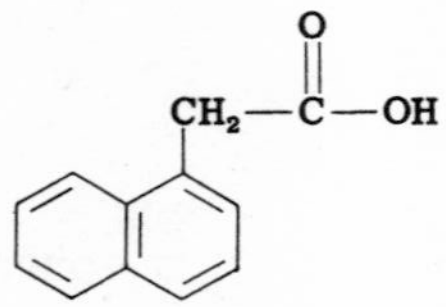	282 230	270, 305	327	Water (pH 11)
Naphthalene-acetamide	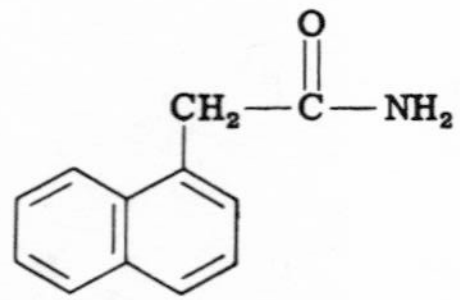	286 230	270, 305	327	Water (pH 11)

[a] After Hornstein.[31]

[b] Excitation wavelengths at which fluorescence intensity decreases to 50% of peak value while fluorescence wavelength is kept constant at its maximum value.

[c] This solution must stand for 0.5 hour before readings are taken. Fluorescence is caused by a hydrolysis product.

Bickoff *et al.*[28] The compound possesses an intense native fluorescence (blue). Less than 0.01 μg./ml. can be detected. It is readily extracted from alkaline tissue preparations into isooctane. Fluorescence is measured in the solvent before and after treatment with a permanganate solution. The latter selectively destroys the antioxidant.

Antibiotics in foods and milk are not only undesirable for health purposes, but they can also be undesirable for other reasons. They are a nuisance to cheese makers by preventing the action of bacterial starters. Antibiotics can be determined by the methods reported in Chapter 13. Hargrove *et al.*[29] have suggested that with antibiotic preparations intended for intramammary infusion in cattle a visibly fluorescent marker be used as a rapid and sensitive means for detecting the antibiotic in milk. A combination of uranine and oil soluble fluorescein (Fluoral 7 Ga.) when injected along with penicillin could be detected in the milk as long as positive tests for penicillin were obtained. The parallelism between fluorescence and antibiotic concentration was found to be very good.

Seed wheat is frequently treated with mercuric compound to control seed-borne diseases. To make certain that such mercury-coated wheat does not contaminate grain supplies, many mercury preparations are prepared with a red dye. Contamination with mercury-treated seed wheat endows grain with a pink color. Only a trace of such a color is sufficient to condemn a whole carload. The dye used is also fluorescent and permits more rapid and more sensitive detection. When unpigmented mercury preparations are used, it is more difficult to detect. A relatively simple and sensitive test for mercury is to place the sample on a metal surface in front of a fluorescent screen. A mercury source (254 mμ) is directed on this screen to make it fluoresce. When the metal is heated, mercury is volatilized and the vapors rise. As they do so, they absorb the light and cast a shadow on the fluorescent screen. The test is highly sensitive.[30]

Pesticides are being used more widely than ever in agriculture, necessitating control measures to make certain that they do not get passed on to humans. Since some are known to be extremely toxic and others have not been tested in man, extremely low limits of tolerance have been set. Fluorescence, where applicable, is certainly the method of choice. Hornstein[31] has surveyed a large number of pesticides with a fluorescence spectrometer and has found those shown in Table IV to possess native fluorescence. Excitation and fluorescence spectra (uncorrected) are reported as well as the media in which maximal fluorescence is obtained. Other compounds investigated which failed to exhibit native fluorescence at a concentration of 1 μg./ml. are D.D.T., methoxychlor, *p*-chlorophenoxyacetic acid, Diazinon, rotenone, aramite, hexachlorocyclohexane, toxaphene, aldrin, dieldrin, chlordan, and heptachlor. Assay procedures for those with fluorescence will, of course, require the development of suitable isolation procedures.

Of course, pesticides which are not themselves fluorescent can be converted to fluorophors by suitable chemical means. Giang[32] has developed a highly sensitive fluorometric method for the insecticide O,O-diethyl O-naphthalimido phosphorothioate (Bayer 22,408) which involves oxida-

Bayer 22,408

tion to a fluorescent derivative. Although the compound itself is fluorescent, this is not sufficiently intense to satisfy the requirements of sensitivity. The method for the estimation of Bayer 22,408 residues in milk, butter fat, and plant samples involves the following steps: extraction of the insecticide from the sample with a mixture of diethyl ether and n-hexane; evaporation of the solvent on a steam bath; removal of fats and waxes by dissolving the residues in n-hexane and extracting with acetonitrile; removal of even traces of lipids and plant pigments by chromatography on Florex xxx using chloroform for elution; evaporation of the chloroform to dryness; and dissolving the residue in 2 ml. of 0.1 N methanolic sodium hydroxide, then adding 10 ml. of a dioxane solution of hydrogen peroxide (1 ml. of 30% hydrogen peroxide in 100 ml. of dioxane). Before fluorometric assay, the samples are filtered through about 1 g. of anhydrous sodium sulfate. The oxidization product, which has not yet been identified, has an excitation maximum at 372 mμ and a fluorescence maximum at 480 mμ (uncorrected). With a commercial fluorescence spectrometer, as little as 0.01 μg./ ml. of Bayer 22,408 can be measured in the final extracts. There is no interference by milk, butter fat, and plant materials not treated with the insecticide. Of a large number of other insecticides tested, only two related compounds yielded a small amount of fluorescence with this method. They would not produce interference unless they were present in very large amounts. One of these related compounds, CoRal [O-(3-chloro-4-methyl-umbelliferone) O,O-diethylphosphoro-thioate-Bayer' 21,199] can be assayed fluorometrically by a method devised by Anderson et al.[33]

IV. MEDICAL DIAGNOSIS

Fluorescence emitted under ultraviolet light has also been used as a diagnostic aid in medicine. It has been used in ophthalmology[34] in a variety of ways. It is possible to diagnose the fungal infection tinea capitis by the characteristic yellow-green fluorescence of the infected hairs under 360-mμ light.[35] Similar diagnoses can be made of ringworm diseases in animals. Fluorescence has been used as a tool for investigating the structure of teeth.[36, 37] There have been many other medical applications of fluorescence in ultraviolet light. Additional references to such applications may be obtained from Ultra-Violet Products Inc.*

One widely used technique in medical diagnosis makes use of fluorescein as a tracer in the living patient. The fluorescence of fluorescein is so intense that it can be detected at enormous dilutions. The dye does not give any evidence of toxicity when injected into animals or humans. When several milliliters of fluorescein solution (10–15%) are injected into an arm vein, the dye is quickly transported to other regions of the body. Within a few seconds the bright-green fluorescence can be observed in the eyes, mucosa of the buccal cavity, and the lips.[38] The same principle can be used to determine circulation time[39] and to detect areas of poor circulation. Möller and Rorsman[40] have utilized the fluorescein method to study vascular permeability in man and have used it in studies on the effects of drugs on vascular permeability.[41] Kirshner and Proud[42] have reported using fluorescein to detect cerebrospinal fluid fistulas. When a fluorescein solution is injected into the spinal fluid, fistulas are made apparent by the appearance of fluorescein-tagged (yellow fluorescing) spinal fluid in various compartments, the nasopharynx, middle ear, etc. With this technique it has been possible to detect and repair surgically a number of lesions which would otherwise have remained undetected.

The following application of the fluorescent tracer technique in humans, which indirectly comes under the heading of Public Health and Sanitation, was carried out by a police department of one of the larger cities in the United States.[43] "The head of the Vice Squad . . . instructed one of his men to take two dollars and go to a known house of prostitution on North —— Street. Before entering, the officer put invisible fluorescent powder on his hands. As he entered the room, he left invisible marks from his hands along the wall and on the door jamb so that the officers could identify the room he entered. He handed her the money which had been treated with invisible powder. As the case progressed, the officer left traces of the fluorescent powder from his hand on the suspect's body. Meantime, other officers in the Vice Squad followed his trail with a portable, battery-operated Mineralight and easily identified the room the officer had entered. They were able thus to make quick entry into the proper room, identify the money given to the girl by the officer through its fluorescence, examine

---

* Ultra-Violet Products Inc., 5114 Walnut Grove Avenue, San Gabriel, California.

her hands and see that she had received it, and have the matron strip the girl and examine the fluorescent powder markings with the ultraviolet lamp to show where the officer's hands had been. In court the officer's testimony as to the marked money, established the transaction beyond doubt and the conviction was easily obtained." It is quite obvious from this and previous reports that fluorescence offers revolutionary techniques in all branches of public health.

REFERENCES

1. Przibram, K., *Nature* **183**, 1848 (1959).
2. Inglett, G. E., Miller, R. R., and Lodge, J. P., *Mikrochim. Acta No.* **1**, 95 (1959).
3. Chaikin, S. W., Symposium on Analytical Methods and Instrumentation in Air Pollution, New York, 1954, Division of Analytical Chemistry, 126th Meeting, American Chemical Society.
4. Chaikin, S. W., and Glassbrook, T. D., *Research for Industry*, Stanford Research Institute, Palo Alto, Calif. **5**, 2 (1953).
5. Schachter, D., *J. Lab. Clin. Med.* **54**, 763 (1959).
6. Parker, C. A., and Barnes, W. J., *Analyst* **85**, 3 (1960).
7. Waller, R. E., *Brit. J. Cancer* **6**, 8 (1952).
8. Van Duuren, B. L., *J. Natl. Cancer Inst.* **21**, 623 (1958).
9. Sawicki, E., Hauser, T. R., and Stanley, T. W., *Intern. J. Air Pollution* **2**, 253 (1960).
10. Sawicki, E., Elbert, W., Stanley, T. W., and Fox, F. T., *Anal. Chem.* **32**, 810 (1960).
11. Sawicki, E., Elbert, W., Stanley, T. W., Hauser, T. R., and Fox, F. T., *Intern. J. Air Pollution* **2**, 273 (1960).
12. Goulden, F., and Tipler, M. M., *Brit. J. Cancer* **3**, 157 (1949).
13. Van Duuren, B. L., *J. Natl. Cancer Inst.* **21**, 1 (1958).
14. Van Duuren, B. L., and Kosak, A. I., *J. Org. Chem.* **23**, 473 (1958).
15. Gehauf, B., and Goldenson J., *Anal. Chem.* **29**, 276 (1957).
16. Cherry, R. H., Foley, G. M., Badgett, C. O., Eanes, R. D., and Smith, H. R., *Anal. Chem.* **30**, 1239 (1958).
17. Goldenson, J., *Anal. Chem.* **29**, 277 (1957).
18. Carpenter, J. H., Tracer for Circulation and Mixing in Natural Waters. *Public Works*, reprint (1960).
19. Pearce, J. A., and Thistle, M. W., *Can. J. Research* **27**, 73 (1949).
20. Tarassuk, N. P., and Simonson, H. D., *Food Technol.* **4**, 88 (1950).
21. Fraenkel-Conrat, J., Cook, B. B., and Morgan, A. F., *Arch. Biochem. Biophys.* **35**, 157 (1952).
22. Simonson, H. D., and Tarassuk, N. P., *J. Dairy Sci.* **35**, 166 (1952).
23. Jenness, R., and Coulter, S. T., *J. Dairy Sci.* **31**, 367 (1948).
24a. Vannier, S. H., and Stanley, W. L., *J. Assoc. Official Agr. Chemists* **41**, 432 (1958).
24b. Ciusa, W., and Nebbia, G., Bolletino Scientifico della Facoltà di Chimica Industriale Bologna **18**, 113 (1960).
25. Radley, J. A., *Food* **20**, 84 (1951).
26. Auerbach, M. E., Eckert, H. W., and Angell, E., *Cereal Chem.* **26**, 490 (1949).
27. Kulberg, L. M., and Matveev, L., *J. Gen. Chem. U.S.S.R.* (Eng. Transl.) **17**, 452 (1947).
28. Bickoff, E. M., Guggolz, J., Livingston, A. L., and Thompson, C. R., *Anal. Chem.* **28**, 376 (1956).

29. Hargrove, R. E., Lehman, R. J., and Matthews, C. A., *J. Dairy Sci* **41,** 617 (1958).
30. Leighton, W. G., and Leighton, P. A., *J. Chem. Educ.* **12,** 139 (1935).
31. Hornstein, I., *J. Agr. Food Chem.* **6,** 32 (1958).
32. Giang, P. A., *J. Agr. Food Chem.* **9,** 42 (1961).
33. Anderson, C. A., Adams, J. M., and McDougall, D., *J. Agr. Food Chem.* **7,** 256 (1959).
34. Hague, E. B., *Am. J. Opthalmol.* **23,** 317 (1940).
35. Wilson, J. W., *A. M. A. Arch. Dermatol.* **63,** 771 (1951).
36. Dickson, G., Forziati, A. F., Lawson, M. E., Jr., and Schoonover, I. C., *J. Am. Dental Assoc.* **45,** 661 (1952).
37. Losee, F. L., Jennings, W. H., Lawson, M. E., Jr., and Forziati, A. F., *J. Dental Research* **36,** 911 (1957).
38. DeMent, J., Fluorobiology, Bull. No. 6, Ultra-Violet Products Inc., San Gabriel California.
39. Fishback, D., *J. Lab. Clin. Med.* **26,** 1966 (1941).
40. Möller, H., and Rorsman, H., *Acta Dermato-Venereol.* **38,** 233 (1958).
41. Möller, H., and Rorsman, H., *Acta Dermato-Venereol.* **39,** 12 (1959).
42. Kirchner, F. R., and Proud, G. O., University of Kansas School of Medicine, Kansas City, Kansas.
43. Black Light Case History, Bull. No. 15, Ultra-Violet Products Inc., San Gabriel, California.

APPENDIX I

FLUORESCENT INDICATORS

Fluorescence indicators can be used in chemistry in the same way as colored indicators. The principles are the same, the indicator changing in fluorescence intensity or color as it changes from the ionized to the unionized form. A good, fluorescent acid-base indicator is one whose fluorescence is intense and highly dependent upon the pH of the solution. For practical reasons the indicator should fluoresce in the visible range. Titration curves for several compounds which can be used as fluorescence indicators are shown in Figs. 1 and 2. 2-Naphthol can be used as an indicator in the range of pH from 8 to 9 (365-mμ excitation). Salicylic acid is useful from pH 2 to 4 and m-hydroxybenzoic acid from pH 9 to 11. Kavanagh and Goodwin[3] have presented pH-fluorescence curves for a large number of biologically important compounds. Some of these are shown in Fig. 3.

Many others have studied the variation of fluorescence intensity and wavelength with pH and there is a vast literature on the subject. Lists of fluorescent indicators and their pH ranges from 0 to 13 have been compiled by several authors. Some of these are discussed by Radley and Grant[4] and by Tomicek.[5] One of the most recent lists of fluorescent indicators was compiled at the Research Laboratories of the Eastman Kodak Company.[6] This is reproduced in Table I. Fluorescent indicators prove advantageous when dealing with colored or turbid solutions where visual color changes are not readily perceptible. Extracts of many foodstuffs,

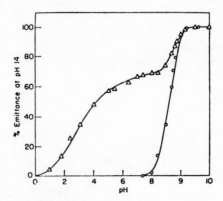

FIG. 1. Variation of fluorescence intensity of 2-naphthol with pH. O — O, 365-mμ excitation; △ — △, 313-mμ excitation (after Hercules and Rogers[1]).

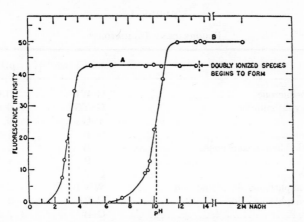

FIG. 2. Plots of pH vs. fluorescence for: curve A, salicylic acid at 410 mμ; curve B, m-hydroxybenzoic acid at 420 mμ. Excitation was in both cases at 314 mμ (after Thommes and Leininger[2]).

plants, and animal tissues and fluids are colored and difficult to titrate with colored indicators. Red wines, for example, can be conveniently titrated with fluorescent indicators or potentiometrically. Fluorescent indicators are also an advantage when dealing with solutions which attack or interfere with the glass electrode. It is also possible to use them in titrations in nonaqueous media.

As with colored indicators, the indicator change can be detected by eye. This can be done simply with a portable ultraviolet lamp (see Chapter 2) in a dark room or in a darkened cabinet. For more exact titrations, fluorometers can be used for the titration. When the titration is carried out over a period of time it is necessary to use a stable instrument as the detector.

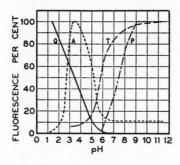

FIG. 3. The pH-fluorescence curves of quinine (Q), —; thiochrome (T), — — —, anthranilic acid (A), - - -; and 4-pyridoxic acid lactone (P), — — — — — (after Kavanagh and Goodwin[3]).

TABLE I

Fluorescent Indicators[a]

Indicator	Color Change[b]	pH range
4-Methylumbelliferone	G–WB	0.0–2.0
3,6-Dihydroxyphthalimide	G–YG	0.0–2.5
Benzoflavin	Y–G	0.3–1.7
Ethoxyacridone	G–B	1.2–3.2
3,6-Tetramethyldiaminoxanthone	G–B	1.2–3.4
Esculin	C–B	1.5–2.0
Eosin yellowish	C–Y	2.0–3.5
7-Amino-1,3-naphthalenedisulfonic acid	WB–B	2.0–4.0
1-Hydroxy-2-naphthoic acid	C–B	2.5–3.5
Salicylic acid	C–B	2.5–4.0
2′,4′,5′,7′-Tetrabromofluorescein sodium salt	C–G	2.5–4.5
2-Naphthylamine	C–V	2.8–4.4
Erythrosin	C–B	3.0–4.0
1-Naphthylamine	C–B	3.0–4.5
3-Hydroxy-2-naphthoic acid	B–G	3.0–6.8
o-Phenylenediamine	G–C	3.1–4.4
p-Phenylenediamine	C–OY	3.1–4.4
5-Aminosalicylic acid	C–G	3.1–4.4
o-Methoxybenzaldehyde	C–G	3.1–4.4
Phloxine	C–Y	3.4–5.0
Fluorescein	WG–G	4.0–5.0
Quininic acid	Y–B	4.0–5.0
4,5-Dihydroxy-2,7-naphthalenedisulfonic acid disodium salt	WB–B	4.0–6.0
2′,7′-Dichlorofluorescein	WG–G	4.0–6.0
Resorufin	C–O	4.0–6.0
β-Methyl esculetin	C–B	4.0–6.2
Acridine	G–B	4.5–6.0
3,6-Dihydroxyxanthone	C–BG	5.4–7.6
3,6-Dihydroxyphthalic acid	B–G	5.8–8.2
Quinine	B–V	5.9–6.1
5-Amino-2,3-dihydro-1,4-phthalazinedione	B–WB	6.0–7.0
3,6-Dihydroxyphthalonitrile	B–G	6.0–8.0
2-Naphthol-6-sulfonic acid sodium salt	C–B	6.0–8.0
Brilliant diazo yellow	C–B	6.5–7.5
Thioflavin	C–G	6.5–7.6
4-Methylumbelliferone	WB–B	6.5–8.0
Umbelliferone	O–B	6.5–8.0
Orcinaurine	C–G	6.5–8.0
2-Naphthol	WB–B	7.0–8.5
2-Naphthol-6,8-disulfonic acid dipotassium salt	WB–B	7.0–8.5
Morin	WG–G	7.0–8.5
1-Naphthol	C–BG	7.0–9.0
2-Naphthol-3,6-disulfonic acid disodium salt	WB–B	7.0–9.0

TABLE I (*Continued*)

Indicator	Color change[b]	pH range
trans-o-Hydroxycinnamic acid	C–G	7.2–9.0
1-Naphthol-4-sulfonic acid sodium salt	DB–LB	8.0–9.0
1-Naphthol-2-sulfonic acid sodium salt	DB–LB	8.0–9.0
Coumarin	WG–G	8.0–9.5
Acridine orange	WYG–Y	8.0–10.0
Naphthol AS	C–YG	8.2–10.3
Ethoxyphenylnaphthastilbazonium chloride	G–O	9.0–11.0
6,7-Dimethoxyisoquinoline-1-carboxylic acid	Y–B	9.5–11.0
Eosin BN	C–Y	10.5–14.0
1-Napthylamine	B–WB	12.0–13.0
7-Amino-1,3-naphthalenedisulfonic acid	B–WO	12.0–13.0
Cotarmine	Y–C	12.0–13.0
β-Naphthionic acid	B–V	12.0–13.0
4-Amino-1-naphthalenesulfonic acid	B–G	12.0–14.0

[a] See reference 6.

[b] C = colorless, B = blue, G = green, V = violet, O = orange, Y = yellow, D = dark, L = light, W = weak.

Fluorescent indicators can be used for other types of determination too. They have been used as adsorption indicators in halide titrations, in inorganic analysis when colored ions are used, and as oxidation reduction indicators. References to such applications can be found in the books by Radley and Grant[4] and by Tomicek[5] and in the report from the Eastman Kodak Research Laboratories.[6]

REFERENCES

1. Hercules, D. M., and Rogers, L. B., *Anal. Chem.* **30,** 96 (1958).
2. Thommes, G. A., and Leininger, E., *Anal. Chem.* **30,** 1361 (1958).
3. Kavanagh, F., and Goodwin, R. O., *Arch. Biochem.* **20,** 315 (1949).
4. Radley, J. A., and Grant, J., "Fluorescence Analysis in Ultra-Violet Light," 4th ed. Chapman and Hall, London, 1954.
5. Tomicek, O. "Chemical Indicators." Academic Press, New York, 1951.
6. *Org. Chem. Bull.* **29,** No. 4 (1957).

APPENDIX II

FREQUENCY, WAVELENGTH, AND ENERGY INTERRELATIONSHIPS
OF RADIATION, AND ENERGIES OF SOME CHEMICAL BONDS

Frequency, ν cm.$^{-1}$	Wavelength A	Energy electron volts	Energy kcal./mole	Bond
—100,000	—1000			
		—12	—280	
			—270	
—90,000	—1100	—11	—260	
			—250	
	—1200		—240	
—80,000		—10	—230	
	—1300		—220	
		—9	—210	
—70,000	—1400		—200	—Na Cl
	—1500	—8	—190	
	—1600		—180	
—60,000	—1700		—170	—N≡N
	—1800	—7	—160	
	—1900		—150	—C=O,C≡N —H-F
—50,000	—2000	—6	—140	
			—130	
			—120	—C≡C
—40,000	—2500	—5	—110	—O-H —C-C aryl
			—100	—C=C —O=C H-CL,C=S
	—3000	—4	—90	—C-H, H-Br
—30,000			—80	—N-H
	—4000	—3	—70	—C-O, H-I —C-CL
—20,000	—5000		—60	—C-C —C-Br, C-S
	—6000	—2	—50	—C-N —C-I
	—7000		—40	
	—8000			
—10,000	—9000 —10000		—30	
				—H-bond

Visible (↕)

Kindly prepared by Drs. Joseph E. Hayes Jr. and Robert L. Bowman.

Appendix III

CORNING GLASS FILTERS

Colored-glass filters are important components in fluorometry and are even used in monochromator instruments such as fluorescence spectrometers. Of the many types of filters which are available, the Corning filters are most widely used. Information concerning these filters can be found in Bulletin CF-1, Corning Glass Works, Corning, New York. However, the selection of proper filters is attended by some confusion since two different systems have been used for their classification. Until fairly recently, Corning filters were described in the literature in terms of the Glass Number, which is assigned to the original molded filter blank. More recently, Color Specification (C.S. No.) has been reported. The latter pertain to the polished filters. Although the C.S. No. is now used most frequently, the Glass No. also appears from time to time, and is present in all the older literature. The following table for interconversion of Corning C.S. No. with Glass No. should be helpful in the proper use and description of filters in fluorometry.

CORNING GLASS FILTERS

C.S. No.	Glass No.	Description
9-54	7910	Ultraviolet transmitting filters
0-53	7740	
0-54	0160	
0-52	7380	
0-51	3850	
9-30	7905	
7-54	9863	Transmit in the ultraviolet and absorb in the visible
7-51	5970	
7-60	5840	
7-59	5850	
7-37	5860	
7-39	5874	
5-57	5030	Blue filters
5-56	5031	
5-59	5433	
5-60	5543	
5-61	5562	
5-58	5113	
1-64	5330	

CORNING GLASS FILTERS (*Continued*)

C.S. No.	Glass No.	Description
4-70	4308	
4-71	4305	
4-72	4303	Blue-green filters
4-94	4784	
4-96	9782	
4-76	9780	
4-97	9788	
4-74	4445	
4-67	4060	Green filters
4-64	4010	
4-65	4015	
4-68	4084	
4-77	9830	
		Cutoff filters; less than 0.5% transmission below
2-64	2030	620 mμ
2-58	2403	617
2-59	2404	608
2-60	2408	599
2-61	2412	590
2-62	2418	579
2-63	2424	568
2-73	2434	558
3-66	3480	547
3-67	3482	524
3-68	3484	507
3-69	3486	493
3-70	3384	466
3-71	3385	441
3-72	3387	411
3-73	3389	391
3-74	3391	—
3-75	3060	—
7-56	2540	Infrared transmitting-visible
7-57	2550	absorbing filters
7-69	2600	
1-56	3961	
1-57	3962	Infrared absorbing-visible trans-
1-58	3965	mitting filters
1-59	3966	
1-69	4600	

APPENDIX IV

The references in this section represent selected papers published mainly from January 1962 through February 1964. A few earlier papers which were omitted from the first and second printings are also included. The references are listed by subject under headings which are, with one or two exceptions, the same as are used for chapter headings. Some papers are listed under more than one heading. Under the final heading, "General," are listed papers, books, and symposia of a review nature covering more than one area of interest; also included are papers which could not appropriately be included in the other categories.

PRINCIPLES OF FLUORESCENCE

Birks, J. B., and Aladekomo, J. B., Excimer fluorescence spectra of aromatic liquids, *Spectrochim. Acta* **20,** 15 (1964).

Bridges, J. W., and Williams, R. T., Fluorescence of some substituted benzenes, *Nature* **196,** 59 (1962).

Crowell, E. P., and Varsel, C. J., Spectrophotofluorometric studies of some aromatic aldehydes and their acetals, *Anal. Chem.* **35,** 189 (1963).

Döller, E., and Forster, Th., Der konzentrationsumschlag der fluoreszenz des naphthalins, *Z. physik. Chem. (Frankfurt)* **31,** 274 (1962).

Hameka, H. F., Resonance fluorescence in gases and molecular crystals, *J. Chem. Phys.* **38,** 2090 (1963).

Jablonski, A., Decay and polarization of fluorescence of solutions, *Proc. Intern. Conf. on Luminescence Org. Inorg. Materials, New York, 1961* p. 110 (Publ. 1962).

Ketskemety, I., Intermolecular energy transitions in fluorescent solutions, *Z. Naturforsch.* **17a,** 666 (1962).

Mataga, N., Dawasaki, Y., and Torihashi, Y., Solvent effects on the fluorescence spectrum of the ion-pair (Naphtholtriethylamine System), *Bull. Chem. Soc. Japan* **36,** 358 (1963).

Mataga, N., and Torihashi, Y., The electronic structure of carbazole in the fluorescent state, *Bull. Chem. Soc. Japan* **36,** 356 (1963).

Muel, B., Excitation d'une fluorescence retardée par des photons d'energie inférieure à celle des photons émis, *Compt. rend. acad. sci.* **255,** 3149 (1962).

Mukai, T., and Tebbens, R., A geometrical aspect of fluorescent spectroscopy, *Natl. Cancer Inst. Monograph No.* **9,** 127 (1962).

Parker, C. A., Sensitized P-type delayed fluorescence, *Proc. Roy. Soc.* **A276,** 125 (1963).

Parker, C. A., Delayed fluorescence of 3:4 benzpyrene solutions, *Nature* **200,** 331 (1963).

Parker, C. A., and Hatchard, C. G., Triplet-singlet emission in fluid solutions, *Trans. Faraday Soc.* **57,** 1894 (1961).

Parker, C. A., and Hatchard, C. G., Sensitized anti-Stokes delayed fluorescence, *Proc. Chem. Soc.* p. 386 (1962).

Parker, C. A., and Hatchard, C. G., Delayed fluorescence from solutions of anthracene and phenanthrene, *Proc. Royal Soc.* **A269,** 574 (1962).

Parker, C. A., and Hatchard, C. G., Delayed fluorescence of pyrene in ethanol, *Trans. Faraday Soc.* **59**, 1 (1963).

Parker, C. A., and Rees, W. T., Fluorescence spectrometry—a review, *Analyst* **87**, 83 (1962).

Sandros, K., and Almgren, M., The relative yields of fluorescence and phosphorescence of biacetyl in fluid solutions, *Acta Chem. Scand.* **17**, 552 (1963).

Van Duuren, B. L., Effects of the environment on the fluorescence of aromatic compounds in solution, *Chem. Rev.* **63**, 325 (1963).

INSTRUMENTATION

Ainsworth, S., and Winter, E., An automatic recording spectrofluorimeter, *Appl. Opt.* **3**, 371 (1964).

American Instrument Co., Silver Spring, Maryland, Fluoro-microphotometer, Bull. No. 2390 (1963).

Baird-Atomic, Inc., Cambridge, Massachusetts, Fluorispec—fluorescence spectrophotometer model SF-1, Bull. VCA 10M/264 (1964).

Beckman Instruments Inc., Fullerton, Calif., DK/Universal, a double-beam dual monochromator instrument (1963).

Brewer, L., James, C. G., Brewer, R. G., Stafford, F. E., Berg, R. A. and Rosenblatt G. M., Phase fluorometer to measure radiative lifetimes of 10^{-5} to 10^{-9} sec., *Rev. Sci. Instr.* **33**, 1450 (1962).

Burton, R. A., and Butterworth, K. R., A semi-automatic fluorimetric method of determining adrenaline and noradrenaline, *J. Physiol.* **166**, 4P (1962).

Chance, B., and Legallais, V., A spectrofluorometer for recording of intracellular oxidation-reduction states, *I.E.E.E. Trans. Bio-Med. Electronics* **BME-10**, 40 (1963).

Glick, D., and von Redlich, D., Fluorometer cuvette adapters for measurements on 0.05 or 0.01 milliliter volumes, *Anal. Biochem.* **6**, 471 (1963).

Iwai, K., and Kozawa, S., A new spectrofluorophotometer and its application to the determination of biological substances, *Abstr. Pittsburgh Conf. on Anal. Chem. and Appl. Spectroscopy* p. 67 (1964).

Johnson, P., and Richards, E. G., A simple instrument for studying the polarization of fluorescence, *Arch. Biochem. Biophys.* **97**, 250 (1962).

Laikin, M., A new double beam fluorometer, *Federation Proc.* **21**, 476 (1962).

Loeser, C. N., In vivo localization of injected carcinogen by fluorescence quartz microscopy, *Cancer Research* **20**, 415 (1960).

Mekshenov, M. I., and Andreitsev, A. P., Universal recording equipment for spectrofluorimetry, *Biophysics (USSR) Engl. Transl.* **6**, 88 (1961).

Muller, R. H., Instrumentation for spacecraft, *Anal. Chem.* **36**, 115A (1964).

Parker, C. A., and Hatchard, C. G., The possibilities of phosphorescence measurement in chemical analysis: tests with a new instrument, *Analyst* **87**, 664 (1962).

Peticolas, W. L., Goldsborough, J. P., and Riekhoff, K. E., Double photon excitation in organic crystals (laser induced fluorescence), *Phys. Rev. Letters* **10**, 43 (1963).

Rockenmacher, M., and Farr, A. F., Modification of the fluorescence attachment for the Beckman Model DU spectrophotometer, *Clin. Chem.* **9**, 554 (1963).

Shimadzu Seisakusho Ltd., Kyoto, Japan, Spectrofluorometer accessory set type GF-16E, Catalogue No. P63-171.

Taketomo, Y., Modification of spectrophotofluorometer for scanning paper strips, *Nature* **190**, 1094 (1961).

Turner Associates, Palo Alto, California, Turner Model 252 Process Stream Fluorometer; designed to continuously measure fluorescence in product pipelines (1963).

Turner, G. K., A precision recording absolute spectrofluorometer—Turner Model 210, Turner Associates, Palo Alto, California, *Abstr. Pittsburgh Conf. on Anal. Chem. and Appl. Spectroscopy* p. 74 (1964).

Vladimirov, Yu. A., and Litvin, F. F., Use of the photomultiplier to measure fluorescence spectra of model systems and living objects, *Biofizica* **3**, 606 (1958).

Vladimirov, Yu. A., and Litvin, F. F., Investigation of luminescence of very low intensity in biological systems, *Biofizika* **4**, 601 (1959).

West, S. S., Loesser, C. N., and Schoenberg, M. D., Television spectroscopy of biological fluorescence, *IRE Trans. Med. Electronics* **ME-7**, 138 (1960).

PRACTICAL CONSIDERATIONS

Argauer, R. J., and White, C. E., Fluorescent compounds for calibration of excitation and emission units of spectrofluorometer, *Anal. Chem.* **36**, 368 (1964).

Burdick and Jackson Laboratories, Muskegon, Michigan, Distilled in glass solvents, Bull. BJ-13a (June 21, 1963).

Chapman, J. H., Förster, Th., Kortüm, G., Parker, C. A., Lippert, E., Melhuish, W. H., and Nebbia, G., Proposal for standardization of methods of reporting fluorescence emission spectra, *Appl. Spectroscopy* **17**, 171 (1963).

Hartmann-Leddon Co., Inc., Philadelphia, Pennsylvania, Reagents for fluorometric clinical procedures, Harleco Insert No. 68.

King, R. M., and Hercules, D. M., Correction for anomalous fluorescence peaks caused by grating transmission characteristics, *Anal. Chem.* **35**, 1099 (1963).

Melhuish, W. H., Calibration of spectrofluorimeters for measuring corrected emission spectra, *J. Opt. Soc. Am.* **52**, 1256 (1962).

Parker, C. A., Spectrofluorometer calibration in the ultraviolet region, *Anal. Chem.* **34**, 502 (1962).

Van Duuren, B. L., Effect of the environment on the fluorescence of aromatic compounds in solution, *Chem. Revs.* **63**, 325 (1963).

AMINO ACIDS, AMINES, AND THEIR METABOLITES

Allgén, L. G., Funke, K. E., and Naukhoff, B., Fluorometric determination of tryptamine in urine with the Zeiss ultraviolet spectrophotometer-fluorometer, *Scand. J. Clin. Lab. Invest.* **13**, 390 (1961).

Andén, N. E., Roos, B. E., and Werdinius, B., On the occurrence of homovanillic acid in the brain and cerebrospinal fluid and its determination by a fluorometric method, *Life Sciences* **7**, 448 (1963).

Bischoff, F., and Torres, A., Determination of urine dopamine, *Clin. Chem.* **8**, 370 (1962).

Brunjes, S., Wybenga, D., and Johns, V. J. Jr., Fluorometric determination of urinary metanephrine and normetanephrine, *Clin. Chem.* **10**, 1 (1964).

Burton, R. A., and Butterworth, K. R., A semi-automatic fluorimetric method of determining adrenaline and noradrenaline, *J. Physiol.* **166**, 4P (1962).

Carlsson, A., Falck, B., and Hillarp, N. Å., Cellular localization of brain monoamines, *Acta Physiol. Scand.* **56**, Suppl. No. 196 (1962).

Carlsson, A., and Lindquist, M., A method for the determination of normetanephrine in brain, *Acta Physiol. Scand.* **54**, 83 (1962).

Chin, L., Picchioni, A. L., and Childs, R. F., Light—an essential factor in the trihy-droxyindole-spectrofluorometric assay of norepinephrine, J. Pharm. Sci. **52**, 907 (1963).

Cooper, J. R., The fluorometric determination of acetylcholine, Biochem. Pharmacol. **13**, (1964).

Cornog, J. L., Jr., and Adams, W. R., The fluorescence of tyrosine in alkaline solution, Biochim. Biophys. Acta **66**, 356 (1963).

Corrodi, H., and Hillarp, N. Å., A fluorescence method for the histochemical identification of catecholamines, Helv. Chim. Acta **46**, 2425 (1963).

Falck, B., Observations on the possibilities of the cellular localization of monoamines by a fluorescence method, Acta Physiol. Scand. **56**, Suppl. No. 197 (1962).

Gally, J. A., and Edelman, G. M., The effect of temperature on the fluorescence of some aromatic amino acids and proteins, Biochim. Biophys. Acta **60**, 499 (1962).

Goldenberg, H., and White, D. L., A new fluorometric reaction for determination of the 3-O-methyl catecholamines, Clin. Chem. **8**, 453 (1962).

Harrison, W. H., Detection of intermediate oxidation states of adrenaline and noradrenaline by fluorescence spectrometric analysis, Arch. Biochem. Biophys. **101**, 116 (1963).

Hsia, D. Y., Litwack, M., O'Flynn, M., and Jakovcic, S. Serum phenylalanine and tyrosine levels in the newborn infant, New Engl. J. Med. **267**, 1067 (1962).

McCaman, M. W., and Robins, E., Fluorimetric method for the determination of phenylalanine in serum, J. Lab. Clin. Med. **59**, 885 (1962).

Nagatsu, T., and Yagi, K., Identification of the main ethylenediamine condensate of noradrenaline with that of catechol, Nature **193**, 484 (1962).

Noah, J. W., and Brand, A., Simplified micromethod for measuring histamine in human plasma, J. Lab. Clin. Med. **62**, 506 (1963).

Pomeranz, Y., and Shellenberger, J. A., A fluorescent spot test for sulfhydryl compounds, Anal. Chim. Acta **26**, 301 (1962).

Saunders, P. P., and Parks, L. W., Some observations on the fluorescence analysis of anthranilic acid, Anal. Biochem. **3**, 354 (1962).

Sharman, D. F., A fluorometric method for the estimation of 4-hydroxy-3-methoxy phenylacetic acid (homovanillic acid) and its identification in brain tissue, Brit. J. Pharmacol. **20**, 204 (1963).

Smith, E. R. B., and Weil-Malherbe, H., Metanephrine and normetanephrine in human urine: method and results, J. Lab. Clin. Med. **60**, 212 (1962).

Spector, S., Melmon, K., Lovenberg, W., and Sjoerdsma, A., The presence and distribution of tyramine in mammalian tissues, J. Pharmacol. Exptl. Therap. **140**, 229 (1963).

Vanable, J. W., Jr., A ninhydrin reaction giving a sensitive quantitative fluorescence assay for 5-hydroxytryptamine, Anal. Biochem. **6**, 393 (1963).

Wiegand, R. G., and Scherfling, E., Determination of 5-hydroxytryptophan and serotonin, J. Neurochem. **9**, 113 (1962).

Zachariae, H., Histamine in human blood, Scand. J. Clin. Lab. Invest. **15**, 173 (1963).

PROTEINS AND PEPTIDES

Andersen, S. A., Characterization of a new type of cross-linkage in resilin, a rubber-like protein, Biochem. Biophys. Acta **69**, 249 (1963).

d'Antona, D., and Mannucci, E., Fluorescent antibodies in laboratory diagnosis, Ann. Sclavo **3**, 37 (1961).

Brand, L., and Shaltiel, S., Appearance of fluorescence on treatment of histidine residues with *N*-bromosuccinimide, *Biochim. Biophys. Acta* **75**, 145 (1963).

Bulen, W. A., and LeComte, J. R., Isolation and properties of a yellow-green fluorescent peptide from *Azotobacter* medium, *Biochem. Biophys. Research Communs.* **9**, 523 (1962).

Churchich, J. E., The polarization of fluorescence of reoxidized muramidase (lysozyme), *Biochim. Biophys. Acta* **65**, 349 (1962).

Cowgill, R. W., Fluorescence and the structure of proteins. I. Effects of substituents on the fluorescence of indole and phenol compounds, *Arch. Biochem. Biophys.* **100**, 36 (1963).

Fox, K. K., Holsinger, V. H., and Pallansch, M. J., Fluorimetry as a method of determining protein content of milk, *J. Dairy Sci.* **46**, 302, (1963).

Frieden, C., Glutamate dehydrogenase, V. The relation of enzyme structure to the catalytic function, *J. Biol. Chem.* **238**, 3286 (1963).

Goldman, M., and Carver, R. K., Microfluorimetry of cells stained with fluorescent antibody, *Exptl. Cell Research* **23**, 265 (1961).

Gray, R. W., and Hartley, B. S., A fluorescent end-group reagent for proteins and peptides, *Biochem. J.* **89**, 59P (1963).

Hall, C. T., and Hansen, P. A., Chelated azo dyes used as counterstains in the fluorescent antibody technic, *Zentr. Bakteriol., Parasintenk. Abt. I, Orig.* **184**, 548 (1962).

Hiraoka, T., and Glick, D., Measurement of protein in millimicrogram amounts by quenching of dye fluorescence, *Anal. Biochem* **5**, 497 (1963).

Kaplan, A., and Johnstone, M. A., The fluorometric determination of albumin in cerebrospinal fluid and serum, *Clin. Chem.* **9**, 505 (1963).

Klinman, N., and Karush, F., Assay of disulfide and sulfhydryl content of proteins and peptides by fluorescence quenching, *Federation Proc.* **22**, 348 (1963).

Kulangara, A. C., and Pincus, G., Disappearance of fluorescent labelled gonadotropin from the blood in the rabbit, *Endocrinol.* **71**, 179 (1962).

McKinney, R. M., Spillane, J. T., and Pearce, G. W., Determination of purity of fluorescein isothiocyanates, *Anal. Biochem.* **7**, 74 (1964).

Millar, D., Minzghor, K., and Steiner, R., Enhancement of fluorescence of a conjugate of soybean inhibitor accompanying interaction with trypsin, *Biochim. Biophys. Acta* **65**, 153 (1963).

Nairn, R. C., "Fluorescent Protein Tracing," Williams & Wilkins, Baltimore, Maryland, 1962.

Newmark, M. Z., and Wenger, B. S., Determination of millimicrogram amounts of protein, *Federation Proc.* **22**, 478 (1963).

Partridge, S. M., Elastin, ultraviolet absorption and fluorescence, *Advances in Protein Chem.* **17**, 290 (1962).

Reske, G., and Stauff, J., Fluorimetrischer nachweis einer photo-reaktion von 3,4-benzpyren mit β-lactoglobulin unter beiteilung von molekularem sauerstoff, *Z. Naturforsch.* **18b**, 774 (1963).

Searcy, F. L., Korotzer, J. L., and Bergquist, L. M., A new fluorometric technique for measuring high- and low-density lipoproteins, *Clin. Chim. Acta* **8**, 148 (1963).

Steiner, R. F., and Edelhoch, H., Effect of thermally induced structural transitions on the ultraviolet fluorescence of proteins, *Nature* **193**, 375 (1962).

Steiner, R. F., and Edelhoch, H., The ultraviolet fluorescence of proteins, *Biochim. Biophys.* **66**, 341 (1963).

Tengerdy, R. P., Quantitative immunofluorescein titration of human and bovine gamma globulins, *Anal. Chem.* **35**, 1084 (1963).

Wold, F., and Weber, G., Fluorescence quenching of phenols by carboxylic acids; a model for tyrosine-carboxylate interaction, *Federation Proc.* **22**, 348 (1963).

Young, D. M., and Potts, J. T., Structural transformations of bovine pancreatic ribonuclease in solution: a study of polarization of solution, *J. Biol. Chem.* **238**, 1995 (1963).

VITAMINS, COENZYMES, AND THEIR METABOLITES

Brown, E., and Clark, D. L., Simplified method for the determination of reduced TPNH by means of enzymic cycling, *J. Lab. Clin. Med.* **61**, 889 (1963).

Chen, P. S., Jr., Terepka, A. R., and Lane, K., Sensitive fluorescence reaction for vitamins D and dihydrotachysterol, *Anal. Biochem.* (1964).

Netrawali, M. S., Radhakrishnamurty, R., and Sreenivasan, A., A new fluorometric method for the estimation of citrovorum factor, *Anal. Biochem.* in press (1964).

Ogawa, S., Fluorescent reaction of vitamin C, *Bull. Natl. Hyg. Lab., Tokyo* **70**, 77 (1952).

Rindi, G., and Perri, V., Separation and determination of thiamine and pyrithiamine in biological materials by chromatography on polyethylene powder, *Anal. Biochem.* **5**, 179 (1963).

Toepfer, E. W., Polansky, M. M. and Hewston, E. M., Fluorometric pyridoxamine assay conversion to pyridoxal cyanide compound, *Anal. Biochem.* **2**, 463 (1961).

Uyeda, K., and Rabinowitz, J. C., Fluorescence properties of tetrahydrofolate and related compounds, *Anal. Biochem.* **6**, 100 (1963).

Yagi, K., Kondo, H., and Okunda, J., A simplified measurement of flavins on filter paper by lumiflavin fluorescence method, *J. Biochem. (Tokyo)* **51**, 231 (1962).

METABOLITES, GENERAL

Belman, S., The fluorimetric determination of formaldehyde, *Anal. Chim. Acta* **29**, 120 (1963).

Bourne, B. B., A study of the Momose fluorometric determination of blood glucose, *Clin. Chem.* **9**, 502 (1963).

Chance, B., Legallais, V., and Schoener, B., Metabolically linked changes in fluorescence emission spectra of cortex of rat brain, kidney and adrenal gland, *Nature* **195**, 1073 (1962).

Garay, E., Rodriquez, A., and Argerich, T. C., Fluorometric determination of biliverdin in serum, bile and urine, *J. Lab. Clin. Med.* **62**, 141 (1963).

Gouterman, M., and Stryer, L., Fluorescence polarization of some porphyrins, *J. Chem. Phys.* **37**, 2260 (1962).

Harris, R. A., and Gambal, D., Fluorometric determination of total phospholipids in rat tissues, *Anal. Biochem.* **5**, 479 (1963).

Hess, H. H., Fluorometric assay of sialic acid, *Federation Proc.* **20**, 342 (1961).

Maitra, P. K., and Estabrook, R. W., A fluorometric method for the enzymatic determination of glycolytic intermediates, *Anal. Biochem.* in press (1964).

Mendelsohn, D., and Antonis, A., A fluorometric micro glycerol method and its application to the determination of serum triglycerides, *J. Lipid Research* **2**, 45 (1961).

Sawicki, E., Stanley, T. W., and Johnson, H., Comparison of spectrophotometric and spectrophotofluorometric methods for the determination of malonaldehyde, *Anal. Chem.* **35**, 199 (1963).

Sawicki, E., Stanley, T. W., and Pfaff, J., Spectrophotofluorimetric determination of formaldehyde and acrolein with J-acid; comparison with other methods, *Anal. Chim. Acta* **28,** 156 (1963).

Solov'ev, K. N., Spectroscopic and luminescence relations pertaining to porphin derivatives, *Vestsi Akad. Navuk Belarusk. SSR, Ser. Fiz.-Tekhn. Navuk* **3,** 27 (1962).

Spikner, J. E., and Towne, J. C., Fluorometric microdetermination of alpha-keto acids, *Anal. Chem.* **34,** 1468 (1962).

Towne, J. C., and Spikner, J. E., Fluorometric microdetermination of carbohydrates, *Anal. Chem.* **35,** 211 (1963).

NUCLEIC ACIDS, PURINES, AND PYRIMIDINES

Basu, S., and Loh, L., Absorption and emission spectra of nucleic acids at long wavelengths, *J. Chim. Phys.* **59,** 1031 (1962).

Basu, S., and Loh, L., Absorption and fluorescence spectra of mixtures of nucleosides, *Biochim. Biophys. Acta* **76,** 131 (1963).

Bersohn, R., and Isenberg, I., On the phosphorescence of DNA, *Biochem. Biophys. Research Communs.* **13,** 205 (1963).

Borresen, H. C., On the luminescence properties of some purines and pyrimidines, *Acta Chem. Scand.* **17,** 921 (1963).

Boyle, R. E., Nelson, S. S., Dollish, F. R., and Olsen, M. J., The interaction of deoxyribonucleic acid and acridine orange, *Arch. Biochem. Biophys.* **96,** 47 (1962).

Churchich, J. E., Fluorescence studies on soluble ribonucleic acid labelled with acriflavine, *Biochem. Biophys. Acta* **75,** 274 (1963).

Loeser, C. N., West, S. S., and Schoenberg, M. D., Absorption and fluorescence studies on biological systems: Nucleic acid dye complexes, *Anat. Record* **138,** 163 (1960).

Longworth, J. W., Luminescence of purines and pyrimidines, *Biochem. J.* **84,** 104P (1962).

Schiffer, L. M., Vaharu, T., and Gardner, L. I., Acridine orange as a chromosome stain, *Lancet* p. 1362 (December 1961).

Udenfriend, S., Zaltzman-Nirenberg, P., and Cantoni, G. L., Fluorometric assay of 2-dimethylamino-6-hydroxypurine (dimethylaminoguanine) in the presence of guanine, *Anal. Biochem.* **5,** 258 (1963).

Walaas, E., Fluorescence of adenine and inosine nucleotides, *Acta Chem. Scand.* **17,** 461 (1963).

ENZYMOLOGY

Fasella, P., Turano, C., Giartosio, A., and Hammady, I., Glutamic-oxaloacetate transaminase, spectrofluorimetric study, *Giorn. biochim.* **10,** 174 (1961).

Greenberg, L. J., Fluorometric measurement of alkaline phosphatase and aminopeptidase activities in the order of 10^{-14} mole, *Biochem. Biophys. Research Communs.* **9,** 430 (1962).

Guibault, G. G., and Kramer, D. N., Fluorometric determination of lipase, acylase, alpha- and gamma-chymotrypsin and inhibitors of these enzymes, *Anal. Chem.* **36,** 409 (1964).

Kramer, D. N., and Guilbault, G. G., A substrate for the fluorometric determination of lipase activity, *Anal. Chem.* **35,** 588 (1963).

Pappenhagen, A. R., Use of fluorescein-labelled fibrin for the determination of fibrinolytic activity, *J. Lab. Clin. Med.* **59,** 1039 (1962).

Roth, M., Dosage fluorimétrique de la trypsine, *Clin. Chim. Acta* **8,** 574 (1963).

Rotman, B., β-D-Galactosidase activity of single enzyme molecules, *Proc. Natl. Acad. Sci. U. S.* **47**, 1981 (1961).

Rotman, B., Zderic, J. A., and Edelstein, M., Fluorogenic substrates for β-D-galactosidases and phosphotases derived from fluorescein (3,6-dihydroxyfluoran) and its monomethyl ether, *Proc. Natl. Acad. Sci. U.S.* **50**, 1 (1963).

Strässle, R., Methods for the fluorimetric determination of fibrinolytic activity, *Thrombosis Diath. Haemorrhag.* **8**, 112 (1962).

Sturtevant, J. M., The fluorescence of α-chymotrypsin in the presence of substrates and inhibitors, *Biochem. Biophys. Research Communs.* **8**, 321 (1962).

Tytell, A. A., and Schepartz, A. I., Fluorometric estimation of β-D-galactosidase in cell cultures, *Federation Proc.* **21**, 161 (1962).

Verity, M. A., and Brown, W. J., Spectrofluorometric study of hepatic and cerebral lysosomal phosphomonoesterases, *Exptl. Cell Research* in press (March, 1964).

Verity, M. A., Capter, R., and Brown, W. J., Spectrofluorometric determination of β-glucuronidase activity, *Arch. Biochem. Biophys.* in press (1964).

STEROIDS

Brown, W. R., Evaluation of procedures for urinary 17-hydroxycorticosteroids, *Clin. Chem.* **9**, 468 (1963).

Bruinvels, J., and van Noordwijk, J., Fluorescence of aldosterone in sulfuric acid, *Nature* **193**, 1260 (1962).

De Moor, P., Osinski, P., Deckx, R., and Steeno, O., The specificity of fluorometric corticoid determinations, *Clin. Chim. Acta* **7**, 475 (1962).

Epstein, E., and Zak, B., Stationary phase as color reagent in glass paper chromatography of estogens, *Clin. Chem.* **9**, 70 (1963).

Goldzieher, J. W., Fluorescent spectra, *in* "Physical Properties of the Steroid Hormones" (L. L. Engel, ed.) Pergamon Press, London, 1964.

Ichii, S., Forchielli, E., Perloff, W. H., and Dorfman, R. I., Determination of plasma estrone and estradiol-17-beta, *Anal. Biochem.* **5**, 422 (1963).

Levin, J. J., and Johnston, C. G., Fluorometric determination of serum total bile acids, *J. Lab. Clin. Med.* **59**, 681 (1962).

Mahesh, V. B., and Greenblatt, R. B., Isolation of dehydroepiandrosterone and 17α-hydroxy-Δ⁵-pregnenolone from the polycystic ovaries of the Stein-Leventhal syndrome, *J. Clin. Endocrinol. Metab.* **22**, 441 (1962).

Short, R. V., and Levett, I., The fluorometric determination of progesterone in human plasma during pregnancy and the menstrual cycle, *J. Endrocrinol.* **25**, 239 (1962).

Steinetz, B., Beach, V., Dubnick, B., Meli, A., and Fujimoto, G., Fluorescence of some 17α-substituted steroids in concentrated HCl, *Steroids* **1**, 395 (1963).

PLANTS

Butler, W. L., Effect of red and far-red light on the fluorescence yield of chlorophyll *in vivo*, *Biochim. Biophys. Acta* **64**, 309 (1962).

Butler, W. L., and Norris, K. H., Lifetime of the long-wavelength chlorophyll fluorescence, *Biochem. Biophys. Acta* **66**, 72 (1963).

Döhler, G., Über den Einfluss verschiedener Plasmolytika auf Photosynthese und Chloroplasten-Fluoreszenz von *Chlorella vulgaris*, *Beitr. Biol. Pflanz.* **39**, 147 (1963).

Karapetyan, N. V., Litvin, F. F., and Krasnovskii, A. A., Investigation of the phototransformations of chlorophyll by differential spectrophotometry, *Biofizika* **8**, 191 (1963).

Olson, R. A., Oriented molecules and the structure of chloroplasts, *in* "Photosynthesis Mechanisms in Green Plants," *Natl. Acad. Sci.—Natl. Research Council Publ. No. 1145* (1963).

Olson, R. A., Butler, W., and Jennings, W. H., The orientation of chlorophyll molecules *in vivo*: evidence from polarized fluorescence, *Biochim. Biophys. Acta* **54**, 615 (1961).

Teale, F. W. J., Pigment interaction in chloroplast fluorescence, *Biochem. J.* **85**, 14P (1962).

Tomita, G., and Rabinowitch, E., Excitation energy transfer between pigments in photosynthetic cells, *Biophys. J.* **2**, 483 (1962).

Vennesland, B., Gattung, H. W., and Birkicht, E., Fluorometric measurement of the photoreduction of flavin by illuminated chloroplasts, *Biochim. Biophys. Acta* **66**, (1963).

Yentsch, C. S., and Menzel, D. W., A method for the determination of phytoplankton chlorophyll and phaeophytin by fluorescence, *Deep-Sea Research* **10**, 221 (1963).

Yin, H. C., Hung, H. C., Shen, Y. K., and Wang, M. C., Studies on photophosphorylation, VIII. Quenching of chloroplast fluorescence in the presence of different photophosphorylation systems, *Sci. Sinica (Peking)* **12**, 1945 (1963).

INORGANIC CONSTITUENTS

Armstrong, W. A., Grant, D. W., and Humphreys, W. G., A fluorometric procedure for the determination of cerous ion, *Anal. Chem.* **35**, 1300 (1963).

Block, J., and Morgan, E., Determination of parts per billion iron by fluorescence extinction, *Anal. Chem.* **34**, 1647 (1962).

Carter, D. A., and Ohnesorge, W. E., Spectrofluorometric study of (2-methyl-8-quinolinolato)zinc(II) chelates in absolute alcohol, *Anal. Chem.* **36**, 327 (1964).

Dye, W. B., Bretthauer, E., Seim, H. J., and Blincoe, C., Fluorometric determination of selenium in plants and animals with 3,3'-diaminobenzidine, *Anal. Chem.* **35**, 1687 (1963).

Hill, J. B., An automated fluorometric method for the determination of serum magnesium, *Ann. N. Y. Acad. Sci.* **102**, 108 (1962).

Kepner, B. L., and Hercules, D. M., Fluorometric determination of calcium in blood serum, *Anal. Chem.* **35**, 1239 (1963).

Kirkbright, G. F., and Stephen, W. I., The screening of metallofluorescent indicators, *Anal. Chim. Acta* **27**, 294 (1962).

Parker, C. A., and Harvey, L. G., Luminescence of some piazselenols; a new fluorimetric reagent for selenium, *Analyst* **87**, 558 (1962).

Przibram, K., Fluorescence of organic traces in inorganic materials and their distribution in nature, *Geochim. Cosmochim. Acta* **26**, 1045 (1962).

Wallach, D. F. H., and Esandi, M. P., Fluorescence techniques in the microdetermination of metals in biological materials. III. Method for complexometric determination of magnesium in small serum samples, *Anal. Biochem.* **7**, 67 (1964).

Wallach, D. F. H., and Steck, T. L., Fluorescence techniques in microdetermination of metals in biological materials. II. An improved method for direct complexometric titration of calcium in small serum samples, *Anal. Biochem.* **6**, 176 (1963).

Wallach, D. F. H., and Steck, T. L., Fluorescence techniques in the microdetermination of metals (Al, Co, Cu, Ni, Zn and alkaline earths) in biological materials, *Anal. Chem.* **35**, 1035 (1963).

Wallach, D. F. H., Surgenor, D. M., Soderberg, J., and Delano, E., Preparation of 3,6-dihydroxy-2,4-bis-[N,N'-di-(carboxymethyl)-aminomethyl] fluoran; utilization for the ultramicrodetermination of calcium, *Anal. Chem.* **31**, 456 (1959).

Watanabe, S., Frantz, W., and Trottier, D., Fluorescence of magnesium, calcium and zinc 8-quinolinol complexes, *Anal. Chem.* **5,** 345 (1963).

Yates, W. B., and Ahesson, N. B., Fluorescent tracers for quantitative micro-residue analysis, Paper presented at Ann. meeting Am. Soc. Agr. Eng., Washington, D. C., June, 1962.

DRUGS AND TOXIC AGENTS

Anderson, C. A., Anderson, R. J., and Yagelowich, M., The photofluorometric determination of Bayer 9002 (*N*-hydroxy-naphthalimide diethyl phosphate) residues in animal tissues, *Am. Chem. Soc. Abstr.* (April, 1964).

Bernstein, H. N., Zvaifler, N. J., Rubin, M., and Sr. Mansour, A. M., The ocular deposition of chloroquine, *Invest. Opthalmol.* **2,** 384 (1963).

Brandt, R., Ehrlich-Rogozinsky, S., and Cheronis, N. D., Spectrofluorometric methods for the microdetection and estimation of morphine and codeine, *Microchem. J.* **5,** 215 (1961).

Chiang, H. C., and Chen, W. F., Fluorometric assay of yohimbine, *J. Pharm. Sci.* **52,** 808 (1963).

Child, K. J., Bedford, C., and Tomich, E. G., Common drugs that may invalidate spectrophotofluorometric assays of blood griseofulvin, *J. Pharm. Pharmacol.* **14,** 374 (1962).

Chirigos, M. A., and Mead, J. A. R., Experiments on determination of Melphalan (L-phenylalanine mustard) by fluorescence: Interaction with protein and various solutions, *Anal. Biochem.* **7,** 259 (1964).

Cohen, E. N., Quantitative determination of *d*-tubocurarine in body tissues and fluids, *J. Lab. Clin. Med.* **67,** 979 (1963).

Cramér, G., and Isaksson, S., Quantitative determination of quinidine in plasma, *Scand. J. Clin. Lab. Invest.* **15,** 553 (1963).

Daudel, P., Muel, B., Lacroix, G., and Prodi, G., Recherche de métabolites liés aux protéines de peau de souris badigeonnée par du 3,4-benzopyrène, *J. chim. phys.* p. 263 (1962).

Davis, B., Dodds, M. G., and Tomich, E. G., Spectrophotofluorometric determination of emetine in animal tissues, *J. Pharm. Pharmacol.* **14,** 249 (1962).

Frerejacque, M., and DeGrave, P., Reactions colorées et réactions de fluorescence des digitaliques, *Ann. pharm. franç.* **21,** 509 (1963).

Giang, P. A., Weissbecker, L., and Schechter, M., Fluorometric method for terephthalic acid residues in chicken tissues, *Am. Chem. Soc. Abstr.* (April, 1964).

Hedrick, M. T., Rippon, J. W., Decker, L. E., and Bernsohn, J., Fluorometric determination of isonicotinic acid hydrazide in serum, *Anal. Biochem.* **4,** 85 (1962).

Ibsen, K. H., Saunders, R. L., and Urist, M. R., Fluorometric determination of oxytetracycline in biological material, *Anal. Biochem.* **5,** 505 (1963).

Jakovljevic, I. M., New fluorometric micromethod for the simultaneous determination of digitoxin and digoxin, *Anal. Chem.* **35,** 1513 (1963).

Jellinek, M., and Willman, V. L., Fractionation and fluorometric quantitation of digoxni and its metabolites from human urine, *Federation Proc.* **22,** 186 (1963).

Koechlin, B. A., and D'Arconte, L., Determination of chlordiazepoxide (Librium) and of a metabolite of lactam character in plasma of humans, dogs and rats by a specific spectrofluorometric micro method, *Anal. Biochem.* **5,** 195 (1963).

Kügemagi, U., and Terriere, L. C., Nematocide residues—The spectrofluorometric determination of *O,O*-diethyl *O*-2-pyrazinyl phosphorothioate (Zinophos) and its oxygen analog in soil and plant tissues, *Agr. Food Chem.* **11,** 293 (1963).

Kupferberg, H. J., Burkhalter, A., and Way, E. L., A sensitive fluorometric assay for morphine, *Federation Proc.* **22**, 249 (1963).

Loeser, C. N., In vivo localization of injected carcinogen by fluorescence quartz rod microscopy, *Cancer Research* **20**, 415 (1960).

MacDougall, D., The use of fluorometric measurements for determination of pesticide residues, *Residue Revs.* **1**, 24 (1962).

Mellinger, T. J., and Keeler, C. E., Spectrofluorometric identification of phenothiazine drugs, *Anal. Chem.* **35**, 554 (1963).

Muel, B., Etudes sur la luminescence du 3,4-benzopyrène et de polycycles aromatiques apparentés, en solution à basse température, Thesis, Fac. Sci. Univ. de Paris, November 27, 1962.

Muel, B., and Hubert-Habart, M., Premier triplet d'hydrocarbures aromatiques polycycliques et activité cancérigène, *in* "Advances in Molecular Spectroscopy," p. 647. Pergamon Press, London, 1962.

Panier, R. G., and Close, J. A., Quantitative fluorometric determination of panthenol in multivitamin preparations, *J. Pharm. Sci.* **53**, 108 (1964).

Ragab, M. T. H., Fluorescent paper chromatographic analysis of several chlorinated organic pesticides, *Am. Chem. Soc. Abstr.* (April, 1964).

Roth, M., and Rieder, J., Spectrofluorimetrische Bestimmung von Hydrazinen. Anwendung zur Bestimmung von Benzylhydrazine in biologischen Proben, *Anal. Chim. Acta* **27**, 20 (1962).

Rubin, M., Bernstein, H. N., and Zvaifler, N. J., Studies on the pharmacology of chloroquine, *A. M. A. Arch. Opthalmol.* **70**, 474 (1963).

Spock, J., and Katz, S. E., Fluorometric determination of chlortetracycline in premixes, *J. Assoc. Offic. Agr. Chemists* **46**, 434 (1963).

Swagzdis, J. E., and Flanagan, T. L., Spectrophotofluorometric determination of low concentrations of amobarbital in plasma, *Anal. Biochem.* **7**, 147 (1964).

Tishler, F., Sheth, P. B., and Giaiomo, M. B., A quantitative fluorometric method for the determination of serpasil (reserpine) in feeds at the micro level, *J. Assoc. Offic. Agr. Chem.* **46**, 448 (1963).

Weissler, A., Ultrasonic hydroxylation in a fluorescence analysis for microgram quantities of benzoic acid, *Nature* **193**, 1070 (1962).

Yates, W. B., and Ahesson, N. B., Fluorescent tracers for quantitative micro-residue analysis, Paper presented at Ann. meeting, Am. Soc. Agr. Eng., Washington, D. C., June, 1962.

PUBLIC HEALTH AND MEDICINE

Berk, J. E., and Kantor, S. M., Fluorescence of gastric sediment in differentiation of gastric lesions, *Geriatrics* **18**, 787 (1963).

Feuerstein, D. L., and Selleck, R. E., Fluorescent tracers for dispersion measurements, *J. Sanit. Eng. Div., Am. Soc. Civil Engrs.* **89**, 1 (1963).

Forziati, A., Kumpula, J. W., and Barone, J. J. Tooth fluorometer, *J. Am. Dental Assoc.* **67**, 663 (1963).

Fox, K. K., Holsinger, V. H., and Pallansch, M. J., Fluorometry as a method of determining protein content of milk, *J. Dairy Sci.* **46**, 302 (1963).

Gracian, J., and Martel, J., The fluorescence of olive oils in the ultraviolet. I. Origin of the compounds responsible for the blue fluorescence observed with a Wood filter, *Grasas Aceites (Seville, Spain)* **13**, 128 (1962).

Hoffman, H. J., and Olszewski, J., Fluorescein in normal brain tissue, a study of blood-brain barrier, *Neurology* **11**, 1081 (1961).

Kanno, Y., and Lowenstein, W. R., Intercellular diffusion (using fluorescein as a fluorescent tracer), *Science* **143,** 959 (1964).

Mori, I., Fluorescence of soap attached to the skin and to cloth, *Kyoto Furitsu Ikadaigaku Zasshi* **70,** 453 (1961).

O'Connell, R. L., and Walter, C. M., Hydraulic model tests of estuarial waste dispersion, *J. Sanit. Eng. Div., Am. Soc. Civil Eng.* **89,** P 51 (1963).

Porro, T. J., Dadik, S. P., Green, M., and Morse, H. T., Fluorescence and absorption spectra of biological dyes, *Stain Technol.* **38,** 37 (1963).

Rice, E. W., and Grogan, B. S., 1960 survey of clinical chemistry procedures used by members of the American Association of Clinical Chemists, *Clin. Chem.* **8,** 181 (1962).

Roe, D. A., The application of fluorescence to the identification of acidic components of psoriatic lesions, *J. Invest. Dermatol.* **41,** 319 (1963).

Rogers, R. L., Kirchner, F. R., and Proud, G. O., The evaluation of eustachian tubal function by fluorescent dye studies, *Laryngoscope* **72,** 456 (1962).

Silber, R., Gabrio, B. W., and Huennekens, F. M., Studies on normal and leukemic leukocytes. III. Pyridine nucleotides, *J. Clin. Invest.* **41,** 230 (1962).

Wong, P., Inouye, T. and Hsia, D. Y., A routine procedure for screening of phenylketonuria in the new-born, *Clin. Chem.* **9,** 476 (1963).

Zussman, H. W., Fluorescent agents for detergents, *J. Am. Oil Chem. Soc.* **40,** 695 (1963).

Zweig, G., ed., "Analytical Methods for Pesticides, Plant Growth Regulators, and Food Additives." Academic Press, New York, 1963.

GENERAL

Barskii, I. Ya., Brumberg, E. M., and Brumberg, V. A., Ultraviolet fluorescence of muscles, *Doklady Akad. Nauk SSSR* **147,** 474 (1962).

Brandt, R., and Cheronis, N. D., Lower limits of organic reactions, *Mikrochim. Acta* p. 467 (1963).

Course on Spectrofluorometric Techniques in Biology, held at Istituto di Ricerche Farmacologiche Mario Negri, Milan, Italy, September, 1963, (to be published).

Freed, S., and Vise, M. H., On phosphorimetry as quantitative microanalysis with application to some substances of biochemical interest, *Anal. Biochem.* **5,** 338 (1963).

International Symposium on Molecular Structure and Spectroscopy, held under the auspices of the Science Council of Japan, Tokyo, Japan, September, 1962.

Obrien, D., and Ibbott, F. A., eds., "Laboratory Manual of Pediatric Ultramicrobiochemical Techniques," 3rd ed. Harper and Row, New York, 1962.

Parker, C. A., and Hatchard, C. G., Triplet-singlet emission in fluid solution, *J. Phys. Chem.* **66,** 2506 (1962).

Parker, C. A., and Rees, W. T., Fluorescence spectrometry—a review, *Analyst* **87,** 83 (1962).

Seiler, N., Werner, G., and Wiechmann, M., Direkte quantitative bestimmung von fluoreszierenden substanzen auf dunnschlichtchromatogrammen, *Naturwissenschaften* **50,** 643 (1963).

Turner, G. K., Assoc., Palo Alto California, Radiation dosimetry employing fluorescent glass dosimeters; an annotated bibliography, August, 1963.

Vickers, T. J., and Winefordner, J. D., Atomic fluorescence spectrometry as a means of chemical analysis, *Anal. Chem.* **36,** 161 (1964).

White, C. E., Fluorometric analysis; review of fundamental developments, *Anal. Chem.* **34,** 81R (1962).

AUTHOR INDEX

Numbers in parentheses are reference numbers, and are included to assist in locating a reference when the author's name is not cited at the point of reference in the text. Numbers in italics indicate the page on which the full reference is listed.

A

Aarons, J. H., 149 (56), *188*

Abelson, D., 367, 368, *371*

Abraham, D., 177 (107), *189*

Adair, F., 433 (58), *443*

Adams, J. M., 466 (33), *469*

Adler, A., 309, *311*

Adler, E., 239, *277*

Ahrendt, M. E., 14 (8), *36*

Airth, R. L., 373 (2a), 374 (2a), *386*

Aizawa, Y., 356, 357, 358 (13), *371*

Alba, R. T., 151 (65), *188*

Albers, R. W., 122 (35), *124*, 287, *310*, 317, *347*, 350, 351, 352, *370*

Alejo, L. G., 251 (56), *278*

Allfrey, V., 269, *279*

Allinson, M. J. C., 184 (114), *190*

Almy, G. M., 289 (22, 23), 299 (9), *310*

Amesz, J., 86 (29), 87, 88, 89, *95*, 246, *278*, 338 (44), 339 (44), 342, *348*

Amiard, G., 283 (8, 9), *310*

Anderson, C. A., 466, *469*

Anderson, D. R., 184 (112a), 185, *189*

Andreae, W. A., 383, *387*

Angell, E., 412, *442*, 461 (26), *468*

Aoyama, M., 232, *277*

Arakawa, T., 235 (14), *277*

Archibald, R. M., 263, 264, *279*, 439, 440, *443*

Armstrong, M. D., 135 (23), *187*

Aronoff, S., 384, *387*

Aronow, L., 104, *123*

Asami, T., 251, *278*

Audus, L. J., 379, 384 (23), *387*

Auerbach, M. E., 412, 442, 461, *468*

Axelrod, J., 151 (64), 159, *188*, *189*, 327, 328, *347*, 433, 434 (60), *443*

Ayres, P. J., 366, 367, *371*

B

Bacon, K., 248 (48), *278*

Badgett, C. O., 457 (16), 458 (16), *468*

Baer, J. E., 97 (2), 101 (2), 102 (2), *123*

Baglioni, C., 261 (74a), 262 (74a), *278*

Baltscheffsky, H., 246, *278*, 340, 342, 343, *348*

Bandier, E., 252, *278*

Bannister, T. T., 376 (9), 377, 378, *387*

Barker, H. A., 270, 271 (90), *279*

Barnes, C. C., 249 (49), *278*

Barnes, W. J., 109 (20), *123*, 394, 396, 397, 399, 445, 446, *468*

Barr, C. G., 291, *310*

Barsel, N., 151 (61), *188*

Bartholomew, R. J., *35*, 80, 82, *95*

Bates, R. W., 354, *370*

Baudisch, O., 288, *310*

Bauld, W. S., 69 (18), *94*, 354 (10), 355, *370*

Bayly, R. J., 285, *310*

Beauchamp, C., 381, *387*

Bedford, C., 412, 413, *442*

Beers, R. F., 302, *311*

Benard, H., 370, *371*

Bergquist, L. M., 226, *229*

Berliner, E., 220 (63), *229*

Bernstein, D. E., 361, 362, *371*

Bertler, A., 136 (28), 138, 139, 140 (28), 142, 147 (28), 148 (28), 149, 159, *187*, *188*, *189*

Besch, P. H., 360, *371*

Bessey, O. A., 234 (11), 237, 238, 240, 242, 251 (56), 268 (86), *277*, *278*, *279*, 318 (10), 336 (10, *347*

Betheil, J. J., 225, 226, *229*

Beye, H. K., 411 (24), *442*

Beyer, R. E., 276, *279*

Bicknell, L., 251 (56), *278*

Bickoff, E. M., 385 (37, 38, 39, 40), 386, *387*, 465, *468*

U

Udenfriend, S., 27 (23), 33 (24), *36*, 68 (14), 69, (15, 16, 17), 73 (14), *94*, 97 (2), 101 (2), 102 (2, 8), 103 (9, 10, 11, 12, 13, 14), 106 (17), 107 (19), 110, 113 (25), 117 (29), *123*, *124*, 125 (2), 130, 131, 132, 133, 134 (20), 135 (24, 25, 26, 27), 140 (27), *158*, 159 (27), 162 (15), 163, 164, 165 (89), 166, 167, 168 (89, 90, 93), 169, 170 (97, 98, 99, 100), 171 (100, 101, 102), 172, 177 (107), 179 (101), 184 (111), *187*, *189*, 191 (1, 2), 197 (16), 232 (1), 254 (1), 265 (1), 267 (1), 269 (1), 271 (1), 273 (1), 274 (1), 276 (1), *227*, 294 (37, 38b), 296 (37, 38b), 297, *311*, 323 (19), 324 (21, 23), 325 (24), 326 (26), 329 (30a), 336 (21), *347*, 353 (6), 354 (6), *370*, 379 (25), 380 (25), *387*, 402 (3, 6, 8), 403 (9), 404 (8, 11), 405 (13), 406 (13), 407 (3), 409, 411 (26), 412 (3), 414 (3), 415 (3), 417, 419 (3), 420 (3), 421 (3), 422, 424 (41), 425 (3), 426 (3), 427 (3), 428 (3), 429 (3), 430 (3), 431 (6, 8, 46), 434 (3), 435 (3), 436 (3), *441*, *442*
Uehleke, H., 222 (77), *227*, *229*
Ulshafer, P. R., 439 (65), *443*
Uphaus, R. A., 169, *189*
Upton, V., 367, *371*

V

Valzelli, L., 85 (28), *95*
Van Duuren, B. L., 451 (8), 455, 456, *468*
Van Gieson, P., 409 (18), 410 (18), *442*
Van Horst, S. H., 92, 94, *95*
Vannier, S. H., 460, *468*
Vannotti, A., 304, *311*
Vasta, B. M., 69 (16), *94*, 117 (29), *124*, 402 (3), 407 (3), 409 (3), 412 (3), 414 (3), 415 (3), 417 (3), 419 (3), 420 (3), 421 (3), 422 (3), 425 (3), 426 (3), 427 (3), 428 (3), 429 (3), 430 (3), 434 (3), 435 (3), 436 (3), *441*
Velick, S. T., 131 (19), 162, 163 (87), *187*, *189*, 200, 204, 205, 209, 210 (22), 211, 212, 214, 215, 223, *228*, 246, *278*
Velluz, L., 283, 286, 299, *310*
Vendsalu, A., 142, 151 (48), *188*
Vennesland, B., 293 (30), *310*, 319 (13), *347*

Virgin, H. I., 373 (2a), 374 (2a), 377 (17), *386*, *387*
Vivanco, F. B., 239, *277*
Vladimirov, Yu. A., 128, *187*, 203, *228*
Vles, F., 201 (23), *228*
Vogel, M., *35*
Vogt, M., 151 (59), *188*
Volcani, B. E., 270 (91), *279*
Von Euler, H., 239, *277*
Von Horst, H., 127, *187*

W

Waalkes, T. P., 33 (24), *36*, 132, 133, 171 (101), 179 (101), *187*, *189*, 191 (2), *227*, 324, *347*
Wadman, W. A., 93, 94 (37), *95*, 127, *187*, 285, 286, *310*
Wagner-Jauregg, T., 243 (28), *277*
Walaas, E., 213, *228*
Walaas, O., 213, *228*
Waldeck, B., 137, 139, *187*, *188*
Walker, A., 266, *279*, 308 (59), *311*
Waller, R. E., 451 (7), 455, *468*
Wallerstein, J. S., 151, *188*
Wallner, A., 334 (41), 335 (41), 336 (41), *348*
Wang, Y. L., *253*
Wanta, R. C., 391 (8), *399*
Warburg, O., 242 (25), 243 (27), *277*, 376, *387*
Ward, G. E., 309 (62), *311*
Wassink, E. C., 374 (4), 376 (4), *386*
Watkinson, J. H., 392, 393, *399*
Watson, C. J., 307 (57, 58), *311*
Wavilov, S. I., 20 (16), *36*
Waxman, H. E., 149 (56), *188*
Weber, G., 13 (4), 17, 18, 19, 20 (18a, 18c), 27 (15), 34 (25), *36*, 118 (34), *124*, 127, 129, 160, *187*, 194, 195, 196, 197, 198, 200, 201, 202, 204, 205, 206, 207, 208, 212 (12), 213, 216, 218, 223, 227, *227*, *228*, 237, 239, 246, 247, *277*, *278*, 294, 295, *311*, 379 (24), *387*
Weil-Malherbe, H., 100, *123*, 136 (29), 151, 152, *153*, 156, 159, *187*, *188*, *189*, 297, *311*
Weimer, E. Q., 75 (21), *95*, 118 (31), 121 (31), *124*

SUBJECT INDEX

A

Absorptiometry errors due to fluorescence, 13

Absorption spectra
  adrenal steroids in polyphosphoric acid, 360
  anthracene, 21
  1:2 benzanthracene, 80
  benzo[a]pyrene, 451
  benzonaphthenedione (glucose assay), 282
  corticosterone (sulfuric acid fluorophor), 355
  estrone (sulfuric acid fluorophor), 355
  flavin-adenine dinucleotide, 239
  flavin mononucleotide, 239
  5-hydroxytryptamine, 81
  phycocyanin, 378
  quinine, 21
  riboflavin-5'-pyrophosphate, 239

Acceptor, in energy transfer, 193
Acetaldehyde, 286
Acetoacetic acid, 292
Acetol, 288
Acetone, fluorescence due to traces of diacetyl, 289
Acetophenone, 287
Acetylcholine, 186
Acetylmethylcarbinol, 289
Acetylpyridine analog of DPNH, complex with dehydrogenases, 210, 212
Acid alizarin garnet R, reagent for aluminum, 388
Aconitic acid, 290
Acridine orange, nucleic acid stain, 301
Acriflavine, see Trypaflavine
F-Actin, detection of interaction with myosin, 219
Actinometer, 118
Activation spectra, see Excitation spectra
Adenine
  fluorescence characteristics, 295
  quantum yield, 295
Adenosine, 295

Adenosine diphosphate, 295
  enzymatic assay, 331
Adenosine triphosphate, 295
  enzymatic assay, 332
  luminescent assay (luciferin), 333
Adenylic acid, 295
  assay, 333
Adrenal cortical steroids, 359–371
Adrenaline, see Epinephrine
Adrenochrome, 141, 151
Adrenolutine (trihydroxyindole) epinephrine fluorophor, 141
Adsorption onto surfaces, 101
  chloroquine, 102
Adulteration, detection by fluorescence, 460
Agmatine, 185
  diamine oxidase substrate, 327
Air
  dust, 444
  fluoride, 445
  oil, 445
  pollen, 445
  pollution, 444–458
Alcohol dehydrogenase, fluorescence spectrum of DPNH complex, 209, 210
Aldehydes, 286
Aldosterone
  assay in urine, 367, 368
  soda fluorescence, 366
  tert-butoxide fluorescence, 367
Alloxan, 439
N-Allylnormorphine, 420, 422
Aluminum
  fluorescence characteristics of complexes, 388
  assay in plants, 390
  reagents for assay, 388
Amethopterin, 414
Amines, aromatic, 32
L-Amino acid decarboxylase, See Aromatic L-amino acid decarboxylase
D-Amino acid oxidase, 213

502